ISIS
INTERNATIONAL SYMPOSIUM
ON INTERDISCIPLINARY SCIENCE

Related Titles from AIP Conference Proceedings

757 Statistical Physics and Beyond: 2nd Mexican Meeting on Mathematical and Experimental Physics
Edited by F.J. Uribe, L. García-Colín S., E. Díaz-Herrera, March 2005, 0-7354-0242-6

742 Experimental Chaos: 8th Experimental Chaos Conference
Edited by S. Boccaletti, B. J. Gluckman, J. Kurths, L. M. Pecora, R. Meucci, and O. Yordanov, December 2004, 0-7354-0226-4

735 Bayesian Inference and Maximum Entropy Methods in Science and Engineering
Edited by R. Fischer, R. Preuss, and U. von Toussaint, November 2004, 0-7354-0217-5

729 Global Analysis and Applied Mathematics: International Workshop on Global Analysis
Edited by K. Taş, D. Krupka, O. Krupková, and D. Baleanu, October 2004, 0-7354-0209-4

718 Computing Anticipatory Systems: CASYS '03 - Sixth International Conference
Edited by D. M. Dubois, August 2004, 0-7354-0198-5

676 Experimental Chaos: 7th Experimental Chaos Conference
Edited by V. In, L. Kocarev, T. L. Carroll, B. J. Gluckman, S. Boccaletti, and J. Kurths, August 2003, 0-7354-0145-4

665 Unsolved Problems of Noise and Fluctuations: UPoN 2002: Third International Conference on Unsolved Problems of Noise and Fluctuations in Physics, Biology, and High Technology
Edited by S.M. Bezrukov, May 2003, 0-7354-0127-6

661 Modeling of Complex Systems: Seventh Granada Lectures
Edited by P.L. Garrido and J. Marro, April 2003, 0-7354-0121-7

658 Modern Challenges in Statistical Mechanics: Patterns, Noise, and the Interplay of Nonlinearity and Complexity; Pan American Studies Institute
Edited by V. M. Kenkre and K. Lindenberg, March 2003, 0-7354-0118-7

To learn more about these titles, or the AIP Conference Proceedings Series, please visit the webpage
http://proceedings.aip.org/proceedings

ISIS
INTERNATIONAL SYMPOSIUM
ON INTERDISCIPLINARY SCIENCE

Northwestern State University,
Natchitoches, Louisiana 6-8 October 2004

EDITORS
Andrei Ludu
Nathan R. Hutchings
Darrell R. Fry
Northwestern State University
Natchitoches, Louisiana

SPONSORING ORGANIZATIONS
Northwestern State University
The IDEAS Program
Richardson Technologies, Inc

Melville, New York, 2005
AIP CONFERENCE PROCEEDINGS ■ VOLUME 755

Editors:

Andrei Ludu
Nathan R. Hutchings
Darrell R. Fry

Northwestern State University
Natchitoches, LA 71497

E-mail: ludua@nsula.edu
           hutchingsn@nsula.edu
           fryd@nsula.edu

L.C. Catalog Card No. 2005922481

ISBN 0-7354-0240-X
ISSN 0094-243X
Printed in the United States of America

CONTENTS

Preface vii

Acknowledgments viii

ISIS Welcome... ix
    A. Ludu

ISIS Opening Address: Complexity versus Simplicity...................... 1
    M. Gell-Mann

Super Heavy Nuclei-Clusters of Matter and Antimatter................... 17
    W. Greiner and T. J. Bürvenich

Heavy-Ion Reactions in Time-Dependent Hartree-Fock 34
    J. A. Maruhn

Chaotic Dynamics of Modulational Instability in Optical Fibers 40
    M. J. Ablowitz and C. M. Schober

Dynamics of Patterns on Elastic Hyper-Surfaces: Part I - Shear
Waves in the Middle Surface 46
    C. I. Christov

Dynamics of Patterns on Elastic Hyper-Surfaces: Part II - Wave
Mechanics of Flexural Quasi-Particles................................. 53
    C. I. Christov

Harmonic Oscillators as Bridges between Theories...................... 61
    Y. S. Kim and M. E. Noz

Fourier-Galerkin Method for Time-dependent Problems of
Interacting Localized Waves 70
    M. A. Christou and C. I. Christov

Generalized Quasilinearization Method for Reaction Diffusion
Equation with Numerical Applications 78
    A. S. Vatsala and J. Yang

2D Solitary Waves of Boussinesq Equation.............................. 85
    J. Choudhury and C. I. Christov

Nonlinear Modeling of 3-D Flagellar Dynamics.......................... 91
    A. Ludu and N. Hutchings

Symmetry Breaking in a Model for Nodal Cilia 107
    C. J. Brokaw

What Organizes the Molecular Ballet that Promotes the Movement of
the Axoneme in Such a Way that Its Molecular Machinery Seems to
Be a Whole.. 117
    C. Cibert

Regulation of Eukaryotic Flagellar Motility........................... 130
    D. R. Mitchell

Flagellar Bend Dynamics in African Trypanosomes 137
    N. R. Hutchings and A. Ludu

MSP Dynamics and Retraction in Nematode Sperm........................ 145
    C. W. Wolgemuth

Divalent Cation Control of Flagellar Motility in African Trypanosomes 153
    A. M. Westergard and N. R. Hutchings

Non-Equilibrium Studies of Voltage-Gated Ion Channels 159
    A. Kargol

Phenomenological Energetics for Molecular Motors 165
    T. Harada

Chemical Master Equation Reduction Methods 172
    R. Zhu and M. R. Roussel

Automated Cell Tracking Tools for Quantitative Motility Studies 177
    C. Zimmer, B. Zhang, S. Blazquez, E. Labruyère, F. Frischknecht,
    R. Ménard, N. Guillén, and J. C. Olivo-Marin

Differential Tethering of Log Phase *Trypanosoma brucei* onto Chemically Distinct Surfaces 185
    L. Archuleta, A. Dunham, J. Rains, and D. Fry

Microbial Biofilms: Persisters, Tolerance and Dosing 190
    N. G. Cogan

Computational Model of Population Dynamics Based on the Cell Cycle and Local Interactions 198
    S. A. Oprisan and A. Oprisan

Extensions of Self-Organizing Maps 204
    M. Trutschl and U. Cvek

Cognitive Radio-Genetic Algorithm Approach 215
    Y. B. Reddy

Medical Imaging and the Human Brain: Being Warped is Not Always a Bad Thing 225
    J. C. Patterson II

Do Ecosystems Ever Converge? Evidence from Faunal Size Distributions of Late Miocene North American Mammals 233
    W. D. Lambert

Interdisciplinary Social Science: An Example of Vertical and Horizontal Integrative Strategies 244
    S. Durlabhji

Conference Summary and Closing Remarks 253
    A. Ludu, N. R. Hutchings, and C. J. Brokaw

Author Contact List 258

Author Index 259

Preface

The present book of proceedings reflects the presentations and discussions at the first ISIS: International Symposium on Interdisciplinary Science, which was held at Northwestern State University in Natchitoches, Louisiana from October 6-8, 2004. The participants were scientists from academic, industrial, and government laboratories from the U.S.A., Canada, Germany, France and Japan. The subjects relate to the dynamics of complex and/or nonlinear physical, biological, chemical and social systems. The selection of symposium topics was based on our assessment of where the intersections between nonlinear physics, cell biology, mathematics, engineering, computer simulations, modeling, and medical research have a great potential to spark new discovery. The symposium was intended to stimulate researchers solving problems such as collective dynamics of populations at different scales from exotic nuclei to ecosystems, functional behavior of complexity, and nonlinear driven waves in living systems, etc. In all such examples of complex scientific questions, the answers are only possible through *interdisciplinary science*, which can empower researchers to deduce the fundamental behavior of matter and energy in both living and nonliving systems.

Although the contributions focus on a wide variety of specific problems, each paper is linked by interdisciplinary connections between important areas of cutting-edge research. The articles are arranged to reveal the overlapping (interdisciplinary) nature of topics that may have been classically treated as unrelated. For example, through these works, one can draw a common thread between the theoretical analysis of waves and solitons to the experimental approaches to modeling flagellar motion; and with a few more degrees of freedom, the reader can relate the social behavior of living populations (cells, organs, and societies) to instabilities within super heavy atomic nuclei. We challenge the reader to creatively search for additional commonalities between each of the articles in order to widen his or her perspective about such problems and to gain insights into approaches that might be applied in his/her own work.

January 5, 2005
Andrei Ludu, Nathan R. Hutchings, and Darrell R. Fry

Acknowledgements

The ISIS Planning Committee would like to thank all those who contributed to the funding, planning, and organization of ISIS. We appreciate your hard work, sacrifice, and dedication to the advancement of interdisciplinary science. Thank you.

Northwestern State University
Mr. Ted Jones
Richardson Technologies, Inc
The McGraw Hill Companies
Impressions by Dunagan
Aramark Catering
The Ramada Inn, Natchitoches
Palmer Air Taxi
Dr. Randall Webb
Dr. Anthony Scheffler
Dr. Priscilla Kilcrease
Dr. Austin Temple
Mr. Tommy Whitehead
Ms. Darlene Williams

The College of Science and Technology
The Department of Biological Sciences
The Department of Chemistry & Physics
Dr. Michael Bodri
Dr. Paul Withey
Ms. Melanie Bedgood
Ms. Leah Jackson
Ms. Louise Martin
Ms. Julie O'Bannon
Ms. Laney LaCour
Mr. Richard Manion
Ms. Melinda Parrie
Mr. Tracy Brown
The NSUAV/IT support staff

We would like to extend our gratitude to Andrei Ludu, Charles Brokaw, Charles Lindeman, Karsten Kruse, Christophe Zimmer, and Walter Greiner for serving as ISIS session chairs. We thank Nathan Hutchings for designing the cover art for the abstract book and the proceedings. We are grateful to the Northwestern State University Enrichment Fund (NEF) grant #NEF PD 04-05 R1-007 to D.F., N.H, and A.L. for providing the publication costs for this collection.

We are especially thankful for the fund raising efforts of Mr. Ted Jones. Without his generous contribution and support, ISIS would not have been possible. Likewise, we are very grateful to Ms. Carla Magness from the McGraw Hill Companies and Mr. Eric Lejune, Mr. Dan Williams, and Mr. Tim Richardson from Richardson Technologies, Inc. for their generous financial contributions to the event.

ISIS Welcome

Andrei Ludu

Interdisciplinary Experimentation and Scholarship (IDEAS) Program
Department of Chemistry and Physics.
Northwestern State University of Louisiana. Natchitoches, Louisiana 71497

Thank you Dr. Scheffler for your introduction. Let me quote here Dr. Vischenevski's (University of Georgia) favorite remark about speeches. He used to say: "A speech is like a lady's evening gown: should be long enough to cover the topics, but short enough to draw the attention."

Ladies and Gentlemen,
My dear colleagues,

On behalf of the IDEAS Program I would like to say good morning and thank you all for your participation in this event. It is a great pleasure and honor for me to open this 1st International Symposium on Interdisciplinary Science, ISIS 2004 in this lecture hall where NSU has accepted to welcome us. We would like to thank all participants for their efforts to attended ISIS by all means, at the risk of temporarily abandon their research, their teaching, or their administrative work. Thank you for being here.

We have for this first year a good number of exceptional participants providing highest scientific quality. So we hope it will be a very stimulating 3 day event, and it will allow many exchanges between the participants. Actually that was the idea: not only is the interdisciplinary approach a new and efficient tool for modern science research, but it is also necessary for everyone that is very focused on a specific activity, from time to time, to have a "look around" for creative breakthroughs and cross disciplinary oversights - for more integrated and refreshing ideas from other fields.

We know some fundamental facts about science, and we all want to know the answer to fundamental questions like: who are we, where are we coming from, and where do we go? In one question: how is it possible to generate adaptive complexity from simplicity?

In the process of doing science, one realizes that science doesn't actually belong to someone specifically. It just moves from mind to mind, from hand to hand. We take it over from someone else's mind or hands, leaves own finger prints, foot prints, mind prints on it, and pass it forward to other mind. It's like trying to rebuild some original primordial cosmic flower vase, somehow created in a very complicated, multi-dimensional manifold, intricate, fuzzy, and topologically twisted, and later on…broken. Out of the debris, we are constantly trying to re-build it, and sometimes we succeed to construct a perfect vase, completely closed, nice…but smaller or simpler than the original one. On the top of it, we (or our inter-disciplinary collaborators) find some debris left. So, we must have the courage to break the vase, and start a new one from the beginning, one larger, more complex, based on the smaller one, but aiming towards the primordial prefect original. In that, there are 2 major tendencies:

1. The "ancient Greek" tendency, based on the unity of knowledge, like: knowing nothing about everything;   and
2. The modern tendency, based on super narrowing the field, like: knowing everything about nothing.

The first tendency, generated by the Aristotelian model of science, tries to answer in a logical deterministic way to all natural issues. It propagated until the Renaissance, towards the ideal *hommo universalis*, Leonardo da Vinci-like scientist. The second tendency, the present way, is a reaction to the constantly increasing value of the ratio between the volume--and the degree of sophistication--of information over human and computer capabilities.   However lately, we notice an emerging revival of the Renaissance scientist, the scientist interested in a multi-disciplinary criticism, in nexialistic dialogues - the inter-disciplinary scientist.

As I was going to St. Ives,
I met a man with seven wives;
Every wife had seven sacks,
Every sack had seven cats,
Every cat had seven kits.
Kits, cats, sacks, and wives,
How many were going to St. Ives?

This is a good illustration of an interdisciplinary approach in education: rhyme and math. Another illustration can be provided by the story of Archimedes, the king of Syracuse, the hypothetic gold manufacturer, the bath tub, and the water. This could be another example of generating complexity from simplicity, through the question: who was the genius: Archimedes who discovered the concept of density, or the king who provided the job and the grant?

In that we invited people from all disciplines, and everybody will speak about everything. Because life is such an important, intriguing and exciting topic, many scientists will speak about life sciences; because physics is the closest logical science to biology we have here a lot of physics talks; and because chemistry is for these two, what the bath tub plus water was for Archimedes, there are also talks in chemistry. And so on…

Among the complex scientific problems there is one of particular interest, namely the dynamics of systems with free boundaries. I can give you lots of examples from the shape of nuclei, to cell division, from neutron star nonlinear tides creating gravitational waves patterns, to population growth, from birds and motile cell swimming to city dynamics, CA or vortexes in clouds. And through these models, we know that by starting from a random distribution of patterns, some units may grow faster than others, and the bigger ones try to swallow the smaller ones and later becoming unstable and break up into parts, or many smaller ones join into a larger one, etc. In one word: the nonlinear balance between competition and cooperation. However, sometimes there are exceptions to this dynamic law. For example, when a small center begins to grow up, and grows, and grows more. It's like in the Tom Thumb story when the bad giant asked him to mow and harvest an infinite field of wheat. So, Tom Thumb found himself tiny and alone in front of this infinite problem, to be solved in one day. So what did he do? Well, the story says he just started to mow, and to harvest and to eat the wheat, and did so for a while, until he identified himself with the wheat, until he became himself wheat. And then, the field wasn't anymore infinite for him. This is again a signature of complexity through simplicity.

One necessary premise to nurture a smaller unit into a larger one is to have internal coherence, to be fed, and to be backed up from inside. It happened, and this is rather the exception than the rule. Internal coherence was destined to halt here at NSU, yet I managed to find it, as a physicist in our IDEAS program together with Nathan Hutchings in biology, and Darrell Fry in Chemistry. The appropriate size of this university and the exquisite support of the administration fed it from inside. And the result is this symposium.

I'd like to mention and to thank Dean Austin Temple, the vice-president of academic affairs, Dr. Anthony Scheffler, and our president, Dr. Randall Webb for their generous support and confidence in the interdisciplinary program. It's my great pleasure to mention here the exquisite and professional help of Tommy Whitehead, and Theodore Jones who secured the primary financial support. It is a particular pleasure for me to thank Ms. Louise Martin, assistant to the president, Ms. Melinda Prarrie and Mr. Richard Manion from Space Science who exquisitely helped us with everything, Ms. Leah Jackson and the whole NSU News team, our students, and last but not least the heads of our three departments.

That's why we devote our energy to build this program (IDEAS) and this ISIS symposium at NSU towards a multi-disciplinary center like the well-known Santa Fe Institute in New Mexico. In that, it is a great honor and pleasure to have with us this morning Professor Murray Gell-Mann who will speak to you in a moment. We are very grateful to him for his attendance. Murray Gell-Mann is Distinguished Fellow of the Santa Fe Institute, and author of the popular science book, The Quark and the Jaguar, Adventures in the Simple and the Complex. In 1969, Professor Gell-Mann received the Nobel Prize in physics for his work on the theory of elementary particles. Besides being a Nobel laureate, Professor Gell-Mann has received the Ernest O. Lawrence Memorial Award of the Atomic Energy Commission, the Franklin Medal of the Franklin Institute, the Research Corporation Award, and the John J. Carty medal of the National Academy of Sciences. He has been awarded honorary doctoral degrees from many institutions, including Yale University, the University of Chicago, the University of Turin, Italy, and Cambridge and Oxford Universities. In 1988 he was listed on the United Nations Environmental Program Roll of Honor for Environmental Achievement (the Global 500). He also shared the 1989 Erice "Science for Peace" Prize. In 1994 he received an honorary Doctorate of Natural Resources from the University of Florida. In closing, I extend warm greetings to all of you, along with my best wishes for a successful and fruitful meeting, expecting from 'ya'll', a very productive dialogue.

ISIS Opening Address: Complexity versus Simplicity

Prof. Murray Gell-Mann

Santa Fe Institute, Santa Fe, New Mexico 87501-8943 USA

Abstract. There are a lot of different concepts to cover all the meanings that we attach intuitively to the word complexity, and to its opposite simplicity. There is one kind of complexity that corresponds best to what is meant by the word complexity in ordinary conversation, and in most scientific dialog. It's what I call *effective complexity*. Roughly, *effective complexity* refers to the length of a very precise description of the regularities of an entity. Not the features that are treated as random or incidental, but the features that are treated as regularities.

Keywords: effective complexity, randomness, phenomenon, simplicity, order, chaos
PACS: 80.0

EFFECTIVE COMPLEXITY

Complexity does not mean randomness. For example if we take a sheet of paper with a lot of dots arranged randomly that is not complexity, in fact it's rather simple. A complex novel, for example, would have many different characters, many different scenes, many different sub-plots, each of them taking quite a long description. Here is Snoopy's take on that. Sally says: "That too many characters in the book, too much going on… I can't keep track of them all." Snoopy agrees, Snoopy says: "I like a book when it's only one character and nothing happens to him." So that would be a simple novel, not a complex novel. The U.S. Tax code is complex. Every rule in the book is a regularity because it's a rule, and there are a great many of them. The U.S. Tax code fills up a large heavy book.

We also see complexity used in the world of advertising. There is an advertisement for a cognac, a brandy. Both the brandy and the young lady shown in the picture are labeled appropriately complex. This is a very good illustration of what people in the advertising business mean by complex, it's slightly sexist unfortunately but here's the young lady rated according to a number of characteristics, instigator, muse, daughter, klutz, accomplish, gardener, slow kisser, journalist, optimist, pessimist, flirt, insomniac and in each one she's rated numerically so there are a great many different traits and a rather lengthy description of the measure of each one. Clearly she is complex and presumably the brandy is complex as well, but you see that they mean the same thing that we mean by effective complexity.

You may notice that I'm defying convention here by not wearing a neck tie. But I brought along a whole packet. Suppose we look at the patterns of these various ties. Here is one that is relatively fashionable two or three years ago among people who

CP755, ISIS: International Symposium on Interdisciplinary Science
edited by A. Ludu, N.R. Hutchings and D.R. Fry

wear ties. You see it is very simple, especially for those of you in the back. You see a rather coarse gray version of the tie, a dark blue stripe, a thin white stripe, a red stripe and the same pattern repeating over and over and over again. So it could be described very simply and in a sort description. For those of you in the front, perhaps you can see inside each red stripe there is some thin blue lines and one red seam. And inside the blue there is also some lines with a red thick. That makes the description slightly longer but only very slightly. It's still a simple neck tie. We can look at one that's somewhat more complex. This one has a chain and some other decorations besides the stripes and a little bit more color, but is relatively simple, too. This one it's rather simple, especially for those of you in the back, and needs a very short description. Now here's one that's genuinely complex, this was designed by Jerry Garcia, when he was still a visual artist. See now, we take a long time to describe adequately the regularities of this tie. So this is definitely a complex one. Now you notice we've concentrated on the pattern of the ties and asked whether they're simple or complex but we've been ignoring the soup stains, the wine stains, the milk stains and so on, on the ties. We just looked at the patterns. But how do we know what is important? Is it the pattern, or the stain? Suppose you're a dry cleaner, then you're probably interested mainly in the stain and not very much in all the patterns. So what is the regularity it depends some how on the kind of the judge. A judge of what is important and what isn't important, not directly a judge of regularity and randomness but a judge of least what is to be treated as important and what is to be treated as unimportant. That judge doesn't have to be human, doesn't even have to be alive. But it has to be something to make it distinction between what is treated as important and what is not. We see that understanding regularity and randomness is the key to understanding the distinction between the simple from the complex. Everything we see around us displays this delicate interplay of regularity and randomness. The universe does, and as we get to smaller and smaller and smaller things in our experience, the same is true: delicate interplay of regularity and randomness, delicate interplay of law and chance.

We see that the distinction between the regular and the random is often context-dependent and can even be subjective. When we listen to music on the radio and there's a lot of static, a lot of noise on the radio we describe the static as random and the music as regular. But in the 1930s, at the Bell Telephone laboratories in New Jersey, Dr. Janski and his associate Dr. Bailey were instructed to try understand something about where static came from. And what they discovered way back then in 1930s was that there were important sources of static in the sky located in particular constellations in the night sky and corresponding places in the daytime sky. What were they? Well, they were signals from distant stars in a particular distant galaxy, and that is what gave rise to radio astronomy. So static is not just noise, it contains important regularities, actually it is the basis for the whole science of radio astronomy. And we saw in a similar way that neck ties can exhibit regularities not only in the pattern but also in the stain. Some people might be more interested in the stains then in the pattern. Now in science we search for regularities. Natural phenomenon tend to obey laws and as Sir Issac Newton said "It is the business of natural philosophy to find them out" to find the little regularities in nature, the non-random phenomenon in nature, science is a search. In the 17th Century science was called usually natural philosophy. As in that sentence we just quoted from Newton. It was natural philosophy distinct from what we might call armchair philosophy. Natural philosophy involves comparing of nature, getting ones ideas by looking at nature, by making theories, testing the theories by observation of nature and so on. Constantly going back to the natural world to check on ones ideas whether they're right or they're relevant. This is the opposite of armchair philosophy.

There has been some significant controversies about regularities: Are there regularities of a certain kind or are there not? Take one example: for many years there were Wall Street, and else where in the financial world so called chartists who made charts of prices in financial markets prices of stocks, for example. Graphs of stock prices versus time or prices and commodities versus time or whatever. And then claim they could from the fluctuations in these graphs deduce something about future fluctuations and thereby make money for themselves or their clients. Of course they could not deduce the future regularities. All they could do would be to supply probabilities for certain kinds of fluctuations in the future. But the point is they could extract non-random results and apply them, or so they claim. That there were lots of famous neo classical economists saying that the fluctuations around fundamentals were nothing but a random walk. In fact one of them wrote a book called "A Random Walk Down Wall Street". After a random walk you can not extract any useful information: it's just random. By looking at the historical data, past behavior of markets, one can see that there is a non-trivial correlation between price fluctuations of one time and price fluctuations of another time. So from a historical record you could show that the random walk down Wall Street people were simply mistaken.

What gives rise to complexity, where does complexity come from in nature? Well, the fundamental law of physics which govern the behavior of all matter in the universe and of the universe itself. They seem to be simple. You can write them down very concisely as far as we know. Of course we do not exactly have these laws yet, but we are getting closer, and all the indications are that we will get the laws that govern the behavior of elementary particles, the basic building block of all matter, and the boundary condition at the beginning of the expansion of the universe. When we will get these two fundamental laws they will be simple. They will be describable in a brief message. The first thing is the unifying quantum theory of all the elementary particles everywhere in the universe. And the other one is the initial condition of the universe beginning, near the beginning of its expansion around 13 billion years ago. As we said both of these laws seem to be coming out simple, so where does effective complexity come from? It doesn't come, we believe from the fundamental laws. Well let's ask this question, if we know the exact fundamental laws of physics, the theory of the elementary particles and the initial condition of the universe, can we then in principal (not of course in practical) predict the behavior of everything in the universe? A hundred years ago many people would have said yes. If you really know the laws and you know the initial condition, you can predict everything in principal. But it is not true, because we know that the universe is governed by quantum-mechanical theory, and in quantum-mechanical theory all you get is a set of probabilities for various alternative histories of the universe. You do not get a prediction of a particular history of the universe but a set of probabilities for many different alternatives. So the fundamental laws are probabilistic and not fully deterministic. And the history of the universe is co-determined by the fundamental laws which we believe to be simple and an unimaginable long sequence of chance events which we call accidents which are governed by probabilities. There are a number of different outcomes for each accident or chance event and each one has a probability. But you do not know in advance which one you are going to get.

Let us take a simple laboratory example. Take a radioactive nucleus that emits alpha-particles that is helium nuclei. In advance of the emission, before the radioactive nucleus disintegrates, you have no idea in what direction the alpha-particle will come out. After it comes out of course you would know, but before hand there is no way to tell. In this case the probabilities are all equal. That is not always true, but in this example the probabilities are all equal, every direction is equally probable, all directions are equally probable. Only afterwards can you specify in what direction the alpha-particle went. That is just one simple example of what accidents are chance invested. Theorists may be coming close to a description of the fundamental unified theory of all the elementary particles. That is the research on a particular model of superstring theory and search for a possible generalization which is named m-theory. This kind of research may be coming very close to the unified theory of the elementary particles and their interactions, in other words, a complete theory of the behavior of matter in terms of its fundamental constituents. Already this body of theory based on so-called superstrings has scored a remarkable triumph. Some of this was done actually in my group from Cal Tech, when I was still there, although I didn't do it, but I brought the people there who did. And what they found was the superstring theory, which may be a part of the true underlined theory of elementary particles. In the superstring theory one can derive Einstein's famous general-relativistic theory of gravitation, and further more, one can derive it within quantum-mechanics and without the absurd infinite corrections that have plagued all previous attempts to reconcile general relativity with quantum-mechanics. Here they are fully reconciled and there are no infinite corrections. Everything looks perfectly ok. Well that triumph seems to be the indication that the people who are looking for the correct unified theory starting from superstring theory maybe on the right track.

The other fundamental law is the initial commission of the universe near the beginning of its expansion around 13 billion years ago. In that, it is a simple initial condition, and that is ultimately responsible for the so called arrow of time. That distinguishes the past from the future. If you see a film of a lot of little bits of egg, yolk, white, shell and so on. Starting out scattered around and gradually assembling to form an egg. You would conclude that was a film being shown backwards. Nobody has ever seen a situation where a lot of little bits of egg, yolk, white, and shell assemble to form an egg. But many times you dropped an egg and seen it come apart into little bits of yolk, white, and shell, and you can easily tell which is the movie going forward and which is the movie going backwards. That's the so called arrow of time or at least one of the arrows of time. It's related of course to the second law of thermodynamics, a little fancy expression which means that the average disorder has the tendency to increase in a closed system. So the egg breaking is an example of increased disorder. The egg reassembling would be an example of disorder decreasing in a closed system, which we don't see. So the second law of thermodynamics depends crucially on this orderly state at the beginning of the expansion of the universe 13 billion years ago. Now you can say you observed the phenomenon of the egg in the laboratory, or in your house. What does it have to do with the universe? Well, you can trace all these phenomena in different parts of the universe, back to a fundamental reason why time goes forward, and that has to do with the universe. The universe goes forward in time, therefore the various parts go forward in time, therefore people for example live forward in time, you can not remember what happened tomorrow, and so on. Therefore when you drop an egg, you drop it forward in time and not backwards. So the arrow of time for the universe is ultimately responsible for the little local arrows of time that we see in parts of the universe.

Now, as we take into account all these probabilistic situations, and all these accidents, all these random events we see the alternative possible histories of the universe to be represented as a branch in a tree. You have one accident with various possible results, different probabilities, and another accident with different probabilities, another accident with different probabilities. This way you get a branching tree for all the alternative histories of the universe with a probability, with probabilities in all the branches and with a probability for each history in the tree. Jorge Luis Borges had a story called "El jardin de senderos que se bifurcan" ("The Garden of Forking Paths") it's about a man who left behind a mysterious garden with some message involved, people couldn't figure out at first what it was but then they realized it was a map of the alternative histories of the universe represented as forking paths in the garden. Now some people refer to the still sought unified theory of the elementary particles and their interactions even when it's supplemented by the initial condition of the universe and the other fundamental law. They refer to it as the theory of everything. You may have heard that expression used often by very distinguished scientist, but it's a stupid name, because most things have to do with accidents. There is very little that depends only on the fundamental laws. Elementary particle physics and cosmology are the two sciences that depend on the fundamental laws, but all the other sciences depend on accidents as well. Geology for example depends on the existence of the solar system and the planets, the various histories of the planets all depending on numerous accidents. Biology depends on even more accidents, all the accidents that have taken place in the course of biological evolution in the last 3.9 billion years or whatever it is since life started on Earth. Even chemistry, some of which is derivable directly from elementary particle physics, depends to some extent on accidents because you have chemistry only when conditions of temperature and pressure and so on are such to permit atoms to exist, atoms and molecules then you get chemistry. In the center of the sun for example there is essentially no chemistry, it's too hot. You do have nuclear physics in the center of the sun, but not much chemistry. So, all the other sciences, except elementary particle physics and cosmology, depend on accidents. Think of all the accidents that produced the various people in this room. Some nasty little quantum fluctuation that gave rise to our galaxy long ago billions of years ago, and then gave rise to the evolution of all the stars in the galaxy, including this very ordinary star that we call the Sun. And along with the sun, when the matter condensed, a lot of planets were formed, too, and all that depended on a large number of accidents. One of those was the third planet from the sun, the Earth. And it underwent all sorts of changes, many of which depended on accidents. Then life came about in a particular way on Earth almost 4 billion years ago. But that involved accidents too then, in the course of biological evolution there were enormous numbers of accidents that took place, all though of course things like natural selection created regularities among those accidents, but they were still accidents. Think of all the other things that had to happen to produce us, two people meeting, sperm meeting egg and so on in a particular manner determining our various genomes, but it's not only the genomes that determine the person. Identical twins have the same genomes, but they're different people, different experiences in the womb, different experiences in early childhood, and different experiences in adulthood, and so on. So, there were enormous numbers of accidents that produced all of us the way we are today.

Some of the chance events, accidents, or branching in the tree produce more future regularities than others, in finite regions of space and time, and we call those *frozen accidents*. They must be the main source of effective complexity because the fundamental laws are thought to be simple. So it is the frozen accident that creates most of the regularities we see over and above the fundamental laws. Take some examples of life on Earth, many right handed molecules claim working roles while the corresponding left handed molecules do not. For example, left handed amino acids and right handed sugars are very important in biology where its' mirror image molecules play very little role. Now it has been possible to explain why the sugars would all have the same handedness and why the amino acids had the opposite handedness. People have succeeded in giving theoretical explanations of that, but why it does it a particular way that it does it, left handed amino acids and right handed sugars, nobody has explained that, it seems to be an accident. People have tried very hard to contribute it to the fact that for matter is opposed to antimatter, that we interact with left handed but they never succeeded in doing that. So, it looks as if this is an accident, an accident of the earliest form of life on Earth, and one that has been propagated through the whole process of biological evolution and all living things, the characteristics of all living things today. We can also look at accidents in human history, and ask this what consequences they had, where they important frozen accidents.

Now of days, so many distinguished historians are much more tolerant than they used to be, asking what if so and so had happened instead of so and so. It's not something you can easily test of course, but that kind of speculation has proved to be interesting and historians are more and more tempted by it. They call it Counter Factual History or Contingent History and many of them like to talk about an incident that took place in 1889. While Buffalo Bill's Wild West Show was touring Europe and made a stop in Berlin, the star of the show was the famous female marksman, Annie Oakley. Annie would ask for a male volunteer from the audience to smoke a cigar and have her stand there with the cigar in his mouth while Annie shot the ash off the end of the cigar. Normally there were no male volunteers, then her husband who was himself a distinguished marksman, would step up, he would smoke the cigar, leave some ash at the end, and have his wife Annie shoot the ash of the end of the cigar. But on this one occasion in 1889 in Berlin, there was a male volunteer from the audience, the Kaiser, William the second, who had just ascended the throne a year before on the untimely death of his father. And there he was all dressed up in a uniform, this very elegant uniform, he took out an expensive Havana cigar and clipped off the end and took off his band, lit it and waited for some ash to accumulate at the end of it and then stood at attention on the stage waiting for Annie to shoot the ash of the end of the cigar. Well Annie was worried, she had been drinking heavily the night before, and her husband was one thing but the Kaiser was another. But it all went off all right, she shot the cigar and not the Kaiser, we know that because otherwise we would have read about it in history and other places. But what if she had actually killed the Kaiser? William the Second was a difficult character as most you know. He canceled the reinsurance treaty with Russia, he engaged in naval competition with Britain, which Bismarck advised against, fired Bismarck, who was trained to construct the stable order in Europe, and his work led in many ways to the First World War. Would the First World War have been quite different if he had been killed in 1889? Would it perhaps never have happened… we don't know, but people can speculate about it. Anyway it may well be an example of a very important frozen accident. And historians argue about whether the results of major accidents like deaths of prominent figures are eventually healed by grand historical forces or whether they continue into the future to have very important effects for a very long time.

In any case since the fundamental laws are thought to be simple most effective complexity anywhere in the universe can be traced to frozen accidents. And as time goes on in many domains of experience, we see that entities of greater and greater effective complexity come into being. Why is that? Well, any given entity can become less complex. People die, civilizations die, and they become less complex. But the point is that the envelope of effective complexity keeps getting pushed out as more and more complex things come into existence, even though many things decline in complexity. The boundaries of complexity keep being expanded. The envelope of effective complexity is pushed out when the accumulation of the results of frozen accidents outstrips the forgetting or erasure of the results of frozen accidents. And as time goes on you get this accumulation. It doesn't contradict the famous second law of thermodynamics that average disorder in a closed system tends to increase because mechanisms of self-organization can cause local order to increase while order is decreased elsewhere, like for example your refrigerator makes ice cubes which are very regular and very ordinary but if you go around to the back of a refrigerator there's a huge amount of heat coming out which is very disorderly and which makes up for the order that's being created in the freezing compartment, more than makes up for it. That's how local order can create spiral arms of galaxies, different forms of snowflakes and so on. As many theories believe, protons disintegrate with a half-life of 10^{33}-10^{34} years. Then, after the 10^{36}-10^{37} years there'll be almost no matter left of the kind that we know, made of atoms and molecule, and so on, instead there'll just be a soup of electrons, positrons, photons, neutrinos, and antineutrinos. Very few regularities as they are now conceived. At this point the envelope of effective complexity might shrink, but this is not something to worry about right away, 10^{36} years is a long time.

Complexity versus simplicity

Above, we talked about effective complexity, the length of a very concise description of the regularities, and about where it comes from, since the fundamental laws of nature seem to be simple. And we found it comes mostly from accumulating accidents, accumulating frozen accidents. What is a lengthy concise description? What do we mean by regularities? Is there some mathematical way of distinguishing regularities from random futures? We pointed out that everything around us exhibits mixture of regularities and randomness including the whole universe. And the job of the scientist is to locate the regularities and understand how they interact with randomness. Now since we're in the age of computers we'll have to talk about the bit string, the language that computers use. So a bit string is a string of zeroes and ones, that's what information is these days. And if we're talking about some entity and we want to talk about its effective complexity then we represent it by a bit string. And in order to do that we have to specify various characteristics of the entity we were talking. One thing is the level of detail, which we are calling coarse-grained. In physics it's usually called coarse graining at which it's being described. I think the phrase coarse graining probably comes from photography, you have a very grainy photograph then you're seeing only certain features out of many, if you have a much finer grain photo you're seeing much more features and so on and so forth. Besides that we need the language in which the entity is being described, and then most important--and often neglected--, the knowledge and understanding of the world that is assumed.

Imagine for instance being an explorer running across a new hitherto un-contacted Indian tribe in the Amazonian jungle. It is un-contacted, but it speaks a language that's the same as that of a neighboring group that has been contacted. And you have actually learned that language, so you can talk to these people in their own language even though they have not encountered people from the outside directly before. Now your job is to explain to these Indians in their own language what a tax manual's mutual fund is. You would have to give a lot of background and that is the idea here, the knowledge and understanding of the world is assumed very much influences the characters of the description. Finally, there is the system of coding from the language from which the entity is being described to bit string. Given all that, you can represent the entity by bit string and it is one of many different possible bit strings and one of many possible bit strings of the same length. Now this quantity called Algorithmic Information Content of a bit string or the like of the entity that the bit string is describing. It's the length of the shortest program that will call a given universal computer U to print out the bit sting and then halt. Universal computer of course is either capable of doing any calculation, and has infinite memory, or more plausibly, it has the capacity to create memory when ever needed in order to solve problems. So it can solve any problem, although it may take a very long time. It's something of an idealization in universal computers. When I say solve any problem, it means solve any solvable problem. And of course often it takes so long that it's not practical. So here it is again, they talk about the entity, e, and the bit string, s_e, and we call the A.I.C. the Algorithmic Information Content $K_u(s_e)$ or $K_u(e)$, where K means the Algorithmic Information Content and the u is the particular universal computer. As this gets larger and larger and larger, it becomes more and more independent of the particular universal computer that's in use, so it kind of covariance problems. Now our job is describing effective complexity. Remember, effective complexity was defined roughly as the length of a very concise description of a regularity, not the features treated as random or incidental. So our job is to split K, the A.I.C. of the entity, or the bit string that describes it, into two pieces. The A.I.C. of the regularities and the A.I.C. of the rest of the features treated as random or incidental. Then first part would be the effective complexity, the second we can call the random information. Now we are using A.I.C. here as a way to represent the idea of concise description. What do we mean by the length of a concise description? Well, we can say the A.I.C. Otherwise we might deal with a lot of redundancy. For example, in my book I quote the story of a school teacher, who assigns to her primary school class writing a three hundred word essay on something that recently happened in your household. Now let us assume that there is a student who did what I would have done in those circumstances years ago. Which is to fool around outside all weekend and finally scribble something on Monday morning to please the teacher. So what this ridiculous student wrote was yesterday the neighbors had a fire in their kitchen and I leaned out of the window and yelled: "Fire, fire, fire, fire…" Now that is an example of something that can be compressed. If the teacher had not insisted on a three hundred word essay, he could have said I leaned out of the window and yelled fire two hundred and eighty-three times, the equivalent, but much shorter. So in the same spirit, we're using Algorithmic Information Content to represent the length of a concise description of what we're talking about.

Let us give a couple of examples at both ends of the spectrum so to speak. Lets take a bit string that is perfectly regular, like 1111111111 or 00000000 it has very little A.I.C. because it's so regular. The regularities have very little A.I.C., all they have to say is it's all ones or it's all zeroes and the length. The length of the bit string is about the only thing that is important or otherwise the regularity takes a trivial time to describe, all ones or all zeroes. So it's effective complexity if very low, the effective complexity is the A.I.C. of the regularity. At the other end of the spectrum, we take a bit string with no regularities, an incompressible string. It has maximum A.I.C. for its life because it's incompressible. But the A.I.C. of the regularities is again very low because it doesn't have any regularities, be about just the length and nothing else. So in both cases, at both ends of the spectrum so to speak the A.I.C. of the regularities, the effective complexity is very low. So you can draw the diagram in Figure 1.

The A.I.C. is very high in the middle, that's a string with no regularities at all. Close to the origin O there is a string with very low A.I.C. just all ones or all zeroes. At both ends the A.I.C. of the regularities, the effective complexity has to be very small, very close to zero at both ends. In the middle it can get bigger. So we see that effective complexity can be there in a considerable amount only in the middle, only in this intermediate range between perfect order and perfect disorder. That region is the region where you can have a lot of effective complexity. I say that here, can be high only in the intermediate region between perfect order and perfect disorder. It is not a particular place; it is a sort of anywhere in here.

Now, how do we represent the regularities in the first place, what do we mean mathematically by regularities as opposed to features that are treated as random or incidental? Remember that there is always a judge, not a necessarily human, not necessarily alive, but some kind of judge that makes a distinction between the important and the unimportant. Only then do we define regularities and randomness.

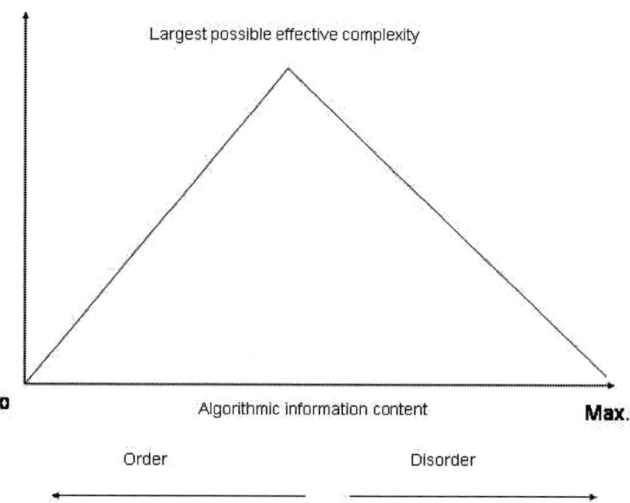

Figure 1. Algorithmic Information Content (A.I.C) and effective complexity

Now we can borrow from statistical mechanics, this way of treating regularities. Regularities of an entity e are represented by embedding the real entity in a set of things, an ensemble Ẽ the rest of which are just imagined. Our entity is real, we talked about it, describing it, but we bury it in a mass of things, all the rest of which have just been made up for the purpose. And we assign probabilities to all the members of the set, making it what's called an ensemble. An ensemble is just a set with probabilities assigned for all members of the set. Here this is written out, to describe the regularities attributed to an entity, embedded it conceptually in a set of entities with probabilities for the member that makes it an ensemble. It's very important though that the entity itself should be a typical member of the ensemble, otherwise the ensemble is hardly describing the entity we're talking about. It has to be a typical member, that is one does not have an abnormally low probability for that set. The probability distribution that will reflect the regularities, that's the way we do it in statistical mechanics, physics, and it seems to be something that quiet generalizing.

This idea of burying the thing we want to talk about, embedding it in a set of things, the rest of which are imaginary and imagined. It's very familiar in the arts, for example in fiction or drama and in many kinds of poetry. What we're doing is constructing an alternative world that we set along side our own. And if there are many works of fiction, many dramas, many poems then it's a whole set of different worlds. You could have imaginary families, but sometimes the novelist or the draftsman makes them so real, that we debate what they would do in various conditions and so we feel we know those people, they represent additional families beyond the ones we actually encounter in the real world. The characters represent additional people that we don't find in the real world, but we find in these books or in the dramas that we see them perform. So it's not an unfamiliar idea even outside of science to understand more the regularities of something by embedding it in a set of things, the rest of which are made up. So how do we do that? Well in statistical mechanics, we always do that. For example, suppose you have a sample of gas, sample of oxygen gas for example. It is a sizable sample, microscopic sample; it may have 10^{24} molecules of oxygen. 10^{24} is a lot: a trillion trillion, trillion trillion molecules in this sample of gas, to specify exactly the state, even classical. Classical you don't have the uncertainty principal to worry about but you have to specify the position in three dimensions of every single molecule. So that there are new molecules in the sample, that's six new coordinates that we have to specify. Up to some accuracy they were good depending on the coarse graining. Where could we possibly get all that information? How could we acquire that information, even with the crude approximation, for the values of the positions and momentum? It is an unbelievable amount of information. If we ever acquired it , where could we store it? If we ever stored it, how could we read it? And if we ever read it, how do we make use of it? So, clearly we don't try to do that. What we do in fact is, in Statistical Mechanics is embed this sample of gas in a huge set of samples, the rest of which are imaginary. And we specify some probability distribution for this set and that's our description then of gas. In this way we can define temperature, we can define pressure, we can define a whole bunch of other variables if we want them and describe the system by specifying these properties only. What we can do then in general, that is a very general procedure. So here I have written it again in Fig. 2. Regularities of an entity can be described embedding e in an ensemble E, as a typical member, and the various elements of this ensemble exhibit the variations in Hypothetical Entities sharing the regularities attributably.

For example, the various states of the gas at certain temperatures. Then we can look at the A.I.C. of the ensemble of the probabilities distributions $\{p_r\}$ which depends for example on how many parameters we are specifying. If it is just temperature and pressure this would be rather low. If we are specifying a lot of things it will be high. So it is a length of a very concise description of the ensemble, in other words, the members and their coarse-grained probabilities. Then we define Y to be the A.I.C. of the ensemble, that's our way of giving a very concise description of the length. So it is a candidate for the effective complexity. Why do I call it a candidate? Because we have not yet specified the ensemble. When we do, Y will be the effective complexity. Now we think the other quantity we are interested in is the ignorance or information or intrepid or whatever we want to call it. It has various names. If you have various outcomes of things and they all have the same probability then that probability is one over the number of possible outcomes, $1/\{p_r\}$. The number of bits of information in it is $\log_2(1/p_r)$, the number of possibilities. So if you have three flips of the fair coin, for example you have eight possibilities that are the same probability and the number of bits is three. If you have five flips of the fair coin then the number of bits is five and so on and so forth. It's the law of the base 2 of number of possibilities and therefore the law of the base 2 of 1/probability, the common probability. Now if the probabilities are different, then we have the average. I, the ignorance, is the average over the probabilities of the law: $I=\Sigma \, p_r \, \log_2(1/p_r)$. It is ignorance, but can also be given as information. Ignorance and information are exactly the same thing. The ignorance of a letter is exactly the same thing as the information you derive from when you open a letter. It just depends whether you are talking about the time before you open the letter or the time after you open the letter. Before you open it, it's ignorance. After you open it, it's knowledge, information. So ignorance or information: fortunately they both begin with the same letter I. Now of days people are experimenting very much with alternative formula patterns, but I am not going to go into that here today. Mostly this is the right formula, it is certainly the one used in campus. So we have then two quantities we're concerned with, the A.I.C. in the ensemble, and that is the effective complexity once we choose the right ensemble, and the other is the ignorance or information, I, which will be the random information once we have chosen the correct ensemble (Fig. 2).

Now suppose that what we are doing is looking at a set of data and trying to make up a theory to account for those data, that is something many of us do, those of us that who at least part time theoreticians do that all the time. Now to explain a given set of data, we can go to two extremes, though we try to avoid those extremes, one is to make the theory have a huge number of parameters and a huge number of bells and whistles so that it really specifies those data and not any other data. Well that is very good in zeroing in on the data, but it is not very good as a theory because it has a huge number of adjustable parameters… used to say that way you could fit in an elephant. So we do not want that to be, we do not want the Y to be to large. If the ensemble is very complicated with lots of bells and whistles and lots of parameters, Y would be much too large. We also do not want I to be to large, because I is the ignorance, and we do not want a huge amount of ignorance. In other words we do not want the theory to predict the data that we see, but numbers of other sets of data at the same time. We do not want the theory to allow almost any set of data, which would not be very good either. But we can trade of between I and Y because what we can do if the theory has a lot of ignorance, we can add enough bells and whistles so as to get rid of the ignorance but then we have a huge amount of Y, which we do not want.

The other way around also, if we lots and lots of bells and whistles and extra parameters, we can get rid of a lot of them but often at the cost of increasing the ignorance, in other words not zeroing it very well on the particular data that we have in hand. Since there are all these possible trade offs, what we really need is to minimize the sum of the two. Try to make K=Y+I as small as possible. Later we can worry about the trade offs. Occasionally the theorist is in a very fortunate situation, a win win situation, where he's reducing both I and Y together. For example, James Clark Maxwell, when he formulated Maxwell's equation of the electromagnetic field was doing a great job both ways, he was simplifying the equation and he was also zeroing in better on the data, the electromagnetic fields. So occasionally we can do that, but also we have to make use of trade offs. In any case what we want to do is minimize the sum of the two, the ignorance plus the Algorithmic Information Content, plus the effective complexity. Now if the ensemble consisted of just the entity itself, we aren't burying it in the midst of a whole bunch of other things, we're just looking at the entity itself, one element, it's a one element ensemble, with probability one for that element and zero for anything else. Then Y would just be K, where K is the Algorithmic Information Content of the entity because that's all there is in the ensemble. So the Algorithmic Information Content of the ensemble is the same as that of the entity, it's just K, and I is zero in that case, because there is just one entity and that's all we have. So for the probability 1, p is just zero. Now it turns out that K, which is equal to Y+I in this very special case, is actually the minimum possible value of this of Y+I. We couldn't have a smaller value. But there are many different ensembles that have obtained that minimum least roughage. We have to decide which one to use.

So here we are, we have information or ignorance which is also proportionate to entropy, we have the Algorithmic Information Content of the ensemble which is the candidate for effective complexity and we have the, this is the candidate for the random information if we choose the right ensemble. This is the candidate for effective complexity if we choose the right ensemble. We talked about the sum of the two, which we can call total information. We want that to be its minimum and we already said what the minimum is, it's just K. So here's what we do, to choose the ensemble, we're finally getting down to choosing it, so we're actually describing our entity. We keep the total information equal to its minimum value, which is K the A.I.C. of the entity itself, and we maximize I and minimize Y, while keeping fixed particular other quantities beside this, the ones that are judged to be important. And what those are, are the average values, the ensemble averages of various quantities. For example, in statistical mechanics we usually keep fixed the ensemble average of the energy, and if we do that we define the absolute temperature. It's just one parameter in the Maxwell-Boltzmann distribution, E to the minus the energy, that's a constant and in that case the only quantity judged to be important besides I+Y is just the energy, the average energy. Fix the average energy, otherwise we maximize I and minimize Y and we get the maximum of both sides. We can, if we choose to treat a lot of things as important and keep them all fixed while maximizing the ignorance, the measure of ignorance, which is maximizing the entropy. Then we are proceeding just the way we do in statistical mechanics, we are maximizing the ignorance and keeping certain important quantities fixed. And we have the average energy held fixed, and we get this reduced distribution, the Maxwell distribution. So here is what we have in a diagram, the final diagram, describing what we are doing. I am plotting Y versus I, where Y is the candidate for the effective complexity, I is the candidate for the random information (the measure of ignorance). We want to hold Y+I=K, that is this straight line here called minus one fixing Y plus I equals K. As we go down to very small I and very large Y, this line ends in the allowable space of ensemble, one resolves here, this line one resolves here toward the maximum. So this is as far as we can go, we want to minimize or maximize I while staying on this line; we have to come down to this point. Provide we don't keep anything else fixed, but if we keep other things fixed like the average energy, the average this and the average that and so on, we can gradually move up this line further, further and further attributing more and more and more properties to our ensemble and knowing more and more and more therefore about our energy. The ignorance is being reduced, effective complexity is being increased, because we have a more and more complex description of the entity but it involves less and less ignorance, we go up here so. Now the problem with being down here at the bottom not specify nothing besides Y+I equals K, then maximizing Y and minimizing I going to this point. The problem is: this point lies very low. We have shown, many

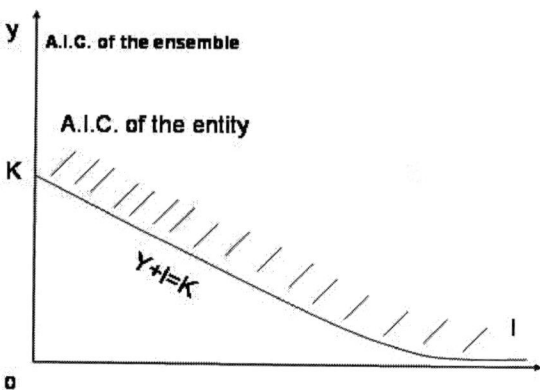

Figure 2. Ingorance, Information, and Effective complexity.

people have shown that this point lies very, very low, which means the effect complexity if very low here. Everything is simple, everything would come out simple, and it's not really satisfactory, what's the use of a theory of simplicity and complexity where everything is simple. So we do not want that, in fact we keep a number of quantities fixed and move up this line to some reasonable place like this. Now the Great Russian mathematician Kolmogorov, one of the people who invented A.I.C., fell into this trap. He called it the *minimum sufficient statistics*, but one of his students, David, who is now a famous professor of computer science at Boston University, told him that it is wrong: "not good everything simple", but Kolmogorov would not listen. So he kept talking about this minimum sufficient statistic here, where as in fact we need to keep a lot of important quantities fixed, while maximizing I and minimizing Y and keeping this on this straight line. We have to keep a number of things fixed and then we move up here. We have a more complex description than with one of less ignorance. So here we have the mathematics of describing effective complexity.

We just maximize the measure of ignorance while holding this at its minimum value and Y always comes out tiny, that is that point with the curve departs from the straight line, everything comes out simple. In one case out of the universe, it is ok. But in every other situation it is no good. Here we are, say this once more, just for emphasis. More regularities recognized, the entity is allowed more individuality instead of being treated merely a statistic, move up the straight line just a bit, your characterizing the number of properties of the entity not just say the temperature but the number of those. Now another way look this whole matter which may appeal to many of you, another way to look at Y and I as that in terms of the program, you remember A.I.C. was the length of the shortest program, it causes the universal computer to print out the bit string under consideration and that's all. But we can think of a program in two parts, a basic program and then another program that supplies the data for the basic program. We can divide it into two parts, a more fundamental program and then data that you fed into it. You can think of Y as the length of the basic program and I is the length of the program that feeds in data to the system. So we have a general program that can accommodate various data sets and I will describe that message to be condensed. For most people that's a more congenial way of talking about this. Now the judge can come up with all sorts of different ideas about what is important and what is not. Suppose you're a anthropologist for example, then the kind of complexity you're interested in is social complexity. How many roles are there in society, how many different professions are there in this society and over. How complicated are those professions, do they have lawyers in that society, do they have doctors in that society, do they have witch-doctors in that society. How complicated is the stuff the lawyers deal with, or the witch-doctors deal with. That's social complexity. And of course a lot depends on the data that are available. Lets take Mars for example, the planet Mars, before the voyages by NASA and then after them. In the old days there was an astronomer looking through a telescope, peering through a telescope hoping for a night of good seeing, when the boiling of the atmosphere wasn't so bad and trying to get a clear picture of Mars and understand what kind of features there were on the surface of Mars. The astronomer Perceval Loel, who worked in Arizona, but obviously his ancestors were from Boston, who claimed he could on nights of especially good seeing spot lines on Mars, straight lines. Now back in 1877 they have seen lines on Mars also, and he called them *canali*, channels in Italian. But Perceval Loel interpreted them as canal dubbed by intelligent creatures. How could you possibly get a straight line on Mars, if there were not intelligent organisms some how engineering these straight lines? They must be canals and there must be intelligent Martians. NASA photos even the early ones, which were not very fine grained, showed clearly that this was all imaginative; this was just the human mind seeking patterns, which it does as you know. Perceval Loel was just looking for a pattern, and he could see these straight lines, which the earliest NASA pictures show was not there. They were just pieced together with bits of other things. Now ever since the time of Linden Johnson, the Vice President of The United States, who has been interested in the space program and has been chairman for the space council and when Dan Quail became vice president, he assumed that role also and became very enthusiastic about manned voyages to Mars. And he was invited to open a NASA meeting on that subject, give a little talk at the beginning of the meeting. So somebody in NASA prepared a nice two page speech for Dan Quail. And the first page was about the old Mars, Perceval Loel, the canal, the intelligent Martians and so on. The second page was about all the things that NASA had discovered to be actually true of the surface of Mars. The old stuff turned out to be wrong but unfortunately Dan Quail lost the second page. So his entire speech was about the old Mars, the canals, the intelligent Martians and so on and so forth. You see how much difference greater course graining and greater fine graining can be. Now the latest wrinkle of course is astrobiology which is very exciting and has to do with what life or something like it would look like on other planets and evidence could possible be collected to bear about that. It seems extremely likely to me that there must be life on other planets, because there are so many planets in the universe. The number of stars is gigantic as you know and planets are not unusual. When I was a undergraduate, the usual theory then was that planets had to be created in a rare triple collision star, but very shortly afterwards scientist went back to the old time investigatable hypotheses of the late eighteenth century, which was a condensation of dust, the big condensation given to the star and the little condensation to the planet, very simple, that's what people believe today. And it seems to be very common, now that people are actually able to see large planets, discovering a lot of which of course theory predicted along time ago and when they are able to see smaller planets they will find a lot of others too. Life does not seem to be anything very special, once you have the right conditions as we did on the earth some four billion years ago, life presumably springs up, or something very life like, we don't know how to exactly define it and it's up to the astrobiologist to define it as that. So it is difficult but not crazy to think about what form, how different life could be. How much depends on accidents, which we discussed so much this morning and how much depends on the fundamental laws of physics. How much of biochemistry is rooted in physics and how much is the result of a historical accident. People at the Santa Fe Institute are very anxious, very eagerly pursuing that question and it's crucial for astrobiology. The most interesting question is about intelligence life, that is, if you define us to be intelligent, which I am not sure I would. And it seems to me there too, that there is likely intelligent life because we know evolution works toward more and more complex things, usually and solve more and more difficult problems, provided the relevant creatures don't kill themselves. So I think very, very, likely in an enormous number of planets in the universe, there are others that contain what we would call intelligent life. But Pogo said the last word in this subject, Pogo is the cartoon character called Kelly who lived in the Okey Ponokey Swamp on the border of Florida and Georgia. Along with a lot of other swamp creatures, alligators, bugs and so on. And one day Pogo was talking to one of the other swamp creatures and said, "Out there on some planet orbiting some other star, there maybe entities that are more intelligent than we are, we humans or on the other hand, maybe we humans are the most intelligent entities in the universe. Either way it's a mighty sobering thought.

Superheavy Nuclei – Clusters of Matter and Antimatter

Walter Greiner[*] and Thomas J. Bürvenich[†]

[*]Institut für Theoretische Physik, Goethe Universität, D–60054 Frankfurt am Main, Germany
[†]Theoretical Division, Los Alamos National Laboratory, Los Alamos, New Mexico 87544, USA

Abstract. The extension of the periodic system into various new areas is investigated. Experiments for the synthesis of superheavy elements and the predictions of magic numbers with modern meson field theories are reviewed. Different channels of nuclear decay are discussed including cluster radioactivity, cold fission and cold multifragmentation Furthermore, we present the vacuum for the e^+-e^- field of QED and show how it is modified for baryons in nuclear environment. Then we discuss the possibility of producing new types of nuclear systems by implanting an antibaryon into ordinary nuclei. The structure of nuclei containing one antiproton or antilambda is investigated within the framework of a relativistic mean-field model. Self-consistent calculations predict very enhanced binding and considerable compression in such systems as compared with normal nuclei. We present arguments that the life time of such nuclei with respect to the antibaryon annihilation might be long enough for their observation. A perspective for future research is given.

Keywords: Periodic system, Superheavy elements, meson field theory, QED, Cold fission, Cluster radioactivity
PACS: 25.75.-q, 23.70.+j, 25.85.Ca

INTRODUCTION

The elements existing in nature are ordered according to their atomic (chemical) properties in the **periodic system** which was developped by Mendeleev and Lothar Meyer. The heaviest element of natural origin is Uranium. Its nucleus is composed of $Z = 92$ protons and a certain number of neutrons ($N = 128 - 150$). They are called the different Uranium isotopes. The transuranium elements reach from Neptunium ($Z = 93$) via Californium ($Z = 98$) and Fermium ($Z = 100$) up to Lawrencium ($Z = 103$). The heavier the elements are, the larger are their radii and their number of protons. Thus, the Coulomb repulsion in their interior increases, and they undergo fission. In other words: the transuranium elements become more instable as they get bigger.

In the late sixties the dream of the superheavy elements arose. Theoretical nuclear physicists around S.G. Nilsson (Lund)[1] and from the Frankfurt school[2, 3, 4] predicted that so-called closed proton and neutron shells should counteract the repelling Coulomb forces. Atomic nuclei with these special **"magic" proton and neutron numbers** and their neighbours could again be rather stable. These magic proton (Z) and neutron (N) numbers were thought to be $Z = 114$ and $N = 184$ or 196. Typical predictions of their life times varied between seconds and many thousand years. Fig.1 summarizes the expectations at the time. One can see the islands of superheavy elements around $Z = 114, N = 184$ and 196, respectively, and the one around $Z = 164, N = 318$.

The important question was how to produce these superheavy nuclei. There were

CP755, *ISIS: International Symposium on Interdisciplinary Science*
edited by A. Ludu, N.R. Hutchings and D.R. Fry
© 2005 American Institute of Physics 0-7354-0240-X/05/$22.50

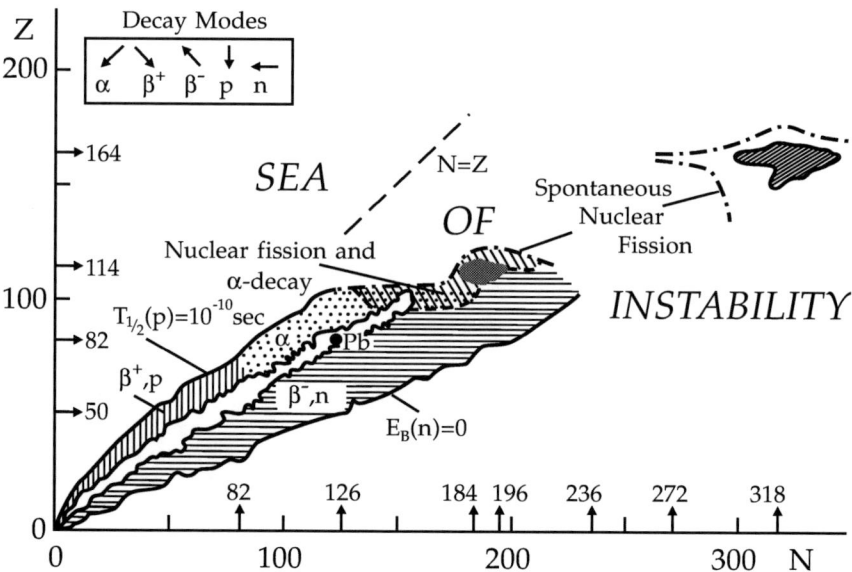

FIGURE 1. The periodic system of elements as conceived by the Frankfurt school in the late sixties. The islands of superheavy elements ($Z = 114$, $N = 184$, 196 and $Z = 164$, $N = 318$) are shown as dark hatched areas.

many attempts, but only little progress was made. It was not until the middle of the seventies that the Frankfurt school of theoretical physics together with foreign guests (R.K. Gupta (India), A. Sandulescu (Romania))[6] theoretically understood and substantiated the concept of bombarding of double magic lead nuclei with suitable projectiles, which had been proposed intuitively by the russian nuclear physicist Y. Oganessian[7]. The two-center shell model, which is essential for the description of fission, fusion and nuclear molecules, was developped in 1969-1972 together with U. Mosel and J. Maruhn[8]. It showed that the shell structure of the two final fragments was visible far beyond the barrier into the fusioning nucleus. The collective potential energy surfaces of heavy nuclei, as they were calculated in the framework of the two-center shell model, exhibit pronounced valleys, such that these valleys provide promising doorways to the fusion of superheavy nuclei for certain projectile-target combinations (Fig. 8). If projectile and target approach each other through those **"cold" valleys**, they get only minimally excited and the barrier which has to be overcome (fusion barrier) is lowest (as compared to neighbouring projectile-target combinations).

COLD VALLEYS IN THE POTENTIAL

In this way the correct projectile- and target-combinations for fusion were predicted. Indeed, Gottfried Münzenberg and Sigurd Hofmann and their group at GSI [9] have followed this approach. With the help of the SHIP mass-separator and the position

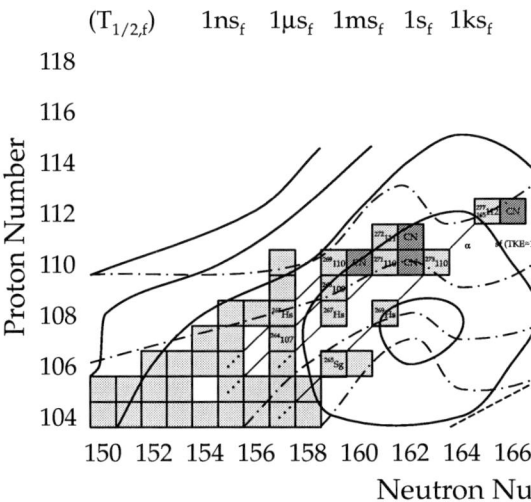

FIGURE 2. The $Z = 106 - 112$ isotopes were fused by the Hofmann–Münzenberg (GSI)–group. The two $Z = 114$ isotopes and the $Z = 116$ isotope were produced by the Dubna–Livermore group. It is claimed that three neutrons are evaporated. Obviously the lifetimes of the various decay products are rather long (because they are closer to the stable valley), in crude agreement with early predictions [3, 4] and in excellent agreement with the recent calculations of the Sobicevsky–group [12].

sensitive detectors, which were especially developped by them, they produced the pre-superheavy elements $Z = 106, 107, \ldots 112$, each of them with the theoretically predicted projectile-target combinations, and only with these. Everything else failed. This is an impressive success, which crowned the laborious construction work of many years. Very recently the Dubna–Livermore–group produced two isotopes of $Z = 114$ element by bombarding ^{244}Pu with ^{48}Ca and also $Z = 116$ by ^{48}Ca + ^{248}C m.(Fig. 2). Also these are cold–valley reactions (in this case due to the combination of a spherical and a deformed nucleus), as predicted by Gupta, Sandulescu and Greiner [10] in 1977. There exist also cold valleys for which both fragments are deformed [11], but these have yet not been verified experimentally.

SHELL STRUCTURE IN THE SUPERHEAVY REGION

Studies of the shell structure of superheavy elements in the framework of the meson field theory and the Skyrme-Hartree-Fock approach have recently shown that the magic shells in the superheavy region are very isotope dependent [5, 14] (see Fig. 3). According to these investigations $Z = 120$ being a magic proton number seems to be as probable as $Z = 114$. Additionally, recent investigations in a chirally symmetric mean–field theory result also in the prediction of these two magic numbers[20, 21]. The corresponding magic neutron numbers are predicted to be $N = 172$ and - as it seems to a lesser extend - $N = 184$. Thus, this region provides an open field of research.

The charge distribution of the $Z = 120, N = 172$ nucleus, calulated with mean-field

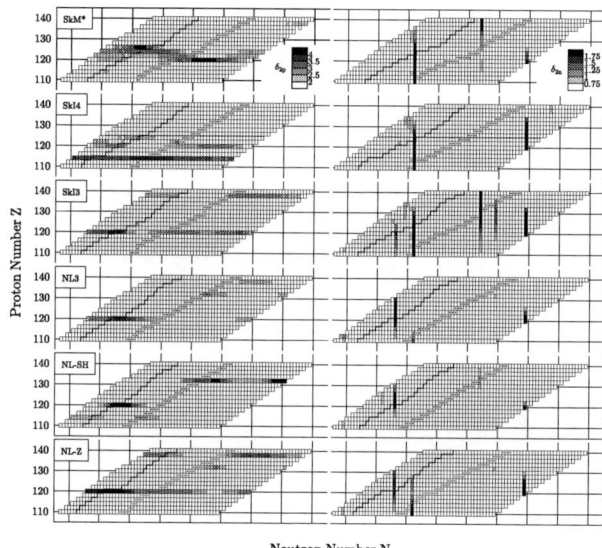

FIGURE 3. Grey scale plots of proton gaps (left column) and neutron gaps (right column) in the N-Z plane for spherical calculations with the forces as indicated. The assignment of scales differs for protons and neutrons, see the uppermost boxes where the scales are indicated in units of MeV. Nuclei that are stable with respect to β decay and the two-proton dripline are emphasized. The forces with parameter sets SkI4 and NL-Z reproduce the binding energy of $^{264}_{156}108$ (Hassium) best, i.e. $|\delta E/E| < 0.0024$. Thus one might assume that these parameter sets could give the best predictions for the superheavies. Nevertheless, it is noticed that NL-Z predicts only $Z = 120$ as a magic number while SkI4 predicts both $Z = 114$ and $Z = 120$ as magic numbers. The magicity depends — sometimes quite strongly — on the neutron number. These studies are due to Bender, Rutz, Bürvenich, Maruhn, P.G. Reinhard et al. [14].

models, indicates a hollow inside. This leads us to suggest that a system with 120 protons and 180 neutrons might essentially be a fullerene consisting of 60 α-particles and one additional binding neutron per alpha. This is illustrated in Fig 5. The protons and neutrons of such a superheavy nucleus are distributed over 60 α particles and 60 neutrons. Such an object could be expected to have interesting decay modes such as multifragmentation, spitting out many α particles. The possible condensation of α particles in light nuclei (in ground-states and exctited states) is a modern topic. It would be fascinating if such condensation could occur also in these superheavy systems. Figure ?? depicts this scenario of a nuclear fullerene structure built of α particles.

One might further ask how collective motions of these spherical superheavy elements might look like. We will take a first look at these aspects in the following section.

VIBRATIONAL MODES IN SPHERICAL SUPERHEAVY NUCLEI

We consider vibrational collective properties of the putative double magic SH nucleus $^{292}120$ as predicted by the RMF axial-symmetric model and compare them with those

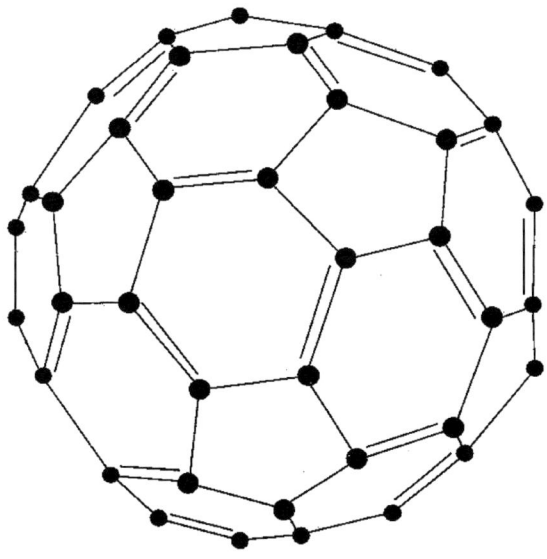

FIGURE 4.

of the well-known double magic heavy nucleus ^{208}Pb [15]. As one can see in Fig.7, the nucleus ^{208}Pb has a pronounced harmonic behaviour, at least for the three vibrational states, i.e. $0^+, 2^+$ and the triplet $0^+, 2^+, 4^+$. In contrast to that the SHE ^{292}120, computed also with the force NL-Z2, exhibits a clear prolate-oblate asymmetry and consequently the sequence of states follows a non-equidistant behaviour. This result was expected because the SHE are less stable (calculations give barriers up to 5 times smaller when the first symmetric barrier of ^{292}120 is compared with that of ^{208}Pb). Therefore the departure of the deformation energy from that of the harmonic oscillator well will be larger.

It is important to stress that in view of the width and height of the potential well in the β_2-coordinate, no more than two phonon states exist. Clearly, the future observation of such β-vibrational states will yield further useful information about the structure of these nuclei. Also the sensitivity of this structure to the underlying effective forces is interesting.

ASYMMETRIC AND SUPERASYMMETRIC FISSION - CLUSTER RADIOACTIVITY

The potential energy surfaces, which are shown prototypically for $Z = 114$ in Fig 8, contain even more remarkable information: if a given nucleus, e. g. Uranium, undergoes fission, it moves in its potential mountains from the interior to the outside. Of course, this happens quantum mechanically. The wave function of such a nucleus, which decays by tunneling through the barrier, has maxima where the potential is minimal and minima

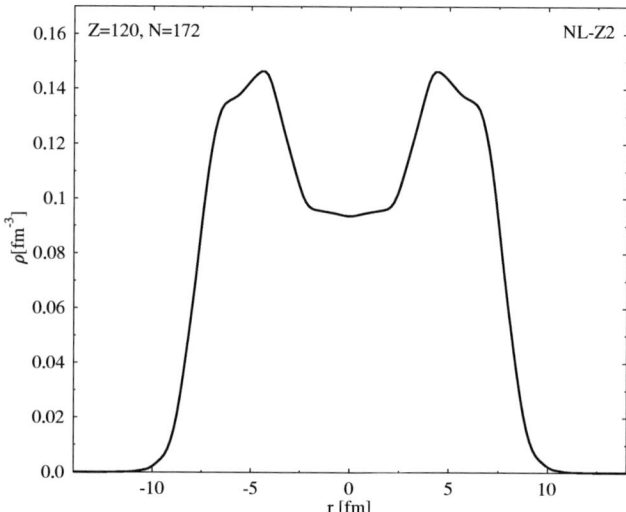

FIGURE 5. Typical structure of the fullerene ^{60}C. The double bindings are illustrated by double lines. In the nuclear case the Carbon atoms are replaced by α particles and the double bindings by the additional neutrons. Such a structure would immediately explain the semi–hollowness of that superheavy nucleus, which is revealed in the mean–field calculations within meson–field theories. The radial density of the nucleus with 120 protons and 172 neutrons, as emerging from a meson-field calculation with the force NL-Z2 is shown on the right side. Note that the semi-bubble structure is mostly pronounced for this nucleus. When going to higher neutron numbers, this structures becomes less and less.

where it has maxima.

The probability for finding a certain mass asymmetry $\eta = \dfrac{A_1 - A_2}{A_1 + A_2}$ of the fission is proportional to $\psi^*(\eta)\psi(\eta)d\eta$. Generally, this is complemented by a coordinate dependent scale factor for the volume element in this (curved) space. Now it becomes clear how the so-called **asymmetric** and **superasymmetric** fission processes come into being. They result from the enhancement of the collective wave function in the cold valleys. And that is indeed, what one observes. For large mass asymmetry ($\eta \approx 0.8, 0.9$) there exist very narrow valleys. They are not as clearly visible in Fig. 8, but they have interesting consequences. Through these narrow valleys nuclei can emit spontaneously not only α-particles (Helium nuclei) but also ^{14}C, ^{20}O, ^{24}Ne, ^{28}Mg, and other nuclei. Thus, we are lead to the **cluster radioactivity** (Poenaru, Sandulescu, Greiner [16]).

By now this process has been verified experimentally by research groups in Oxford, Moscow, Berkeley, Milan and other places. Accordingly, one has to revise what is learned in school: there are not only 3 types of radioactivity (α-, β-, γ-radioactivity), but many more. Atomic nuclei can also decay through spontaneous cluster emission (that is the "spitting out" of smaller nuclei like carbon, oxygen,...). Fig. 9 depicts some examples of these processes.

The knowledge of the collective potential energy surface and the collective masses

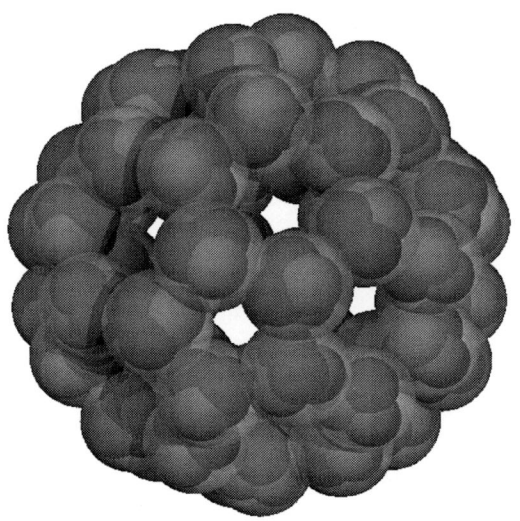

FIGURE 6. An artists view of the nuclear fullerene structure that might occur for the superheavy nucleus $^{300}120_{180}$.

$B_{ij}(R,\eta)$, all calculated within the Two-Center-Shell-Modell (TCSM), allowed H. Klein, D. Schnabel and J. A. Maruhn to calculate lifetimes against fission in an "ab initio" way [17]. The discussion of much more very interesting new physics cannot be persued here. We refer to the literature [18, 22, 23, 24].

The **"cold valleys"** in the collective potential energy surface are basic for understanding this exciting area of nuclear physics! It is a master example for understanding the **structure of elementary matter**, which is so important for other fields, especially astrophysics, but even more so for enriching our "Weltbild", i.e. the status of our understanding of the world around us.

STRUCTURE OF THE VACUUM

It is generally accepted that the physical vacuum has nontrivial structure. This conclusion was first made by Dirac on the basis of his famous equation for a fermion field which describes simultaneously particles and antiparticles. The Dirac equation in the vacuum has a simple form $(i\gamma^{\mu}\partial_{\mu} - m)\Psi(x) = 0$, where $\gamma^{\mu} = (\gamma^0, \vec{\gamma})$ are Dirac matrices, m is the fermion mass and $\Psi(x)$ is a 4-component spinor field. For a plane wave solution $\Psi(x) = e^{-ipx}u_p$ this equation is written as $(\hat{p} - m)u_p = 0$, where $\hat{p} = \gamma^0 E - \vec{\gamma}\mathbf{p}$. Multiplying by $(\hat{p} + m)$ and requiring that $u_p \neq 0$ one obtains the equation $E^2 - \mathbf{p}^2 - m^2 = 0$ which has two solutions

$$E^{\pm}(\mathbf{p}) = \pm\sqrt{\mathbf{p}^2 + m^2} . \tag{1}$$

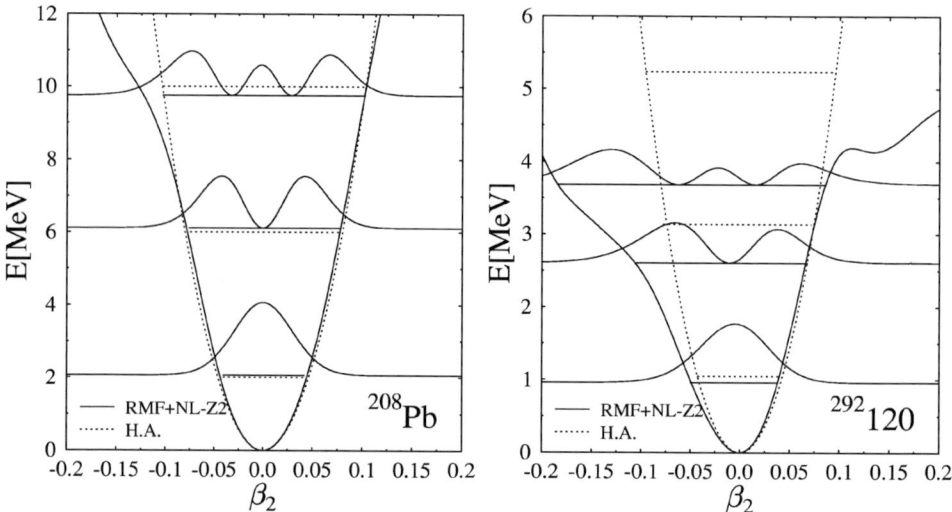

FIGURE 7. Potential well and first three vibrational states of the potential, calculated in the frame of the RMF model with NL-Z2 force (RMF+NL-Z2) and in the Harmonic approximation (HA) for two nuclei. The wave functions of the states are also shown. The left panel represents the case of ^{208}Pb where the harmonic approximation works quite well. The right panel shows the putative double-magic nucleus ^{292}120 for which the anharmonic distortions in the potential are inducing a sensitive departure of the collective level spacing from the equidistant harmonic behaviour.

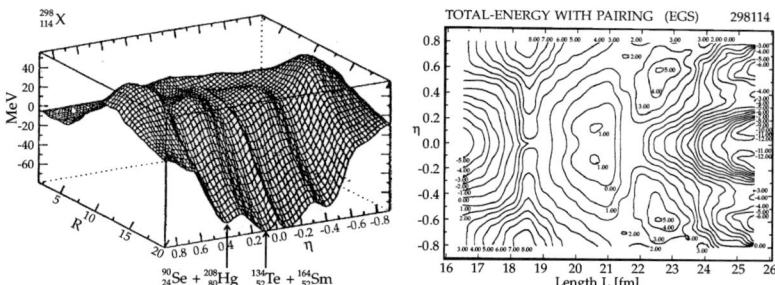

FIGURE 8. The collective potential energy surface of ^{184}114, calculated within the two center shell model by J. Maruhn et al., shows clearly the cold valleys which reach up to the barrier and beyond. Here R is the distance between the fragments and $\eta = \dfrac{A_1 - A_2}{A_1 + A_2}$ denotes the mass asymmetry: $\eta = 0$ corresponds to a symmetric, $\eta = \pm 1$ to an extremely asymmetric division of the nucleus into projectile and target. If projectile and target approach through a cold valley, they do not "constantly slide off" as it would be the case if they approach along the slopes at the sides of the valley. Constant sliding causes heating, so that the compound nucleus heats up and gets unstable. In the cold valley, on the other hand, the created heat is minimized.

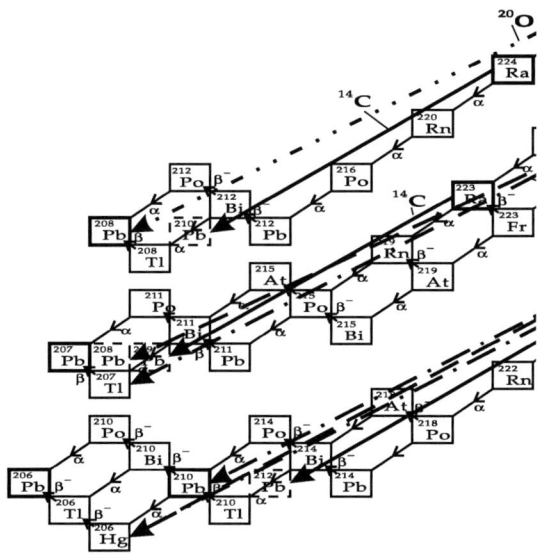

FIGURE 9. Cluster radioactivity of actinide nuclei. By emission of ^{14}C, ^{20}O,... "big leaps" in the periodic system can occur, just contrary to the known α, β, γ radioactivities, which are also partly shown in the figure.

Here the $+$ sign corresponds to particles with positive energy $E_N(\mathbf{p}) = E^+(\mathbf{p})$, while the $-$ sign corresponds to solutions with negative energy. To ensure stability of the physical vacuum Dirac has assumed that these negative-energy states are occupied forming what is called now the Dirac sea. Then the second solution of eq. (1) receives natural interpretation: it describes holes in the Dirac sea. These holes are identified with antiparticles. Their energies are obviously given by $E_{\bar{N}}(\mathbf{p}) = -E^-(-\mathbf{p}) = \sqrt{\mathbf{p}^2 + m^2}$. Unfortunately, the Dirac sea brings divergent contributions to physical quantities such as energy density, and one should introduce a proper regularization scheme to get rid off these divergences. This picture has received numerous confirmations in quantum electrodynamics and other fields.

One of the most fascinating aspects is the structure of the vacuum in QED and its change into charged vacuum states under the influence of strong (supercritical) electric fields [25]. I shortly remind of this phenomenon.

Fig. 10 shows the diving of the deeply bound states into the lower energy continuum of the Dirac equation.

In the supercritical case the dived state is degenerate with the (occupied) negative electron states. Hence **spontaneous** e^+e^- **pair creation** becomes possible, where an electron from the Dirac sea occupies the additional state, leaving a hole in the sea which escapes as a positron while the electron's charge remains near the source. This is a fundamentally new process, whereby the neutral vacuum of QED becomes unstable in supercritical electrical fields. It decays within about $10^{-19}s$ into a charged vacuum. The charged vacuum is now stable due to the Pauli principle, that is the number of emitted

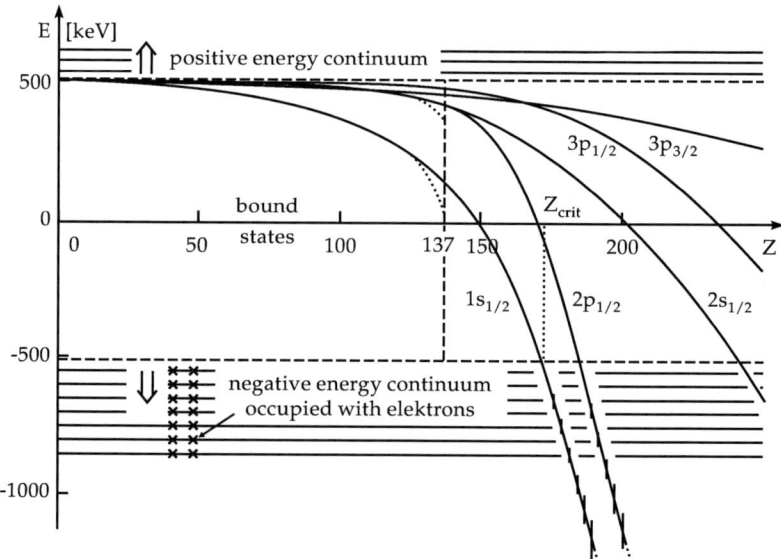

FIGURE 10. Lowest bound states of the Dirac equation for nuclei with charge Z. While the Sommerfeld fine-structure energies (dashed line) for $\xi = 1$ (s states) end at $Z = 137$, the solutions for extended Coulomb potentials (full line) can be traced down to the negative-energy continuum reached at the critical charge Z_{cr} for the 1s state. The bound states entering the continuum obtain a spreading witdth as indicated.

particles remains finite. The vacuum is first charged twice because two electrons with opposite spins can occupy the 1s shell. After the $2p_{1/2}$ shell has dived beyond $Z_{cr} = 185$, the vacuum is charged four times, etc. This change of the vacuum structure is not a perturbative effect, as are the radiative QED effects (vacuum polarization, self-energy, etc.).

The time-dependence of the energy levels in a supercritical heavy-ion collision is depicted in Fig. 11. An electron (or hole) which was in a certain molecular eigenstate at the beginning of the collision can be transfered with a certain probability into different states by the dynamics of the collision. This can lead to the hole production in an inner shell by excitation of an electron to a higher state and/or hole production by ionization of an electron to the continuum. Further possibilities are induced positron production by excitation of an electron from the lower continuum to an empty bound level and direct pair production [26].

It has been noticed already many years ago (see e. g. ref. [27]) that nuclear physics may provide a unique laboratory for investigating the Dirac picture of vacuum. The basis for this is given by relativistic mean-field models which are widely used now for describing nuclear matter and finite nuclei. Within this approach nucleons are described by the Dirac equation coupled to scalar and vector meson fields. Scalar S and vector V potentials generated by these fields modify plane-wave solutions of the Dirac equation as follows pmnuc $E^{\pm}(\mathbf{p}) = V \pm$

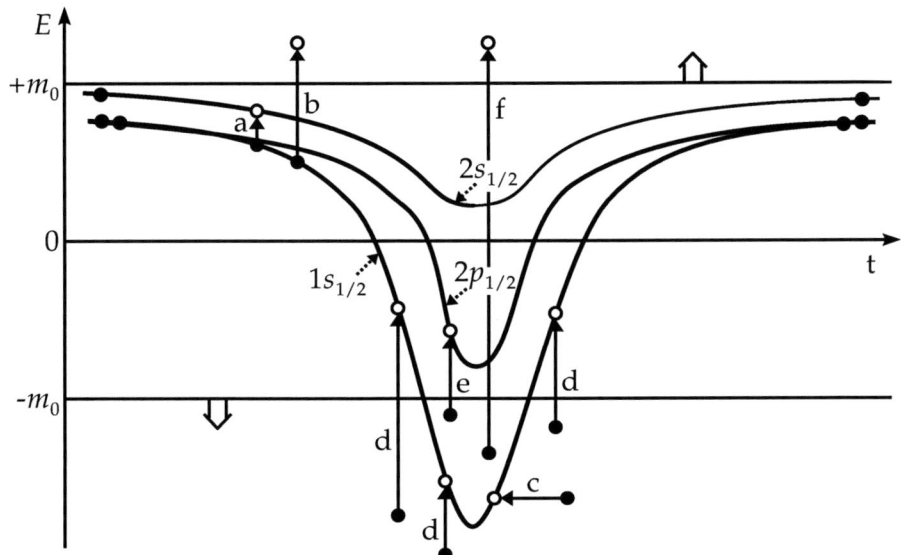

FIGURE 11. Time dependence of the quasi-molecular energy levels in a supercritical heavy ion collision. The arrows denote various excitation processes which lead to the production of holes, i.e., positrons.

$\sqrt{\mathbf{p}^2 + (m-S)^2}$. Again, the + sign corresponds to nucleons with positive energy $E_N(\mathbf{p}) = V + \sqrt{\mathbf{p}^2 + (m-S)^2}$, and the $-$ sign corresponds to antinucleons with energy $E_{\bar{N}}(\mathbf{p}) = -E^-(-\mathbf{p}) = -V + \sqrt{\mathbf{p}^2 + (m-S)^2}$. It is remarkable that changing sign of the vector potential for antinucleons is exactly what is expected from the G-parity transformation of the nucleon potential. As follows from eq. (**??**), in nuclear environment the spectrum of single-particle states of the Dirac equation is modified in two ways. First, the mass gap between positive- and negative-energy states, $2(m - S)$, is reduced due to the scalar potential and second, all states are shifted upwards due to the vector potential. These changes are illustrated in Fig. 1.

It is well known from nuclear phenomenology that good description of nuclear ground state is achieved with $S \simeq 350$ MeV and $V \simeq 300$ MeV so that the net potential for nucleons is $V - S \simeq -50$ MeV. Using the same values one obtains for antinucleons very a deep potential, $-V - S \simeq -650$ MeV. Such a potential would produce many strongly bound states in the Dirac sea. However, if these states are occupied they are hidden from the direct observation. Only creating a hole in this sea, i.e. inserting a real antibaryon into the nucleus, would produce an observable effect. If this picture is correct one should expect the existence of strongly bound states of antinucleons with nuclei.

Below we report on our recent study of antibaryon-doped nuclear systems [28, 29].

27

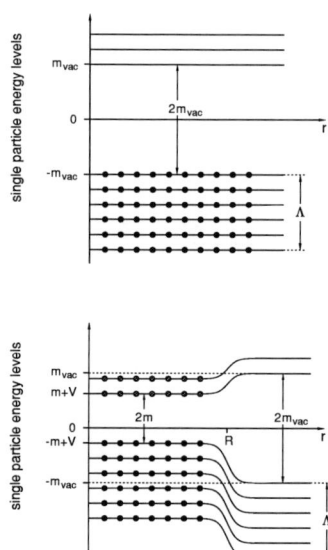

FIGURE 12. Schematic spectrum of Dirac equation in vacuum (upper panel) and in a nucleus of radius R (lower panel). A divergent contribution of negative-energy states is often regularized by introducing a cut-off momentum Λ

NUCLEI CONTAINING ANTIMATTER - COLD COMPRESSION

Unlike some previous works, we take into account the rearrangement of nuclear structure due to the presence of a real antibaryon. The structure of such systems is calculated using several versions of the relativistic mean–field model (RMF): TM1 [30], NL3 and NL-Z2 [31]. Their parameters were found by fitting binding energies and charge form-factors of spherical nuclei from ^{16}O to ^{208}Pb. The general Lagrangian of the RMF model is written as

$$
\begin{aligned}
\mathscr{L} =& \overline{\psi}_j \left(i\gamma^\mu \partial_\mu - m_j \right) \psi_j \\
&+ \frac{1}{2}\partial^\mu \sigma \partial_\mu \sigma - \frac{1}{2}m_\sigma^2 \sigma^2 - \frac{b}{3}\sigma^3 - \frac{c}{4}\sigma^4 \\
&- \frac{1}{4}\omega^{\mu\nu}\omega_{\mu\nu} + \frac{1}{2}m_\omega^2 \omega^\mu \omega_\mu + \frac{d}{4}(\omega^\mu \omega_\mu)^2 \\
&- \frac{1}{4}\vec{\rho}^{\mu\nu}\vec{\rho}_{\mu\nu} + \frac{1}{2}m_\rho^2 \vec{\rho}^\mu \vec{\rho}_\mu \\
&+ \overline{\psi}_j \left(g_{\sigma j}\sigma + g_{\omega j}\omega^\mu \gamma_\mu + g_{\rho j}\vec{\rho}^\mu \gamma_\mu \vec{\tau}_j \right) \psi_j \\
&+ \text{Coulomb part}
\end{aligned}
\tag{2}
$$

Here summation includes valence baryons B, in fact the nucleons forming a nucleus, and valence antibaryons $\overline{B}$ inserted in the nucleus. They are treated as Dirac particles coupled to the scalar-isoscalar (σ), vector-isoscalar (ω) and vector-isovector ($\vec{\rho}$) meson fields.

The calculations are carried out within the mean-field approximation where the meson fields are replaced by their expectation values. Also a "no-sea" approximation is used. This implies that all occupied states of the Dirac sea are "integrated out" so that they do not appear explicitly. It is assumed that their effect is taken into account by nonlinear terms in the meson Lagrangian. Most calculations are done with antibaryon coupling constants which are given by the G-parity transformation $(g_{\sigma\bar{N}} = g_{\sigma N}, g_{\omega\bar{N}} = -g_{\omega N})$ and $SU(3)$ flavor symmetry $(g_{\sigma\bar{\Lambda}} = \frac{2}{3}g_{\sigma\bar{N}}, g_{\omega\bar{\Lambda}} = \frac{2}{3}g_{\omega\bar{N}})$. In isosymmetric static systems the scalar and vector potentials for nucleons are expressed as $S = g_{\sigma N}\sigma$ and $V = g_{\omega N}\omega^0$.

Following the procedure suggested in Ref. [32] and assuming the axial symmetry of the nuclear system, we solve effective Schrödinger equations for nucleons and an antibaryon together with differential equations for mean meson and Coulomb fields. We explicitly take into account the antibaryon contributions to the scalar and vector densities. It is important that antibaryons give a negative contribution to the vector density, while a positive contribution to the scalar density. This leads to increased attraction and decreased repulsion for surrounding nucleons. To maximize attraction, nucleons move to the center of the nucleus, where the antiproton has its largest occupation probability. This gives rise to a strong local compression of the nucleus and leads to a dramatic rearrangement of its structure.

Results for the ^{16}O nucleus are presented in Fig. 2 which shows 3d plots of nucleon density distributions. The calculations show that inserting an antiproton into the ^{16}O nucleus leads to the increase of central nucleon density by a factor 2–4 depending on the parametrization. Due to a very deep antiproton potential the binding energy of the whole system is increased significantly as compared with 130 MeV for normal ^{16}O. The calculated binding energies of the $\bar{p}-^{16}$O system are 830, 1050 and 1160 MeV for the NL–Z2, NL3 and TM1, respectively. Due to this anomalous binding we call such systems super bound nuclei (SBN). In the case of antilambdas we rescale the coupling constants with a factor 2/3 that leads to the binding energy of 560÷700 MeV for the $\bar{\Lambda}-^{16}$O system.

As a second example, we investigate the effect of a single antiproton inserted into the ^{8}Be nucleus. The normal ^{8}Be nucleus is not spherical, exhibiting a clearly visible 2α structure with the ground state deformation $\beta_2 \simeq 1.20$. As seen in Fig. 14, inserting an antiproton in ^{8}Be results in a much less elongated shape ($\beta_2 \simeq 0.23$) and disappearance of its cluster structure. The binding energy increases from 53 MeV to about 700 MeV. Similar, but weaker effects have been predicted [33] for the K^- bound state in the ^{8}Be nucleus.

The calculations have been performed also with reduced antinucleon coupling constants as compared to the G-parity prescription. We have found that the main conclusions about enhanced binding and considerable compression of $\bar{p}$-doped nuclei remain valid even when coupling constants are reduced by factor 3 or so.

LIFETIMES AND SIGNATURES

The crucial question concerning possible observation of the SBNs is their life time. The main decay channel for such states is the annihilation of antibaryons on surrounding nucleons. The energy available for annihilation of a bound antinucleon equals

Sum of proton and neutron densities for ^{16}O (top),
^{16}O with $\bar{\Lambda}$ (bottom left) and ^{16}O with $\bar{p}$ (bottom right)

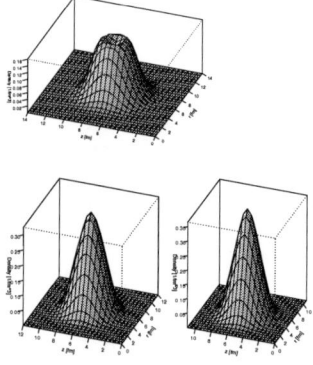

FIGURE 13. Sum of neutron and proton densities for ^{16}O (top), ^{16}O with $\bar{p}$ (bottom right) and ^{16}O with $\bar{\Lambda}$ (bottom left) calculated with the parametrization NL-Z2.

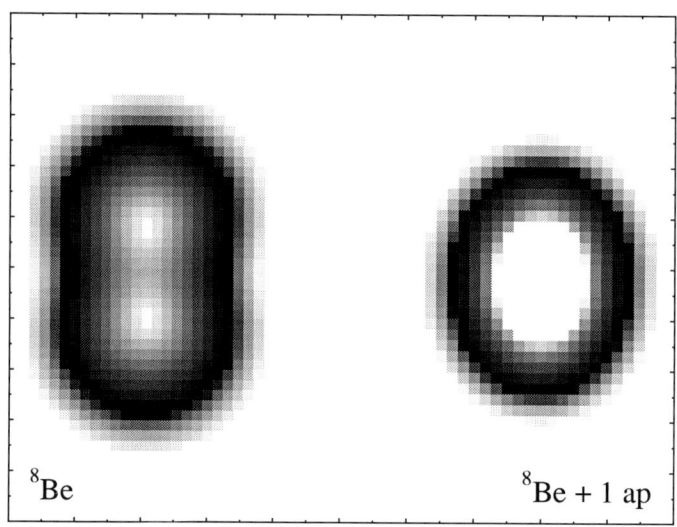

FIGURE 14. Contour plot of nucleon densities for ^{8}Be without (left) and with (right) antiproton calculated with the parametrization NL3.

$Q = 2m_N - B_N - B_{\overline{N}}$, where B_N and $B_{\overline{N}}$ are the corresponding binding energies. In our case this energy is at least by a factor 2 smaller as compared with the vacuum value of $2m_N$. This should lead to a significant suppression of the available phase space and thus to a reduced annihilation rate in medium. We have performed detailed calculations assuming that the annihilation rates into different channels are proportional to the available phase space. All intermediate states with heavy mesons like ρ, ω, η as well as multi-pion channels have been considered. Our conclusion is that decreasing the Q value from 2 GeV to 1 GeV may lead to the reduction of total annihilation rate by factor $20 \div 30$. Then we estimate the SBN life times on the level of 5-25 fm/c which makes their observation feasible. This large margin in the life times is mainly caused by uncertainties in the overlap integral between antinucleon and nucleon scalar densities. Longer life times may be expected for SBNs containing antihyperons. The reason is that instead of pions more heavy kaons must be produced in this case. We have also analyzed multi-nucleon annihilation channels (Pontecorvo-like reactions) and have found their contribution to be less than 40% of the single-nucleon annihilation.

We believe that such exotic nuclear states can be produced by using antiproton beams of multi-GeV energy, e.g. at the future GSI facility. It is well known that low-energy antiprotons annihilate on the nuclear periphery (at about 5% of the normal density). Since the annihilation cross section drops significantly with energy, a high-energy antiproton can penetrate deeper into the nuclear interior. Then it can be stopped there in an inelastic collision with a nucleon, e. g. via the reaction $A(\bar{p}, N\pi)_{\bar{p}}A'$, leading to the formation of a $\bar{p}$-doped nucleus. Reactions like $A(\bar{p}, \Lambda)_{\overline{\Lambda}}A'$ can be used to produce a $\overline{\Lambda}$-doped nuclei. Fast nucleons or lambdas can be used for triggering such events. In order to be captured by a target nucleus final antibaryons must be slow in the lab frame. Rough estimates of the SBN formation probability in a central $\bar{p}A$ collision give the values $10^{-5} - 10^{-6}$. With the $\bar{p}$ beam luminocity of $2 \cdot 10^{32}$ cm^{-2}s^{-1} planned at GSI this will correspond to the reaction rate of a few tens of desired events per second.

Several signatures of SBNs can be used for their experimental observation. First, annihilation of a bound antibaryon can proceed via emission of a single photon, pion or kaon with an energy of about 1 GeV (such annihilation channels are forbidden in vacuum). So one may search for relatively sharp lines, with width of $10 \div 40$ MeV, around this energy, emitted isotropically in the SBN rest frame. Another signal may come from explosive disintegration of the compressed nucleus after the antibaryon annihilation. This can be observed by measuring radial collective velocities of nuclear fragments.

It is interesting to look at the antibaryon-nucleus system from somewhat different point of view. An antibaryon implanted into a nucleus acts as an attractor for surrounding nucleons. Due to the uncompensated attractive force these nucleons acquire acceleration towards the center. As the result of this inward collective motion the nucleons pile up producing local compression. If this process would be completely elastic it would generate monopole-like oscillations around the compressed SBN state. The maximum compression is reached when the attractive potential energy becomes equal to the compression energy. Simple estimates show that local baryon densities up to 5 times the normal nuclear density may be obtained in this way. It is most likely that the deconfinement transition will occur at this stage and a high-density cloud containing an antibaryon and a few nucleons will appear in the form of a multi-quark-antiquark cluster. One may speculate that the whole ^{4}He or even ^{16}O nucleus can be transformed into the quark phase by this mechanism. As shown in ref. [34], an admixture of antiquarks to cold quark matter is energetically favorable. The problem of annihilation is now transferred to the quark level. But the argument concerning the reduction of available phase space due to the entrance-channel nuclear effects should work in this case too. Thus one may hope to produce relatively cold droplets of the quark phase by the inertial compression of nuclear matter initiated by an antibaryon.

CONCLUDING REMARKS - OUTLOOK

For the Gesellschaft für Schwerionenforschung (GSI), which one of the authors (W.G.) helped initiating in the sixties, the questions raised here could point to the way ahead. Working groups have been instructed by the board of directors of GSI, to think about the future of the laboratory. On that occasion, very concrete (almost too concrete) suggestions are discussed — as far as it has been presented to the public. What is necessary, as it seems, is a **vision on a long term basis**. The ideas proposed here, the verification of which will need the **commitment for 2–4 decades of research**, could be such a vision with considerable attraction for the best young physicists. The new dimensions of the periodic system made of hyper- and antimatter cannot be examined in the "stand-by" mode at CERN (Geneva); a dedicated facility is necessary for this field of research, which can in future serve as a home for the universities. The GSI could be such a home for new generations of physicists, who are interested in the **structure of elementary matter**.

REFERENCES

1. S.G: Nilsson et al. Phys. Lett. 28 B (1969) 458
   Nucl. Phys. A 131 (1969) 1
   Nucl. Phys. A 115 (1968) 545
2. U. Mosel, B. Fink and W. Greiner, Contribution to "Memorandum Hessischer Kernphysiker" Darmstadt, Frankfurt, Marburg (1966).
3. U. Mosel and W. Greiner, Z. f. Physik 217 (1968) 256, 222 (1968) 261
4. a) J. Grumann, U. Mosel, B. Fink and W. Greiner, Z. f. Physik 228 (1969) 371
   b) J. Grumann, Th. Morovic, W. Greiner, Z. f. Naturforschung *26a* (1971) 643
5. W. Greiner, Int. Journal of Modern Physics E, Vol. 5 , No. 1 (1995) 1-90. This review article contains many of the subjects discussed here in an extended version, see also for a more complete list of references.
6. A. Sandulescu, R.K. Gupta, W. Scheid, W. Greiner, Phys. Lett. *60*B (1976) 225
   R.K. Gupta, A. Sandulescu, W. Greiner, Z. f. Naturforschung *32a* (1977) 704
   R.K. Gupta, A.Sandulescu and W. Greiner, Phys. Lett. *64*B (1977) 257
   R.K. Gupta, C. Parrulescu, A. Sandulescu, W. Greiner Z. f. Physik A283 (1977) 217
7. G. M. Ter-Akopian et al., Nucl. Phys. A*255* (1975) 509
   Yu.Ts. Oganessian et al., Nucl. Phys. A*239* (1975) 353 and 157
8. D. Scharnweber, U. Mosel and W. Greiner, Phys. Rev. Lett *24* (1970) 601
   U. Mosel, J. Maruhn and W. Greiner, Phys. Lett. *34*B (1971) 587
9. G. Münzenberg et al. Z. Physik A309 (1992) 89
   S.Hofmann et al. Z. Phys A*350* (1995) 277 and 288
10. R. K. Gupta, A. Sandulescu and Walter Greiner, Z. für Naturforschung *32a* (1977) 704

11. A. Sandulescu and Walter Greiner, Rep. Prog. Phys 55. 1423 (1992); A. Sandulescu, R. K. Gupta, W. Greiner, F. Carstoin and H. Horoi, Int. J. Mod. Phys. E1, 379 (1992)
12. A. Sobiczewski, Phys. of Part. and Nucl. 25, 295 (1994)
13. R. K. Gupta, G. Münzenberg and W. Greiner, J. Phys. G: Nucl. Part. Phys. 23 (1997) L13
14. K. Rutz, M. Bender, T. Bürvenich, T. Schilling, P.-G. Reinhard, J.A. Maruhn, W. Greiner, Phys. Rev. C 56 (1997) 238.
15. Ş. Mişicu, T. Bürvenich. T. Cornelius, and W. Greiner, J. Phys. G: Nucl. Part. Phys. 28 (2002) 1441
16. A. Sandulescu, D.N. Poenaru, W. Greiner, Sov. J. Part. Nucl. 11(6) (1980) 528
17. Harold Klein, thesis, Inst. für Theoret. Physik, J.W. Goethe-Univ. Frankfurt a. M. (1992)
     Dietmar Schnabel, thesis, Inst. für Theoret. Physik, J.W. Goethe-Univ. Frankfurt a.M. (1992)
18. D. Poenaru, J.A. Maruhn, W. Greiner, M. Ivascu, D. Mazilu and R. Gherghescu, Z. Physik A328 (1987) 309, Z. Physik A332 (1989) 291
19. P. Papazoglou, D. Zschiesche, S. Schramm, H. Stöcker, W. Greiner, J. Phys. G 23 (1997) 2081; P. Papazoglou, S. Schramm, J. Schaffner-Bielich, H. Stöcker, W. Greiner, Phys. Rev. C 57 (1998) 2576.
20. P. Papazoglou, D. Zschiesche, S. Schramm, J. Schaffner–Bielich, H. Stöcker, W. Greiner, nucl–th/9806087, accepted for publication in Phys. Rev. C.
21. P. Papazoglou, PhD thesis, University of Frankfurt, 1998; C. Beckmann et al., nucl-th/0002046
22. E. K. Hulet, J. F. Wild, R. J. Dougan, R. W.Longheed, J. H. Landrum, A. D. Dougan, M. Schädel, R. L. Hahn, P. A. Baisden, C. M. Henderson, R. J. Dupzyk, K. Sümmerer, G. R. Bethune, Phys. Rev. Lett. 56 (1986) 313
23. K. Depta, W. Greiner, J. Maruhn, H.J. Wang, A. Sandulescu and R. Hermann, Intern. Journal of Modern Phys. A5, No. 20, (1990) 3901
     K. Depta, R. Hermann, J.A. Maruhn and W. Greiner, in "Dynamics of Collective Phenomena", ed. P. David, World Scientific, Singapore (1987) 29
     S. Cwiok, P. Rozmej, A. Sobiczewski, Z. Patyk, Nucl. Phys. A491 (1989) 281
24. A. Sandulescu and W. Greiner in discussions at Frankfurt with J. Hamilton (1992/1993)
25. W. Greiner, B. Müller, J. Rafelksi, Quantum electrodynamics of strong fields, Springer Verlag, 2nd edition, December 1985
26. W. Greiner and J. Reinhardt, Quantum electrodynamics, Springer Verlag, 3rd edition, February 2003
27. N. Auerbach, A.S. Goldhaber, M.B. Johnson, L.D. Miller, A. Picklesimer, Phys. Lett. B182 (1986) 221.
28. T. Bürvenich, I.N. Mishustin, L.M. Satarov, H. Stöcker, W. Greiner, Phys. Lett. B542 (2002) 261.
29. I. N. Mishustin, L. M. Satarov, T. Bürvenich, H. Stöcker, and W. Greiner, nucl-th/0404026
30. Y. Sugahara and H. Toki, Nucl. Phys. A579 (1994) 557.
31. M. Bender, K. Rutz, P.–G. Reinhard, J.A. Maruhn, and W. Greiner, Phys. Rev. C60 (1999) 34304.
32. G. Mao, H. Stöcker, and W. Greiner, Int. J. Mod. Phys. E8 (1999) 389.
33. Y. Akaishi and T. Yamazaki, Phys. Rev. C65 (2002) 044005.
34. I.N. Mishustin, L.M. Satarov, H. Stöcker, W. Greiner, Phys. Rev. C 59 (1999) 3343.

ACKNOWLEDGMENTS

We are grateful to I.N. Mishustin and L.M. Satarov for fruitful discussions and help in the preparation of this talk. Our investigations on antimatter within nuclei did not receive support from the Bundesministerium für Forschung und Technologie (BMBF) because the head of the BMBF advisory board, V. Metag (Gießen) and his theoretical advisors at on the board, K. Goeke (Bochum) and A. Schäfer (Regensburg), denied it.

Heavy-Ion Reactions in Time-dependent Hartree-Fock

J. A. Maruhn

Institut für Theoretische Physik, Universität Frankfurt, Germany

Abstract. Nuclear collisions at low energies present a unique opportunity for the study of degenerate Fermi systems far from equilibrium. Their behavior shows a rich variety: from complete fusion through highly nonelastic collisions to grazing situations with a complicated interplay of collective and single-particle degrees of freedom. The time-dependent Hartree-Fock approximation assumes a dominance of Pauli exclusion, ignoring the two-body collisions and instead using the mean field as the dominant dynamic quantity. For a class of zero-range effective interactions (the so-called Skyrme forces) it is possible to numerically solve the equations of motion to obtain a description of reactions that is parameter-free in the sense that the forces are fitted exclusively to nuclear ground-state properties. In this talk I give an overview of the theoretical issues, ignoring much of the technical detail of nuclear theory involved in such studies, and instead concentrating on the interesting consequences of the nonlinear coupling through the mean field: the spurious interaction between the different exit channels, the "one-body dissipation" mechanism and the essential semiclassical nature of the approach.

Keywords: nuclear theory, heavy-ion reactions, mean-field
PACS: 21.60.Jz, 25.70.-z

INTRODUCTION

The theory of collisions of heavy nuclei has to deal with a quantum many-body problem involving typically 20 to 500 particles. The experimental situation, as well, is quite complicated, because the initial condition cannot be controlled in detail: only the types of nuclei colliding and the energy of relative motion are known, while the impact parameter and, in the case of deformed nuclei, the relative orientation is unknown but highly important for the outcome of the collision. The result of the collision can vary widely: the final kinetic energies, the number of particles produced, and their excitation can be very different.

In this situation theoretical models naturally have to be quite simplified. The most common description of nuclear collisions is based on classical trajectories, describing the motion of the centers of mass and a few other collective coordinates under the influence of classical (usually density-based) forces including friction forces. The coefficients and the functional form of these forces are usually derived by fitting to experiment. Another approach uses a many-body classical description through the Vlasov or Boltzmann equations. The collisions between the particles are governed by laws obtained from nucleon-nucleon scattering properties and/or a mean field. While dissipation comes in naturally in this approach, it is very difficult to describe the ground-state properties of nuclei because of the lack of quantum mechanics and, most noticeably, the Pauli principle.

CP755, *ISIS: International Symposium on Interdisciplinary Science*
edited by A. Ludu, N.R. Hutchings and D.R. Fry
© 2005 American Institute of Physics 0-7354-0240-X/05/$22.50

The time-dependent Hartree-Fock method was used extensively in the 70's and 80's. Its attractiveness lies in the fact that it provides a quantum-mechanical many-body description, incorporating the high-quality description of the nuclear ground states provided by the static Hartree-Fock model. On the other hand, the underlying mean-field approximation has limitations that are more subtle to recognize and this is certainly an interesting aspect for an interdisciplinary audience.

The simplest way to obtain the TDHF equations is from a time-dependent variational principle. Demanding that

$$\delta \int dt \, \langle \Psi(t) \| i\hbar \frac{\partial}{\partial t} - H \| \Psi(t) \rangle = 0 \tag{1}$$

with H a standard Hamiltonian contaning two-body interactions,

$$H = \sum_k \frac{-\hbar^2 \nabla^2}{2m} + \sum_{k<l} v_{kl}, \tag{2}$$

and restricting the wave function $\Psi(t)$ to a single Slater determinant built out of the single-particle wave functions $\phi_k(t)$ leads to the time-dependent Hartree-Fock equations

$$i\hbar \frac{\partial}{\partial t} \phi_k(\vec{r},t) = V(\vec{r},t) \phi_k(\vec{r},t) \tag{3}$$
$$- \sum_k \phi_k(\vec{r},t) \int d^3 r' \phi_l(\vec{r},t) * v(\vec{r},\vec{r}') \phi_l(\vec{r},t)$$

The important ingredient is the mean field V, which in the case of a pure space dependence of the interaction can be written simply as

$$V(\vec{r},t) = \int d^3 r' v(\vec{r},\vec{r}') \sum_k \left(\phi_k(\vec{r}',t) \right)^2, \tag{4}$$

but including a spin or isospin dependence introduces additional densities in the mean field, while a momentum dependence leads to current densities. The full Skyrme force as used in the present calculations of course leads to a complicated expression in the various densities and currents, but this is of interest to specialists only.

The setup for a nuclear collision is quite straightforward: the nuclear ground-state wave functions are obtained from a static Hartree-Fock calculations; the states localized in each fragment are inserted into the computational grid such that the center of mass is located at the desired point, and then the single-particle wave functions are multiplied by a common plane wave phase factor to set them into motion with a specified velocity.

It is very important to realize that these calculations extend a model that was highly successful for describing the properties of nuclei near equilibrium into a new realm of applications without allowing any additional adjustments: there are no friction coefficient, e. g., and the nuclear forces used are fitted purely to ground-state nuclei.

PAST ACHIEVEMENTS AND FAILURES

Upon the original proposal [1] there were quite exaggerated expectations concerning what might be learnt from this method, as even the first one-dimensional calculations showed a rich variety of collisional behaviour in nuclear collisions. Two- (in axial symmetry) and three-dimensional calculations soon followed in a large number of papers; we refer the reader to the review [2, 3, 4]. Then the limitations of the method also became apparent, and it most interesting to note that these are caused by the nonlinearity introduced through the mean field. The mean field in a nuclear collision is highly space and time-dependent and thus to some extent appear as an independent contribution to the dynamics of the system. The limitations recognized are:

1. *Spurious cross-channel correlations:* The crucial feature to understand is the interaction of all wave function configurations in the one mean field generated by the total density. Thus, for example, even though the reseparated fragments in the final state of a nuclear collision do not have fixed nucleon numbers (since the single-particle wave functions are spread out over both fragments), they are located in a potential that is correct for the *average* nucleon number — if $^{16}O+^{16}O$ is the average, the contribution for $^{20}Ne+^{12}C$ is located in ^{16}O potentials, which is obviously energetically unfavorable and leads to a suppression of such a channel. Thus the spread of fragment masses was always found to be quite small. The same is true for the uncertainty in the center of mass, scattering angle, and collective kinetic energy.

2. *Dissipation:* Since the TDHF equations are time-reversal invariant, the TDHF solution is reversible. This does not, however, imply that there is no dissipation from the collective degrees of freedom: certainly the randomization of the single-particle degrees of freedom is very important. In this sense TDHF should be compared to a microscopic simulation like molecular simulations in thermodynamics. On the other hand, the lack of two-body collisions will lead to an underestimation of dissipation.

3. *Symmetries:* The localized mean field seems to violate translational and rotational invariance. This is true in the standard linear quantum-mechanical sense, but a *nonlinear* equations the TDHF equations are perfectly invariant: the mean field as a functional of the density moves together with the wave functions. This explains also why a nucleus boosted by a plane-wave factor behaves as a stable wave packet.

Another successful field of application was that of fusion cross sections. Usually in TDHF calculations at large angular momenta there is a rapid reseparation of the nuclei, but for more central collisions a long-lived compound state may be formed. While TDHF does not allow true fusion, in the sense that the compound nucleus cannot settle down by particle or γ-emission, nevertheless it is highly likely that for such angular momenta fusion will occur.

The description of fusion cross sections based on this concept within an order of magnitude without adjustable parameters, based only on Skyrme interactions whose fit does not include any dynamical properties of nuclei, is a remarkable achievement. Interestingly, in detail the fusion properties seem to crucially depend on the exact Skyrme force used [5].

To summarize, the positive results from TDHF studies were that many qualitative features of heavy-ion reactions can be understood that are difficult to describe otherwise, such as

- the formation and the snapping of the neck,
- the general behaviour of collective dynamics,
- the role of single-particle dissipation
- the gross features of fusion.

One the goals of the present investigations is to check whether a better quantitative description can be achieved by solving the equations under less restrictive assumptions.

THE CODE

The work presented in this paper is based on a new Hartree–Fock code that has the following properties:

- It can solve both the time-dependent (TDHF) equations and determine the self-consistent stationary state of a nucleus.
- It is nonrelativistic and based on Skyrme forces; all modern Skyrme forces can be handled and various pairing treatments can also be included for the static mode.
- The numerical treatment uses a three-dimensional spatial grid, assuming no symmetries for the wave functions. The spatial derivatives are handled using Fourier-transform methods and boundary conditions for the wave functions can be zero or periodic.
- The Coulomb problem is solved exactly for boundary condition at infinity using the well-known grid doubling technique [6].
- The code is written entirely in Fortran 95 and the time-dependent mode can also run in parallel on a message-passing architecture (MPI). The execution times for the calculations discussed below are typicall of the order of hours (light nuclei) to days (heavy nuclei) on a fast PC.

NEW CALCULATIONS

The first applications of the new code in the time-dependent field were devoted to collisions between light deformed nuclei and to an exploration of near-barrier collisions in heavier systems.

We first investigated the sensitivity of the collision dynamics to the relative orientation of the nuclei. In each case two deformed ^{20}Ne nuclei interact centrally at a c.m. energy of 115 MeV. The only difference between the collisions is the different initial orientation. For the "tip-to-belly" configuration (shown in Fig. 1), the most fascinating result is that the orientation of the reseparated nuclei appears to indicate transparency again, in the sense that the nucleus initially in the "tip" position appears to emerge on the other side. For the "tip-tip" configuration, the octupole deformation reverts just at the moment the

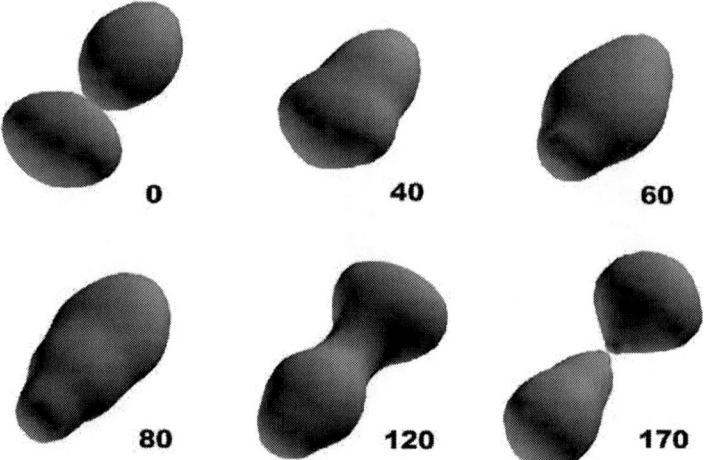

FIGURE 1. Central collision of two deformed Ne nuclei in a "tip-to-belly" configuration. The times corresponding to each plot are given in fm/c. The surface plotted corresponds to $\rho_0/4$.

neck is snapping (see Fig. 2). By contrast, the "belly-to-belly" configuration appears relatively uninteresting.

For an impact parameter of 2 fm and the lower energy of 78 MeV, results showed similar tendencies, but in this case fusion occurred in some cases. Thus the "tip-on-belly" configuration again shows some transparency effect with the "tip" nucleus seeming to try to emerge on the other side; similarly for the "tip-on-tip" configuration at 100 fm/c. The former is the only configuration not leading to fusion in this case, demonstrating that the fusion behaviour is strongly influenced by deformation.

At the present time, such calculations can only have an axploratory character, since to obtain quantitative description of experimental data, one would have to calculate at many different orientations and also to extrapolate the configuration in the computational grid from and to infinity, which is clearly a major problem in its own right for deformed nuclei.

Another situation studied was near-barrier collisions in the systems $^{16}O+^{48}Ca$ and $^{48}Ca+^{208}Pb$. The nuclei were inserted into the numerical grid with surfaces spaced 2 fm apart and then given an initial boost to just overcome the barrier. The resulting behavior was characteristically different in both cases.

For the light system, the development of the density distribution with time shows strong surface oscillations which appear to still show some of the transparency effects seen in light nuclei. The ^{16}O nucleus appears to be "swallowed" by the ^{48}Ca and then to oscillate back and forth.

In contrast, in the heavy system the surface appears to be frozen after about 80 fm/c, once the neck has opened completely. The nuclei appear stuck together and only in the very interior there are strong oscillations apparent only at very high densities. This is true also for noncentral collisions, with a common rotation only superimposed.

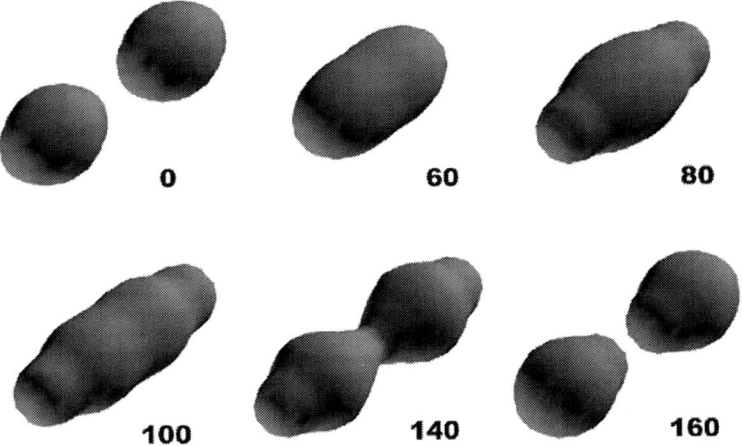

FIGURE 2. Central collision of two deformed Ne nuclei in a "tip-to-tip" configuration. The times corresponding to each plot are given in fm/c. The surface plotted corresponds to $\rho_0/4$.

SUMMARY

The present computational possibilities allow an exploration of nuclear properties and reactions not readily accessible before. Exotic nuclear configurations have been found by a new method (although the states found turned out to be axial after all), and the study of nuclear collisions obviously can be done correctly in TDHF only in full three-dimensional geometry. Such studies will be continued and extended in the future, and the inclusion of pairing in TDHF should provide another important field of activity.

ACKNOWLEDGMENTS

This work was supported by the BMBF, Project No. 06 ER 808.

REFERENCES

1. P. Bonche, S. E. Koonin, and J. W. Negele, Phys. Rev. C13, 1226 (1976).
2. K. T. R. Davies, K. R. S. Devi, S. E. Koonin, and M. R. Strayer, in: *Treatise on Heavy Ion Science*, Vol. 3 (ed. D. A. Bromley), Plenum, New York 1985, p. 3.
3. J. W. Negele, Rev. Mod. Phys. **54**, 913 (1982).
4. J. P. Svenne, Adv. Nucl. Phys. **11**, 179 (1979).
5. J. A. Maruhn, K. T. R. Davies, and M. R. Strayer, Phys. Rev. C31, 1289 (1985).
6. R. W. Hockney, Meth. Comp. Phys. **9**, 135 (1970).

Chaotic dynamics of modulational instability in optical fibers

M. J. Ablowitz[*] and C. M. Schober[†]

[*]Dept. of Applied Mathematics, University of Colorado, Boulder, CO. 80309
[†]Dept. of Mathematics, University of Central Florida, Orlando, FL. 32816

Abstract. The long distance dynamics of modulational instability in optical fibers is studied. A perturbed nonlinear Schrödinger equation which includes higher order dispersive and nonlinear terms as well as damping is investigated. When damping is significant, amplifiers are used to compensate for the fiber losses. Periodically modulated cw waves are shown to evolve chaotically for a variety of physical parameters. A nonlinear spectral analysis of the data shows that the chaotic evolution is characterized by homoclinic transitions in the spectrum.

Keywords: solitons, optical pulse, nonlinear Schrödinger equation, modulational instability
PACS: PACS: 42.65.Tg, 42.65.Wi, 42.82.Et, 63.20.Pw

INTRODUCTION

Modulational instability (MI) is a well known phenomena that occurs in nonlinear optical fibers (cf. [1, 2] and references therein). The first experimental observation of MI in optical fibers was in 1986 [3]. MI is associated with the evolution of quasi-monochromatic wave fields in the anomalous dispersion regime, where the nonlinear Schrödinger (NLS) equation is the leading asymptotic approximation to Maxwells equations [10]. In the usual description of MI the side band frequencies associated with a cw-wave introduced into the fiber become unstable due to Kerr nonlinearity and anomalous dispersion.

In this paper the effects of the higher order terms in a perturbed NLS equation (PNLS) on the MI are studied. The evolution of the following two wave fields are considered: a) periodically modulated (cw) waves and b) solitons. We find that in the former case, the higher order terms have a dramatic impact on the evolution: chaotic dynamics is found to develop. Moreover, as the number of linearly unstable modes in the initial wave form increases, the chaotic dynamics is enhanced. However, when the initial wave is a soliton, the dynamics is coherent and repeatable. In different parameter regimes we find different higher order terms to be primarily responsible for the chaos. For short modulation periods ($T_0 = 200fs$) the main driving term can be traced to the Raman gain term. For longer modulation periods ($T_0 = 3ps$) chaotic dynamics develops when amplifiers are placed in the system, even when the higher order dispersive terms are absent.

In previous work we studied the effect of higher order terms on modulational instability in two different situations. We found that computational chaos associated with numerical models of the NLS ensued because of perturbations due to computational inaccuracies; e.g truncation or round-off errors [4]-[7]. In the other case we studied, both

CP755, *ISIS: International Symposium on Interdisciplinary Science*
edited by A. Ludu, N.R. Hutchings and D.R. Fry
© 2005 American Institute of Physics 0-7354-0240-X/05/$22.50

experimentally and analytically, quasi-monochromatic waves in deep water [8, 9]. Here the asymptotic system is also well known to be the NLS equation with suitable higher order terms. We found that a similar description applies in these cases and it is this scenario that we now find in the case of nonlinear optical fibers. Namely, the higher order terms are responsible for driving waves ("left-or-right" moving waves) to evolve chaotically across a fixed state. The scenario can be understood effectively by appealing to the spectral analysis associated with the NLS equation In the spectral picture there are distinguished eigenvalues (cf. ref. [11, 12]) which are associated with MI in the NLS equation These eigenvalues, while fixed for the "ideal" NLS equation, evolve due to the higher order perturbations in the system. The sensitive evolution of these eigenvalues reflects the true dynamical situation. The evolution of the wave field into chaotic waves is characterized in terms of "homoclinic transitions" [7, 9] which describes the effect of the perturbations on the wave field as it evolves from one side to another of certain bifurcation states.

Thus the phenomena associated with MI that occurs in computational chaos and water waves is now seen to be responsible for the chaotic wave dynamics in nonlinear optical fibers. The phenomena we describe is a "universal" feature associated with MI in a class of problems associated with the NLS equation; the complex dynamics is driven primarily by dispersive higher order terms.

The outline of this paper is as follows: first we give the relevant equations of motion (in both dimensional and non-dimensional form) which govern the evolution of quasi-monochromatic waves in optical fibers, and we briefly describe the phenomena of MI. We then study the evolution of both modulated periodic waves and solitons due to the higher order terms with and without amplification included. We demonstrate that the periodic wave field evolves chaotically whereas the soliton remains coherent in the presence of the higher order terms. We describe the phenomena in terms of the associated spectral analysis of the NLS equation

THE PROPAGATION EQUATIONS

We studied pulse dynamics for short pulses with periods/widths ranging from ~ 200 fs to 3 ps. The governing equation is a perturbed NLS equation [2]:

$$iu_z - \frac{\beta_2}{2}u_{tt} + \gamma|u|^2u = \frac{i\beta_3}{6}u_{ttt} - ic_1(|u|^2u)_t + c_2u(|u|^2)_t - i\alpha u,$$

where β_2 represents the group velocity dispersion (GVD), $\gamma = n_2\omega/(cA_{eff})$ with n_2 the nonlinear coefficient and A_{eff} the effective core area, β_3 and $c_1 = \gamma/\omega$ the higher order linear and nonlinear dispersion, respectively, $c_2 = \gamma T_R$ the self–induced Raman gain, and α the fiber loss, ($\omega = ck$, $c = 3 \times 10^{-4} m/ps$, $k = 2\pi/\lambda$, λ is the pulse wavelength and $\alpha = .0231/km$).

The propagation distance, time and envelope of the electric field are normalized by introducing the dimensionless variables: $z \to z/2z_0$, $t \to t/T_0$ and $u \to u/\sqrt{P_0}$ respectively. The initial pulse period/width, T_0, initial peak power, P_0, and length scale over which both the dispersive and nonlinear effects become important, z_0, are chosen such that $z_0 = T_0^2/|\beta_2|$ and $P_0 = 1/\gamma z_0$. In the anomalous dispersion regime, $\beta_2 < 0$, we obtain

41

the following nondimensionalized perturbed NLS (PNLS) equation:

$$iu_z + u_{tt} + 2|u|^2 u = i\hat{\beta}_3 u_{ttt} - i\hat{c}_1(|u|^2 u)_t + \hat{c}_2 u(|u|^2)_t - i\Gamma u, \tag{1}$$

where $\hat{\beta}_3 = \beta_3/(3T_0|\beta_2|)$, $\hat{c}_1 = 2/(\omega T_0)$, $\hat{c}_2 = 2T_R/T_0$ and $\Gamma = 2\alpha z_0$. The "ideal" NLS equation is recovered when $\hat{\beta}_3 = \hat{c}_1 = \hat{c}_2 = \Gamma = 0$. We note that the higher order dispersive terms break the symmetry of the NLS equation with respect to space translations.

For pulses with $T_0 \sim 3ps$, damping is significant. To compensate for losses we add to the PNLS equation (1) periodic amplification of the form: $i(G-1)\Sigma_{n=1}^{N}\delta(z-nz_a)u$, where z_a is the characteristic amplifier spacing and G is the amplification gain, both normalized. Following [1], we assume that solutions are of the form $u(z,t) = A(z)q(z,t)$. By requiring that the amplification exactly compensates for the total loss over each amplification period, it can be shown that $G = e^{\Gamma z_a/2}$. Further, $q(z,t)$ satisfies

$$iq_z + q_{tt} + 2g(z)|q|^2 q = i\hat{\beta}_3 q_{ttt} + g(z)(-i\hat{c}_1(|q|^2 q)_t + \hat{c}_2 q(|q|^2)_t) \tag{2}$$

where

$$g(z) = A^2(z) = 2\Gamma z_a e^{-2\Gamma(z-nz_a)}/[1 - e^{-2\Gamma z_a}],$$

and $nz_a \le z < (n+1)z_a$, for any integer $n \ge 0$.

MODULATIONAL INSTABILITY AND HOMOCLINIC TRANSITIONS

The MI in optical fibers occurs in the anomalous dispersion regime and, as we show in the numerical experiments, can lead to complicated pulse dynamics. The stability criterion for the periodically modulated waves examined in the numerical experiments is obtained as follows. Consider the steady state solution of the ideal NLS equation, $u_0(z) = ae^{2ia^2 z}$. Let $u(z,t) = u_0(1 + \varepsilon(z,t))$ and linearize for small ε. Assuming $\varepsilon(z,t) = \varepsilon_n(0)e^{iK_n z + i\Omega_n t} + \varepsilon_{-n}(0)e^{iK_{-n} z - i\Omega_{-n} t}$ with $\Omega_n = 2\pi n/L$, K_n and Ω_n satisfy the dispersion relation $K_n^2 = \Omega_n^2(\Omega_n^2 - 4a^2)$. If $0 < (\pi n/L)^2 < |a|^2$, the perturbation $\varepsilon(z,t)$ grows exponentially with z and the steady state solution is unstable. The number of linearly unstable modes, M, is given by $[M] < aL/\pi$ and can be controlled by tuning the amplitude a.

Instabilities can also be examined using Floquet theory. Solutions of the integrable NLS equation, for $u(z,t) = u(z,t+L)$, can be described in terms of the Floquet spectrum of the following linear operator [12]: $\mathcal{L}(u;\lambda)[v_1, v_2]^T = [iv_{1t} - \lambda v_1 - uv_2, -iv_{2t} - \lambda v_2 + u^*v_1]^T$. The fundamental solution matrix M is defined by the conditions $\mathcal{L}(u;\lambda)M = 0$ and $M(t,t;u,\lambda) = I$, and the Floquet discriminant $\Delta(\lambda) = Tr[M(t+L,t;u,\lambda)]$. The spectrum of $\mathcal{L}$ is determined by the condition: $\sigma(\mathcal{L}) = \{\lambda \in \mathbb{C} | \Delta(\lambda) \in \mathbf{R}, -2 \le \Delta(\lambda) \le 2\}$.

The nonlinear spectral transform is used to represent solutions in terms of a set of nonlinear modes whose structure and dynamical stability is determined by the location of the corresponding element of the periodic/antiperiodic spectrum [11, 12]. Multiple points in the spectrum play a particularly important role. Complex multiple points are in general associated with linearized instabilities of the NLS equation and label the orbits homoclinic to the unstable solution [11, 12]. Under a generic perturbation, these multiple points can split. Topological changes in the spectral configuration occur by "homoclinic transitions" and are accompanied by bifurcations in the waveform between left and right running waves (see [7, 9] for a complete description).

NUMERICAL EXPERIMENTS

Equations (1-2) are integrated using an exponentially accurate Fourier pseudo-spectral scheme in time and a fourth order adaptive Runge Kutta discretization in space. Reproducibility of the experiments is studied using a "phase plane" diagnostic: the evolution of $u(0,t)$ is graphed versus that of $u'(0,t) = u(0,t)(1 + \delta r(t))$, $\delta r(t)$ represents random noise ($\delta = .01 - .02$). The dynamics are also investigated by projecting the numerical data onto the nonlinear spectrum of the NLS equation and following its evolution as the pulse propagates. Chaotic evolutions are characterized by irreproducibility (as measured by the phase plane plots) and by changes in the nonlinear spectral configuration.

We consider two classes of initial data: 1) zero velocity solitons, $u(0,t) = \text{sech}(t)$, which are used as the control case and 2) periodic waves, $u(0,t) = a(1 + \varepsilon \cos \Omega_n t)$, where a and L are chosen so that $M = 1, 2, 3$. We have conducted a large body of numerical experiments for a wide variety of wavelengths, pulse widths, fibers (STD and newer fibers) etc. which are typically employed in practice. We varied T_0 from 200 fs to 3 ps, λ from 1350 to 1550, GVD's from 2 to 20ps^2/km, β_3 from .03 for new fiber to .15 for standard fiber etc.

Figures (1) and (2) show typical waveforms and phase plane plots obtained from equation (1) for a soliton and a modulated wavetrain data ($M = 2$), for $0 < z < 100$ (i.e. $z_{max} = .4$ km), with T_0 in the 200 fs pulse regime, respectively. The other parameter values used are: $T_R = 3 fs$, $\lambda = 1550 nm$, $\gamma = 2.51/(W-km)$, $\beta_2 = 20 ps^2/km$, $\beta_3 = .15 ps^3/km \Rightarrow z_0 = .002 km$, $P_0 = 200 W$, $\hat{\beta}_3 = .012$, $\hat{c}_1 = .008$, $\hat{c}_2 = .03$, and $\Gamma = .00008$.

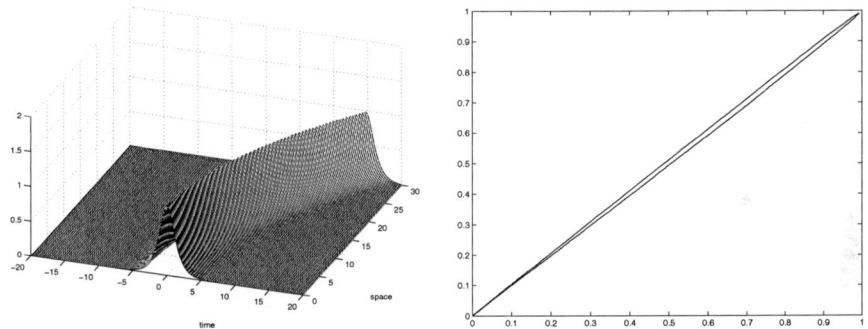

FIGURE 1. The 200 fs wavelength regime. Waveform and the corresponding phase plane plot for $0 < z < 100$ for a soliton.

Similarly, Figures. (3) and (4) show typical waveforms and phase plane plots obtained from equation (2) (with amplifiers) for soliton and periodic wavetrain data ($M = 3$), for $0 < z < 100$ (i.e. $z_{max} = 900$ km) and T_0 in the 3 ps regime, respectively. The other

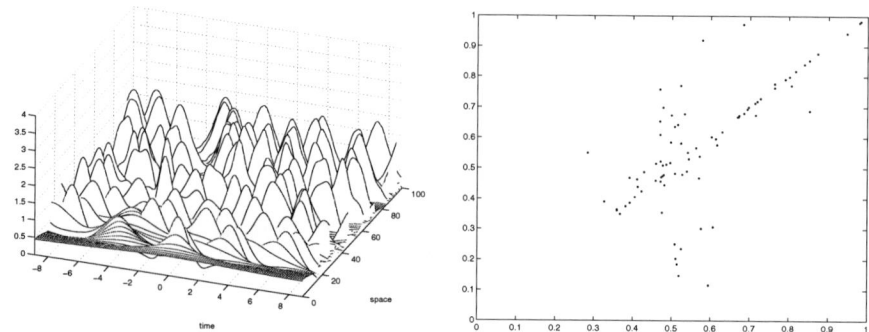

FIGURE 2. The 200 fs wavelength regime. Waveform and the corresponding phase plane plot for $0 < z < 100$ for a modulated cw wave ($M = 2, a = 0.5, L = 4\sqrt{2}\pi$).

parameter values used are: $T_R = 5\,fs$, $\lambda = 1550\,nm$, $\gamma = 2.51/(W-km)$, $\beta_2 = 2\,ps^2/km$, $\beta_3 = .15\,ps^3/km \Rightarrow z_0 = 4.5\,km$, $P_0 = 88\,mW$, $\hat{\beta}_3 = .0084$, $\hat{c}_1 = .00055$, $\hat{c}_2 = .0032$, $\Gamma = .2/km$ and $z_a = .22$.

Figs. (1) and (3) show that the soliton dynamics is regular as the soliton simply develops an $\mathscr{O}(\varepsilon)$ velocity and sheds radiation off the front. The phase plane plot, for $0 < t < 100$, remains close to the 45^0 line (the experiment is reproducible). Although the spectrum is not invariant as in the ideal case, it does not change configuration.

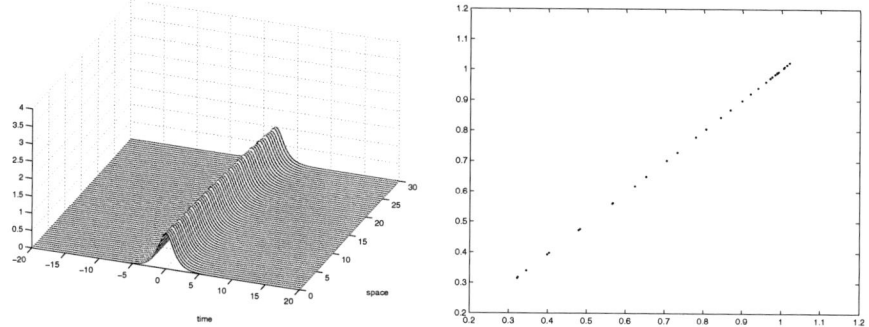

FIGURE 3. The 3 ps wavelength regime. Waveform and the corresponding phase plane plot for $0 < z < 100$ for a soliton.

However, we find that, the periodic wave dynamics is chaotic for $M = M_0 \geq 1$, where M_0 depends on the specific parameter regime. Figs.(2) and (4) display a $M = 2$ and $M = 3$ unstable mode case, respectively. The phase plane plot is spattered significantly away from the 45^0 line. Although we went to $z_{max} = 900$ km, significant chaos is observed as short as $z_{max} = 400$ km. As the pulse propagates down the fiber numerous homoclinic transitions are observed in the spectral decomposition of the numerical data. Each unstable mode bifurcates between left-right traveling waves leading to a very complex waveform. Homoclinic transitions are observed for both the very short 200 fs pulses and for the longer 3 ps pulses. In all the numerical experiments, solitons are stable to the perturbations and evolve regularly whereas the periodic waves can evolve chaotically.

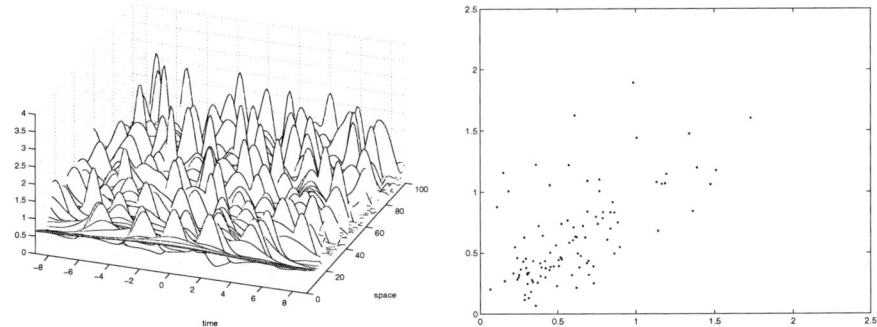

FIGURE 4. The 3 ps wavelength regime. Waveform and the corresponding phase plane plot for $0 < z < 100$ for a modulated cw wave ($M = 3, a = 0.7, L = 4\sqrt{2}\pi$).

The mechanism for the chaotic behavior is the same that we observed for computational chaos and in water waves. We conclude that the long distance chaotic dynamics for MI associated with PNLS due to higher order dispersive terms is universal.

For short modulation periods ($T_0 = 200 fs$) the main driving term is the Raman gain term. The fs pulses are very sensitive and become chaotic for small M with a wide variety of parameters. For longer modulation periods ($T_0 = 3ps$), chaotic behavior is obtained for $0 < z < 100$ with $M = 3$ or else for $M = 2$ when studying longer propagation distances (i.e. $z = 200$). Finally we note that the damping/amplification cycle is sufficient to trigger chaotic behavior. When higher order dispersion is absent ($\hat{\beta}_3 = \hat{c}_1 = \hat{c}_2 = 0$) the dynamics is still chaotic.

ACKNOWLEDGMENTS

This work was partially supported by the NSF, grant numbers DMS-0070792 and DMS-0204714.

REFERENCES

1. A. Hasegawa and Y. Kodama *Opt. Lett.*, **15**, 1443 (1990).
2. G.P. Agrawal. *Nonlinear Fiber Optics*. Academic Press (1995).
3. K. Tai, A. Hasegawa, and A. Tomita. *Phys. Rev. Lett.*, **56**, 135 (1986).
4. M.J. Ablowitz, and B.M. Herbst *Phys. Rev. Lett.*, **62**, 2065-2068 (1989).
5. D.W. McLauughlin and C.M. Schober. *Physica D*, **57**, 447-465 (1992).
6. M.J. Ablowitz, B.M. Herbst, and C.M. Schober. *Phys. Rev. Lett.*, **71**, 2683-2686 (1993).
7. M.J. Ablowitz, B.M. Herbst, and C.M. Schober. *Physica A*, **228**, 212-235 (1996).
8. M.J. Ablowitz, D. Henderson, J. Hammack, and C.M. Schober. *Phys. Rev. Lett.*, **84** (2000).
9. M.J. Ablowitz, D. Henderson, J. Hammack, and C.M. Schober. *Physica D*, **152-153**, 416-433 (2001).
10. A. Hasegawa and F.D. Tappert. *Appl. Phys. Lett.*, **23**, 171 (1973).
11. N. Ercolani, M.G.Forest, and D.W. McLaughlin. *Lect. Appl. Math*, **23**, 149-166 (1986).
12. D.W. McLaughlin and Y. Li. *Comm. Math Phys.*, **162**, 175-214 (1994).

Dynamics of Patterns on Elastic Hypersurfaces. Part I. Shear Waves in the Middle Surface

C. I. Christov

Dept. of Mathematics, University of Louisiana at Lafayette, Lafayette, LA 70504-1010

Abstract. The shear motions in an incompressible elastic continuum are considered and it is shown that, when linearized, the governing equations can be rendered into Maxwell's form. The trace of the deviator stress tensor is analogous to the electric field, while the vorticity (the *curl* of the velocity field) is interpreted as the magnetic field. We show that the analogy can be extended further to incorporate the so-called Lorentz force as the counterpart of the advective nonlinearity of the elastic model. Localized shear dislocations are considered and shown to undergo Lorentz contraction in the direction of motion. Thus an interesting and far reaching analogy between the elastic continuum and the electrodynamics is established.

Keywords: Electrodynamics, Elastic Shear Waves
PACS: 03.50.De, 62.30.+d

INTRODUCTION

Electromagnetic phenomena are an epitome of action at a distance. On its turn the action at a distance is the intrinsic characteristic for any material continuum, and is a consequence of the action of internal stresses. For this reason, it is still a valid avenue of research to attempt to understand the luminiferous *field* as a material continuum in which the *internal stresses* are the transmitter of the long-range interactions.

An all-time candidate for the luminiferous field is the elastic medium because — as shown by Cauchy (see, [1]) — it gives a good quantitative prediction for the shear-wave phenomena (light) and explains quantitatively very well the experiments of Young and Fresnel.

It is only natural to consider the analogy between the shear waves in an elastic continuum and the electromagnetic theory of light based on Maxwell's equations. Building upon our previous work in [2] and [3], we show, in this paper, that the linearized equations of the Hookean elastic continuum admit a Maxwell's form provided that the trace of the stress tensor is understood as the electric field and the *curl* of the velocity vector as the magnetic field. In the present paper we present briefly the derivations pertinent to this analogy.

There is one difference, however. The model of elastic continuum naturally incorporates the Galilean invariance due to the advective part of the material time derivative, while the Maxwell's equations are not Galilean invariant and there is no feasible way to make them invariant if kept in their original form. For this reason, the term connected to the advective part of the time derivative appears in the classical electrodynamics as the Lorentz force on a moving charge. In this sense, the analogy developed here is between the elastic medium and the augmented Maxwellian electrodynamics, in which the

CP755, *ISIS: International Symposium on Interdisciplinary Science*
edited by A. Ludu, N.R. Hutchings and D.R. Fry
© 2005 American Institute of Physics 0-7354-0240-X/05/$22.50

Lorentz force is part of the basic assumptions of the model and not an empirically added term.

We consider localized vortex-like solutions for the displacement field and call them twistons. They can propagate as patterns through the medium without changing shape, and possess topological charge. The kinematics of torsional localized waves is the object of the present short note.

In the second part of the work (see, the present proceedings) we will also consider effects connected with the curvature when the 3D elastic continuum is assumed to be a hypersurface in a 4D geometrical space.

CAUCHY VS MAXWELL

For small velocities, the Lagrangian and Eulerian descriptions of a Hookean elastic medium coincide (see, e.g., [4])

$$\mu_0 \frac{\partial \vec{v}}{\partial t} \stackrel{\text{def}}{=} \mu_0 \frac{\partial^2 \vec{u}}{\partial t^2} = \nabla \cdot \tau, \qquad \tau = \eta \left(\nabla \vec{u} + \nabla \vec{u}^T \right) + \lambda (\nabla \cdot \vec{u}) I, \tag{1}$$

where $\vec{u}, \vec{v}$ are the displacement and velocity vectors; η, λ are Lamé elasticity coefficients, μ_0 is the density of the elastic continuum in material (Lagrangian) coordinates, τ is the stress tensor in the Hooke's law for elastic body, and I stands for the unit tensor. Here the elastic coefficients η, λ and the density μ_0 are constant.

Eqs.(1) govern both the shear and the compression/dilation motions. The phase speeds of propagation of the shear, c, and dilational, c_s, small disturbances are

$$c = \left(\frac{\eta}{\mu_0} \right)^{\frac{1}{2}}, \qquad c_s = \left(\frac{2\eta + \lambda}{\mu_0} \right)^{\frac{1}{2}}, \qquad \delta = \frac{\eta}{2\eta + \lambda}. \tag{2}$$

The interpretation of Cauchy is that c is the speed of light because it is connected to the transverse (shear) waves the latter being the light waves according to Young and Fresnel.

To deal with the second Lamé coefficient, ant thus with the *speed of sound*, c_s, one has two options: to consider an extremely compressible ("volatile") continuum with $c_s = 0$, or an incompressible continuum with $c_s \to \infty$. Cauchy chose the first option and although he succeeded in obtaining quantitative model for the Fresnel observations, his choice left the model of the luminiferous continuum in an unsatisfactory state. The second option refers to an incompressible continuum whose speed of sound is much greater than the speed of light, i.e. $\delta \ll 1$.

Here we examine the limiting case of virtually incompressible continuum when $\lambda \gg \eta$ ($\delta \ll 1$), then eq.(1) can be recast as follows

$$\delta \left(c^{-2} \frac{\partial^2 \vec{u}}{\partial t^2} + \nabla \times \nabla \times \vec{u} \right) = \nabla (\nabla \cdot \vec{u}). \tag{3}$$

Displacement $\vec{u}$ can be developed into an asymptotic power series with respect to δ

$$\vec{u} = \vec{u}_0 + \delta \vec{u}_1 + \cdots . \tag{4}$$

Introducing eq.(4) into eq.(3) and combining the terms with like powers we obtain for the first two terms

$$\nabla(\nabla \cdot \vec{u}_0) = 0, \tag{5}$$

$$c^{-2}\frac{\partial^2 \vec{u}_0}{\partial t^2} + \nabla \times \nabla \times \vec{u}_0 = \nabla(\nabla \cdot \vec{u}_1). \tag{6}$$

From (5) one can deduce

$$\nabla \cdot \vec{u}_0 = \text{const}, \quad \text{or} \quad \nabla \cdot \vec{v}_0 = 0. \tag{7}$$

The preceeding is also a linear approximation to incompressibility condition for an elastic continuum. In the general model of nonlinear elasticity with finite deformations the incompressibility condition is imposed on the Jacobian of transformation from the material to the geometrical variables, but in the first-order approximation in δ, eq.(7) holds true. Henceforth we omit the index "0" for the variable $\vec{u}$ without fear of confusion. We denote formally

$$\varphi \overset{\text{def}}{=} -(\lambda + 2\eta)\nabla \cdot \vec{u}_1, \qquad \vec{A} \overset{\text{def}}{=} \vec{v}_0. \tag{8}$$

It is also convenient to introduce the deviator tensor

$$\tau^0 = \eta\left(\nabla \vec{u} + \nabla \vec{u}^T\right) - 2\eta(\nabla \cdot \vec{u})I.$$

The divergence of the deviator part of the stress tensor τ^0 gives a body force to which the action of the internal shear stresses of the continuum are reduced. We call this body force the "electric field", namely

$$\vec{E} \overset{\text{def}}{=} \nabla \cdot \tau^0 \equiv \eta \nabla^2 \vec{u} = -\eta \nabla \times \nabla \times \vec{u}, \tag{9}$$

where the last equality is obtained by acknowledging the incompressibility condition $\nabla \cdot \vec{u} = const$. Now we can recast the linearized Cauchy balance eq.(6) in terms of $\vec{E}$, as

$$\vec{E} = -\mu_0 \frac{\partial \vec{A}}{\partial t} - \nabla\varphi, \tag{10}$$

which involves $\vec{A}$ and φ. In the same vein we define a "magnetic induction" $\vec{B}$ and "magnetic field" $\vec{H}$ as follows

$$\vec{B} = \mu_0 \nabla \times \vec{A} = \mu_0 \vec{H}, \qquad \vec{H} \overset{\text{def}}{=} \nabla \times \vec{A}. \tag{11}$$

The system of eqs.(10)–(11) is simply the equations of electrodynamics in terms of $\vec{A}$ and φ which play the role of the vector and scalar potentials of electromagnetic field (see, [5]). Note that the density μ_0 of the elastic continuum appears as the magnetic permeability of the Maxwellian field.

Now one can derive the original Maxwell's equations. Taking the operation *curl* of eq.(10) and acknowledging eq.(11) we obtain

$$\nabla \times \vec{E} = -\frac{\partial \vec{B}}{\partial t}, \tag{12}$$

48

which is the first of Maxwell's equations (Faraday's law). Respectively, from eqs.(9), (11), and (8) one obtains

$$\frac{1}{\eta}\frac{\partial \vec{E}}{\partial t} = \nabla \times (\nabla \times \frac{\partial \vec{u}}{\partial t}) \equiv \nabla \times \vec{H}. \tag{13}$$

The last equation is precisely the Maxwell's second equation provided that the shear elastic modulus of metacontinuum is interpreted as the inverse of electric permittivity $\eta = \varepsilon_0^{-1}$. Thus we have shown that the Maxwell's second equation is a corollary of the elastic constitutive relation for the luminiferous continuum and is responsible for the propagation of the shear stresses (action at a distance).

The condition $\nabla \cdot \vec{H} = 0$ (Maxwell's third equation) follows directly from the very definition of magnetic field. Similarly, taking *div* of eq.(9), one gets $\nabla \cdot \vec{E} = 0$.

Thus, we have shown that the linearized equations of elastic continuum admit what can be called *Maxwell's form*. In the framework of such a paradigm, each point of the elastic continuum experiences a body force $\vec{E}$, to which the action of the internal elastic stresses is reduced. We call it the "electric force". The angular momentum of the velocity of a material point is called the "magnetic field".

NONLINEARITY, GALILEAN INVARIANCE, AND LORENTZ FORCE

A far reaching consequence of the previous section is that it gives a clue of how to look for a Galilean invariance of the equations of the luminiferous field. In classical continua the Galilean invariance is connected to the advective nonlinearity of the governing equations. Then the pertinent question here is of what kind of effects are to be expected due to the convective nonlinearity.

Consider the governing equations of an elastic continuum in the Lamb's form [4]

$$\mu_0 \left(\frac{\partial \vec{v}}{\partial t} + \frac{1}{2}\nabla |\vec{v}|^2 - \vec{v} \times \text{rot}\vec{v} \right) + \nabla \varphi = -\vec{E}. \tag{14}$$

where we have already substituted the notations for the above defined scalar potential and electric field.

This form allows one to assess the forces acting at a given material point of the metacontinuum due to the convective accelerations of the latter. The gradient part of the convective acceleration can not be observed independently from the pressure gradient $\nabla \varphi$ in the continuum. In fact one can measure only the quantity $\varphi_1 \equiv \varphi + \frac{1}{2}\vec{v}^2$. Thus the only observable quantity is the last term of the acceleration. By virtue of our definition of magnetic induction (11) the term under consideration adopts the form

$$F_l = \mu_0 \vec{v} \times \vec{H}. \tag{15}$$

Eq.(15) gives a force acting in each material point of the metacontinuum. This force is part of the inertial force which is lost when the equations are linearized and it is

49

analogous to the so-called Lorentz force. To find the exact quantitative coefficient of the Lorentz force one has to integrate the above relation over the spatial extent of a test charge.

The above result suffice to claim that a *Galilean invariant* generalization of the electrodynamics is possible and it incorporates the Lorentz force as an integral part the latter being manifestation of the convective nonlinearity of this more general model.

LOCALIZED SHEAR WAVES — "TWISTONS"

The stationary version of the governing equations of the metacontinuum is the vectorial Laplace equation

$$\nabla^2 \vec{u} \equiv \frac{\partial^2 \vec{u}}{\partial x^2} + \frac{\partial^2 \vec{u}}{\partial y^2} + \frac{\partial^2 \vec{u}}{\partial z^2} = 0, \tag{16}$$

The following stationary vortex-like solution for the components of the displacements

$$u_x = \frac{y}{r^2}, \quad u_y = -\frac{x}{r^2}, \quad r = \sqrt{x^2 + y^2 + z^2}, \tag{17}$$

provides an example of a localized solution. Its divergence is easily proven to be zero, except the origin, where it is undefined, namely

$$\nabla \cdot \vec{u} \equiv \frac{\partial u_x}{\partial x} + \frac{\partial u_y}{\partial y} = 0, \quad \text{for} \quad r \neq 0.$$

The solution (17) has a topological charge ("circulation" of the displacement field) for which a conservation law similar to Thomson's circulation theorem (see, e.g. [4]) holds,

$$\Gamma = \oint u_x dx + u_y dy, \quad \frac{d\Gamma}{dt} = 0,$$

which means that once created, the torsional dislocation (17) cannot be destroyed unless dissipation is introduced in the medium.

In what follows, we call eq.(17) a "twiston". It could serve as a good analogy for the electrical charges. Note that it is only a part of a full fledged charge since there is no component in z-direction. Hence it is a polarized charge. Fig. 1-(a) shows the vector field of the displacement, except for the origin where there is a singularity.

Should the above described "dislocation" be allowed to move, it would not "plow" through the material points of the continuum. It will propagate as a *phase pattern*, much the same way a wave propagates over the water surface.

As already above mentioned, solution in eq.(17) is singular at the origin of the coordinate system. This is the same sort of singularity as the vortex solution in fluid dynamics. The improper behavior can be mitigated if a higher-order elasticity model is considered (see, e. g., the discussion in [6]), such as

$$\frac{\partial^2 \vec{u}}{\partial t^2} = \Delta \vec{u} - \nabla \varphi - \chi \Delta \Delta \vec{u}, \tag{18}$$

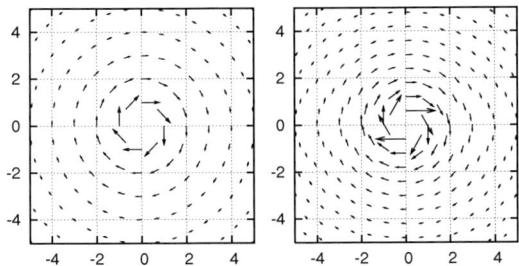

FIGURE 1. The localized torsional dislocation (twiston) in two dimensions for two different phase velocities of propagation. Left panel: $c_1 = 0, c_2 = 0$. Right panel: $c_1 = 0, c_2 = 0.8$

where χ is the coefficient of the higher-order elasticity. It is beyond the scope of present work to give a detailed description of higher-order elasticity. We just mention that upon introducing a "deformation function", $u_x = \frac{\partial \psi}{\partial y}$ and $u_y = -\frac{\partial \psi}{\partial x}$ in 2D, we can reduce eq.(18) for the model 2D case to a scalar equation for ψ; the preceeding admits

$$\chi \Delta\Delta\psi - \Delta\psi = 0, \qquad \psi = K_0(r/\sqrt{\chi}) + \ln(r/\sqrt{\chi}), \qquad (19)$$

where $r = \sqrt{x^2 + y^2}$ and has no singularity at the origin. For the displacement components, this gives

$$u_x = \frac{y}{r\sqrt{\chi}}\left[K_1\left(\frac{r}{\sqrt{\chi}}\right) + \ln\left(\frac{r}{\sqrt{\chi}}\right)\right], \quad u_y = -\frac{x}{r\sqrt{\chi}}\left[K_1\left(\frac{r}{\sqrt{\chi}}\right) + \ln\left(\frac{r}{\sqrt{\chi}}\right)\right]. \quad (20)$$

This form shows that there is a way to tackle the singularity and the vortex-like localized deformation can occur in a real continuum.

Finally, we also note that in the linearized model the amplitude of a twiston remains undetermined unless one considers the nonlinear model of finite elasticity.

FITZGERALD-LORENTZ CONTRACTION OF PATTERNS

Consider now a moving coordinate system, say

$$\xi = x - v_x t, \ \eta = y - v_y t, \ \zeta = z - v_z t, \quad \vec{v} = \frac{\partial \vec{u}}{\partial t} + \vec{V}, \quad \vec{V} = (V_x, V_y, V_z).$$

where $\vec{u}$ are the displacements referred to the Eulerian moving frame. The Galilean invariance means that the last term in the above expression cancels exactly the term in (14) arising from the time derivative in the moving frame. Consider also a stationary solution in the moving frame, when $\vec{u}_t = 0$. Then eq.(16) transforms to the following

$$\left(1 - \frac{V_x^2}{c^2}\right)\frac{\partial^2 \vec{u}}{\partial \xi^2} + \left(1 - \frac{V_y^2}{c^2}\right)\frac{\partial^2 \vec{u}}{\partial \eta^2} + \left(1 - \frac{V_z^2}{c^2}\right)\frac{\partial^2 \vec{u}}{\partial \zeta^2} = 0,$$

It is easily seen that upon scaling the independent variables as

$$\hat{x} = \xi \left(1 - \frac{V_x^2}{c^2}\right)^{-\frac{1}{2}}, \quad \hat{y} = \eta \left(1 - \frac{V_x^2}{c^2}\right)^{-\frac{1}{2}}, \quad \hat{z} = \zeta \left(1 - \frac{V_x^2}{c^2}\right)^{-\frac{1}{2}},$$

one arrives once again to a Laplace equation like eq.(16) in terms of the new variables. Then the solution in eq.(17) is valid for $\hat{\vec{x}}$, thus

$$u_x = \frac{y - c_y t}{r^2 \sqrt{1 - c_y^2}}, \quad u_y = -\frac{x - c_x t}{r^2 \sqrt{1 - c_x^2}}, \quad r = \sqrt{\frac{x^2}{1 - \frac{V_x^2}{c^2}} + \frac{y^2}{1 - \frac{V_y^2}{c^2}} + \frac{z^2}{1 - \frac{V_z^2}{c^2}}}. \tag{21}$$

The lines of equal amplitude $|\vec{u}|$ of displacement for eq.(17) are concentric spheres while the same lines for eq.(21), are ellipsoids. as shown in Fig. 1-(b). The measures of twiston are scaled (shortened) exactly with the respective Lorentz factors. The amplitude of the deformations is increased (proportionally to $[1 - V^2/c^2]^{-\frac{1}{2}}$) in comparison with the twiston at rest, but the total circulation (the charge) remains the same. The increased amplitude means that it is harder to accelerate the steady propagating localized pattern in eq.(21) rather than the pattern at rest in eq.(17). This effect is well known as the increase of mass of moving bodies with the inverse of the Lorentz factor.

The conclusion of this section is that the localized solutions *must* experience contraction in the direction of propagation. The contraction of the localized waves is of the same nature as the Doppler effect for the harmonic waves [7].

CONCLUDING REMARKS

We consider propagation of shear waves in an incompressible elastic medium. We show that the linearized model admits Maxwell's form analogous to the classical electrodynamics. We introduce the notion of a localized shear wave which is a vortex-like structure of the displacement field. The vortex-like localized patterns of displacement field (called twiston) possess topological charges, which are related to the notion of electical charges. The twistons propagating with constant phase speed are shown to experience Lorentz contraction in the direction of motion. We found also that due to the advective nonlinearity of the material time derivative a force arises, which is an exact analogy of the Lorentz force acting on moving charges.

REFERENCES

1. E. Whittaker, *A History of the Theories of Aether & Electricity vol. 1 and 2*, Dover, New York, 1989.
2. C. I. Christov, "Discrete out of continuous: Dynamics of Phase patterns in Continua," in *Continuum Models and Discrete Systems – Proceedings of CMDS8*, edited by K. Markov, World Scientific, Singapore, 1996, pp. 370–394.
3. C. I. Christov, *Annuaire de L'Universite de Sofia*, **95**, 109–121 (2001).
4. L. I. Sedov, *A Course in Continuum Mechanics, vol. I and II*, Walters–Nordhoff, Groningen, 1981.
5. J. D. Jackson, *Classical Electrodynamics*, John Wiley Sons, 1998.
6. C. I. Christov, G. A. Maugin, and M. G. Velarde, *Phys. Rev. E*, **54**, 3621–3638 (1996).
7. T. P. Gill, *The Doppler Effect*, Logos Press, 1965.

Dynamics of Patterns on Elastic Hypersurfaces. Part II. Wave Mechanics of Flexural Quasi-Particles

C. I. Christov

Dept. of Mathematics, University of Louisiana at Lafayette, Lafayette, LA 70504-1010

Abstract. In the first part of this work, the shear wave phenomena in an elastic 3D continuum are investigated and an analogy to Maxwellian electrodynamics is shown. In the present part, the model is extended assuming the continuum to be a momentum-supporting hypersurface in 4D space (a hypershell). The transverse (flexural) deformations of the shell are governed by a Generalized Nonlinear Dispersive Wave Equation (GDWE) of Schrödinger type. The Hamiltonian structure of the model is discussed. The solitary wave solutions are shown to fit the concept of quasi-particles. The concept of pseudo-mass is introduced and the Newtonian mechanics for the centers of quasi-particles/solitons is derived. Numerical examples of the shapes in 2D are presented. The presence of attractive force between the localized elevations of the surface is discussed and shown to depend on the distance between them as $|\vec{x}|^{-2}$.

Keywords: Unified Field Theory, Wave Mechanics
PACS: 03.50.-z, 46.70.De

A NON-PROBABILISTIC APPROACH TO WAVE MECHANICS

De Broglie proposed to explain certain diffraction and interference patterns of electron beams by attaching a wave to a particle. Schrödinger, guided by some heuristic considerations of the principles of wave propagation, derived an equation for the wave function. Schrödinger's equation of wave mechanics is now widely accepted for describing properties of particles (see, for the historical review, [1]). The important characteristic of the Schrödinger equation is that it is not a part of Maxwell's equations. Hence if we are to look for an analogy between the known physical phenomena and the elastic continuum we must transcend the three dimensions because the presence of an additional variable is connected with the existence of an additional dimension. It was first suggested by Hinton [2] that the fourth dimension is not tangible because the 4D thickness of the material world is so minute that it cannot be appreciated.

The derivation of Schrödinger's equation for the purposes of the wave mechanics did not involve physical stipulations of the particular mechanical properties of the field, but Schrödinger [3] did point out in his original paper that the equation at which he had arrived at, was the Euler-Bernoulli equation for the deformation of elastic plates. Since then, the Schrödinger equation proved successful in modeling different atomic phenomena. Now, it is considered as an experimentally established fact that the "master" equation of the wave mechanics is from the genus of the generalized dispersive wave equations (GDWE). In this section we try to come up with a mechanical construct which gives us a fourth order nonlinear dispersive equation, so we can provide arguments for

CP755, ISIS: International Symposium on Interdisciplinary Science
edited by A. Ludu, N.R. Hutchings and D.R. Fry
© 2005 American Institute of Physics 0-7354-0240-X/05/$22.50

the physical basis for the Schrödinger's equation.

Following this line of reasoning, we propose to extend the analogy introduced in Part I. We couple the equations for the laminar components with an equation for the flexural deformation along the fourth dimension. When an observer is situated entirely in the 3D middle surface, the presence of the fourth dimension will be appreciated as some additional forces due to the flexural deformations.

In this paper we assume that the dispersive "master equation" of the wave mechanics is not about an "associated" probability wave, but about a real material wave on the surface of a thin 3D hypersurface (a *hyper*plate or *hyper*shell).

Schrödinger Equation Revisited

To delineate the analogy to plate theory we begin with the linear Schrödinger equation for a complex wave function ψ which reads

$$i\frac{\partial \psi}{\partial t} + \hbar\Delta\psi - \chi\psi = 0, \quad i \equiv \sqrt{-1}, \quad \psi \equiv \psi_1 + i\psi_2, \tag{1}$$

where $\hbar$ is the Plank constant and χ is allegedly connected to the potential of acting forces. In terms of the real (or the imaginary) part of wave function

$$\frac{\partial^2 \psi_1}{\partial t^2} + \hbar^2\Delta\Delta\psi_1 - 2\chi\hbar\Delta\psi_1 + \chi^2\psi_1 = 0. \tag{2}$$

For $\chi = 0$ this is nothing else but the Euler-Bernoulli equation for flexural deformation of an elastic plate with a stiffness proportional to $\hbar^2$. Depending on the sign of χ, different interpretations of Eq. (2) are possible.

Governing Equations for 3D Shells in the 4D Geometrical Space

Consider a very thin elastic structure (a shell) of a 4D material. A 3D momentum supporting hypersurface is a mathematical abstraction for this kind of material structure. Assume also that the shear Lamé coefficient is much smaller than the dilational one (the 4D material is virtually incompressible).

In the middle 3-surface of the shell, the equations for the so-called laminar components are the equations used in Part I to show the analogy to the Maxwell electrodynamics. Consider now the component of the displacement, ζ, in the 4-th dimension. Contrary to the shell theory in technological applications, we consider here the limiting case $L \gg h$ when the deflections are small, the strains (gradients) are of unit-order, and curvatures are large. Such an object is geometrically strongly nonlinear. Under the above assumptions a geometrically nonlinear theory of the very thin but very stiff shells can be derived. We follow in this instance [4, 5, 6, 7], namely

$$\mu_0\frac{\partial^2 \zeta}{\partial t^2} = F^{(4)} + D\left[-\Delta\Delta\zeta - (\Delta\zeta)^3\right] + \sigma_0\Delta\zeta - \frac{a}{R}\zeta. \tag{3}$$

Dimensionless form is obtained by using the scales

$$\zeta = L\zeta', \quad x = Lx', \quad t = Lc_f^{-1}t', \quad c_f = \sqrt{|\sigma_0|\mu_0^{-1}}, \quad \beta = D|\sigma_0|^{-1}L^{-2}, \quad (4)$$

where L is the characteristic length-scale of the waves, μ_0 is the density of the 4D material, σ_0 is the membrane tension and $F^{(4)}$ is a 4D body force (e.g., normal pressure from the fourth dimension). Here $c_f \neq c$.

$$\frac{\partial^2 \zeta'}{\partial t^2} = \beta \left[-\Delta\Delta\zeta' - (\Delta\zeta')^3 \right] + \varepsilon\Delta\zeta' - \kappa\zeta' + F, \quad (5)$$

where $\varepsilon = \pm 1$ and κ is the dimensionless undisturbed curvature of the shell. β is the dimensionless stiffness of the shell. We consider such a deformation for which $\beta \sim O(1)$. for definiteness we set $\beta = 1$.

The linear part of Eq. (5) has the form Eq. (2). The membrane tension and the mean curvature define the potential χ.

Note that the cubic nonlinearity appear in the fourth-order equation for the real part. In this sense the above equation is not exactly what is called NLSE, but the latter has never been derived for the wave mechanics, but only for nonlinear optics. So one can assume that the newly derived GDWE is in very good qualitative agreement with the spirit of Schrödinger's equation.

In order to assess membrane tension in the above model, we can assume that the 3D shell is spherical and subjected to constant normal pressure from the adjacent 4D space outside the shell. Depending on whether the pressure is positive or negative, the resulting membrane tension can be negative or positive. If V_g is the magnitude of the external pressure then, $\sigma_0 = -V_g R^{-1}$, where R is the radius of the shell. When a spherical shell is considered the above model has to be augmented with a liner term with coefficient proportional to the mean curvature R^{-1} (see [8]). A fairly general form of a generic Boussinesq equation describing the model has the dimensionless form

$$\frac{\partial^2 w}{\partial t^2} = -\kappa w + \Delta \left[\varepsilon w - w^3 - \Delta w \right], \quad (6)$$

where w stands for the dimensionless deflection.

Buckling the undisturbed spherical surface (the vacuum state), gives birth to localized creatures – solutions. *Wave Mechanics* is about nonlinear eigenvalue problems. Hence it is by necessity *Quantum*.

Least-Action Principle Formulation

Consider trivial boundary conditions for $w = 0$, $\Delta w = 0$, at $x \in \partial D$ which correspond to an isolated system. Then we prove that the governing equation (5) provides the minimum to the action integral $L \overset{\text{def}}{=} \int_D \mathscr{L} d^3 x$, where

$$\mathscr{L} = -\frac{1}{2} \left[-(\nabla w_t)^2 + \varepsilon(\Delta w)^2 + \kappa(\nabla w)^2 - \frac{1}{2}\beta(\Delta w)^4 + \beta(\nabla\Delta w)^2 \right]. \quad (7)$$

It is readily shown that for asymptotically vanishing wave profiles and for a very large region D, the volume of the wave (sometimes called "wave mass"), and the energy are given by

$$V \equiv \int_D \Delta w \, d^3 x, \quad H = \int_D \left[\frac{1}{2} (\nabla w_t)^2 - \mathscr{L} \right] d^3 x, \tag{8}$$

while for the wave momentum we can derive

$$\vec{P} \equiv \int_D w_t \nabla \Delta w \, d^3 x = - \int_D (\nabla w_t) \cdot (\nabla \nabla w) d^3 x + \oint_{\partial D} w_t \Delta w \, d\vec{s}, \tag{9}$$

where the last integral vanishes due to the asymptotic boundary conditions.

The wave volume and energy are conserved, namely $\dot{V} = 0$, $\dot{H} = 0$. For the wave momentum $\vec{P}$ we have

$$\frac{d\vec{P}}{dt} = \oint_{\partial D} \vec{n} \left[-\frac{1}{2}(\nabla w_t)^2 - \frac{1}{2}(\Delta w)^2 + (\nabla w)^2 - \frac{\beta}{4}(\Delta w)^4 + \frac{\beta}{2}(\nabla \Delta w)^2 \right] ds$$
$$- \oint_{\partial D} \beta (\nabla \Delta w) \frac{\partial \Delta w}{\partial n} d\vec{s} \tag{10}$$

where the r.h.s. is called *pseudoforce*. P is conserved if the pseudoforce is zero. For an isolated system of localized waves, the latter is exactly the case because of the asymptotic conditions at infinity (very large regions D).

SOLITARY WAVES AS QUASI-PARTICLES

Boussinesq Paradigm

A major deficiency of the De-Broglie-Schrödinger wave mechanics is that it tacitly assumes that the wave packets are localized but such a packet is not a solution *per se* of any kind of linear equation. Even in the cases with radial symmetry when localization is observed for the solutions of linear equations, the concept is still not justified because the dispersion will inevitably destroy such a localized wave packet. It turns out that the analogy between waves and particles can be extended farther only if one admits nonlinearity in the model.

Russell [9] discovered a permanent wave (the "Great Wave") on the surface of a shallow water layer (channel). Boussinesq [10] came up with a fundamental idea: the balance between the nonlinearity and dispersion makes the shape of the wave permanent. The equation derived by Boussinesq spawned a variety of models which we call "Boussinesq Paradigm". The next important step was made when Zabusky and Kruskal [11] discovered numerically that the solitary waves retain their shapes after multiple collisions. They introduced the coinage *soliton* in order to emphasize this particle-like behavior of the solitary (localized) waves. Nowadays, the solitons are amply called "quasi-particles". Thus the ground for a non-probabilistic interpretation of the wave mechanics has been set.

Flexural Localized Solutions. The "Flexons"

The flexural localized waves that are solution of Eq. (6) are called in this work "flexons". For numerical example of a shape preserving interaction of *sech*-es in Boussinesq equations we refer the reader to our paper [12] and the references cited therein. Because of the overwhelming numerical problem we were unable until now to come up with stable and fast enough difference scheme in order to examine the full-fledged interaction of two solitons/quasi-particles in 2D. But we have already obtained results for the shapes of stationary propagating solitons for the GDWE Eq. (6) in 2D. In the moving frame, Eq. (6) reads

$$c_1^2 \frac{\partial^2 w}{\partial x^2} + c_2^2 \frac{\partial^2 w}{\partial y^2} + 2c_1 c_2 \frac{\partial^2 w}{\partial x \partial y} = -\kappa w + \Delta \left[-w + w^3 - \Delta w \right], \tag{11}$$

where we have already specified the case with negative membrane tension when the profile of the quasi-particle is more undulated.

Fig 1 presents the result for a soliton propagating along y-line (right panel). Comparing the result to the shape of a resting soliton (left) shows that the moving quasi-particle is contracted in the direction of motion. This is not a Lorentz contraction, however, because the speed of light is not present for the flexural deformations. Yet, the propensity for contraction can be viewed as an universal property of quasi-particles of any GDWE.

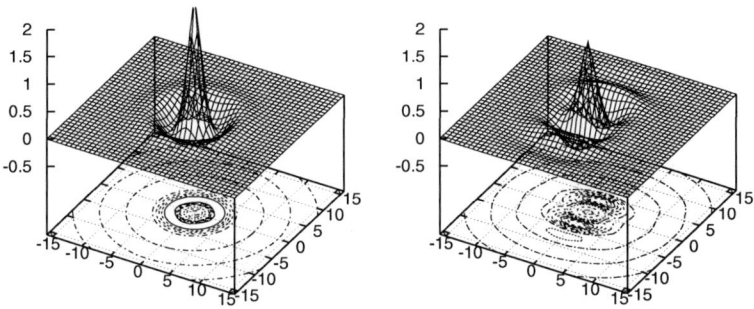

FIGURE 1. Numerical solution for the flexons of (11) with $\kappa = -0.6$. Left panel: flexon at rest. Right panel: flexon propagating with phase velocity (0.,0.6)

Point Particle as a Coarse-Grain Perception of a Localized Wave.

Consider now a single quasi-particle of shape $A(\vec{\xi}, t)$ moving in the field of an external potential $U(\vec{x})$. For the wave amplitude one gets

$$w(\vec{x}, t) = A(\vec{\xi}, t) + U(\vec{x}),$$

where introducing $\vec{\xi} = \vec{x} - \vec{X}(t)$ allows one to acknowledge the moving frame connected with the center of the localized structure. Then

$$\nabla w_t = \nabla A_t - \dot{\vec{X}}(t) \cdot \nabla \nabla A; \quad \nabla^2 w = \nabla^2 A + \nabla^2 U; \quad \nabla \Delta w = \nabla \Delta A + \nabla \Delta U.$$

Then the wave momentum from Eq. (9) $\vec{P}$ gives the following

$$\mathcal{P} = -\int_D (\nabla A_t) \cdot (\nabla \nabla A) d^3 \vec{x} + \frac{d}{dt} \vec{X}(t) \cdot \int_D (\nabla \nabla A) \cdot (\nabla \nabla A) d^3 \vec{x}, \tag{12}$$

which can be called *pseudomomentum* of the geometrical center of the solitary wave when perceived as a point particle. It is well known, however, that the shape of a soliton depends on its phase speed. If we consider the limiting case of slow acceleration/deceleration of the wave, we can neglect the first term in the right hand side and then the above equation gives that the matrix,

$$M = \int_D (\nabla \nabla A) \cdot (\nabla \nabla A) d^n \vec{x},$$

plays the role of a mass. To establish the analogy with the classical particles we mention that the shape $A(\vec{\xi}; \vec{c})$ depends on the phase velocity of the center of the wave or *pseudvelocity* of the quasi-particle (see Fig. 1). When the phase speed is small, the quasi-particle will retain its almost point symmetry and the matrix M will be approximately reduced to a scalar

$$\mathcal{M} = \int_D (\Delta A)^2 d^3 \vec{x}, \tag{13}$$

which we will call *pseudomass* of a single particle in the classical limit. In a sense it is a measure of the resistance (inertia) against changing the measure of the curvature of the shape of pseudo-particle. Then the conservation law for the moment (10), yields the Newton law for the acceleration with $\mathcal{M}$ playing the role of mass, namely $\frac{d}{dt}(\mathcal{M}\dot{X}) = 0$.

For two quasi-particles with shapes A_i and trajectories $\vec{X}_i(t)$, one has

$$w = A_1(\vec{x} - \vec{X}_1(t)) + A_2(\vec{x} - \vec{X}_2(t)) + A_{12}(\vec{x} - \vec{X}_1(t), \vec{x} - \vec{X}_2(t)). \tag{14}$$

For well separated particles, the term F_{12} can be neglected. Then

$$\frac{\partial w}{\partial t} = -\nabla A_1 \cdot \frac{dX_1}{dt} - \nabla A_2 \cdot \frac{dX_2}{dt}, \tag{15}$$

and the term $(\nabla w_t)^2$ yields the following term in the discrete-system Hamiltonian:

$$\sum_{i=1}^{2} \sum_{j=1}^{2} \mathcal{M}_{ij} \cdot \dot{X}_i \dot{X}_j, \quad \mathcal{M}_{ij} = \int (\nabla \nabla A_i(x - X_i)) \cdot (\nabla \nabla A_j(x - X_j)) d^3 x, \tag{16}$$

and the generalization of Newton's law involves the terms

$$\frac{d}{dt} \mathcal{M}_{11} \dot{X}_1 + \frac{d}{dt} \mathcal{M}_{12} \dot{X}_2 = \ldots, \quad \frac{d}{dt} \mathcal{M}_{22} \dot{X}_2 + \frac{d}{dt} \mathcal{M}_{21} \dot{X}_2 = \ldots, \tag{17}$$

where we can call $\mathcal{M}_{11}$ and $\mathcal{M}_{22}$ proper masses of the quasi-particles, while $\mathcal{M}_{12} = \mathcal{M}_{21}$ can be called "cross-mass" of the system of particles. For particles situated far enough from each other, one has $|\mathcal{M}_{12}| \ll \max |\mathcal{M}_{22}|, |\mathcal{M}_{11}|$ for $|X_1 - X_2| \gg 1$. It is interesting to note that the last result means that we have found a kind of Machean model for the quasi-particles in which model the inertial properties of a particle are influenced by the presence of other particles in its vicinity. Naturally, when the particles are far from each other, the mutual influence is negligible.

Thus we have shown that the conservation of the wave energy and momentum one can derive conservation laws for the pseudo-energy and the pseudo-momentum, provided that the pseudomass of the quasi-particle is given by the curvature mass (13).

The Effect of Shell Membrane Tension as Gravitation-like Force

Far from the center of a particle, the shape function w decays to zero and the fourth derivatives and the nonlinear term can be neglected. If we do not consider cosmological scales, we can also disregard the term proportional to $\kappa \ll 1$, the latter being proportional to the inverse of the radius of the hypershell. The linear term plays a stabilizing role preventing long-wave instability on the scale proportional to radius R. Respectively, on short lengths, the stability is ensured by the dispersion term (the fourth derivative). This said we can consider the intermediate scales where the membrane tension is the predominant effect.

$$\frac{\partial^2 w}{\partial t^2} = \pm \Delta w.$$

Quantitatively speaking a *flexon* of shape w will reduce the negative tension (or will introduce positive tension) and the points will experience attractive force proportional to ∇w. The stationary solution for the localized disturbance in 3D gives $w \sim |\vec{x}|^{-1}$ or $\nabla w \sim G|\vec{x}|^{-2}$ for $|\vec{x}| \to \infty$. The shape function of the quasi-particle is the gravitational potential it creates.

An inverse square law (Newton's Inverse-Square Law of Gravitation) means that the shell is *three dimensional*. Please note that for a planar world the asymptotic behavior of w is like $ln|\vec{x}|$ and the attractive force would follow an inverse power of $|\vec{x}|$.

In principle, measuring the attraction between two isolated elementary particles (flexons) could provide the data to identify the shell (anti-)tension ("gravitational") constant.

A final note is due here on the solutions of (11) of type of harmonic waves. They are possible for the model with positive membrane tension, but their phase speed, c_f, is defined in (4) and it differs in general from the phase speed, c, of the shear waves in the middle surface (speed of light in Part I). Even if the gravitational waves exist, they will be highly dispersive and one cannot say that their group speed will have much to do with the phase speed. For the model with negative membrane tension, there is no place for linear harmonic gravitational waves because the problem without the fourth derivative is elliptic in time. Hence the phase speed of propagation of small disturbances will depend strongly on the frequency. And this does not violate the postulate of limiting speed of shear waves (light) in the three laminar dimensions. The flexural waves are in the fourth dimension and should not to be expected to be bound by the limitation from the orthogonal three dimensions.

CONCLUSIONS

We consider a material 3D hypersurface situated in the 4D geometric space. Assuming that it is a mechanical continuum which can support momentum stresses we arrive at the notion of a hypershell. The latter is governed by a nonlinear equation of Boussinesq type whose linear part is the same as for the Euler-Bernoulli and Schrödinger's equations.

The undisturbed shape of the shell can be called "physical vacuum" and the localized flexural waves (solitons or quasi-particles) are the particles. The center of a localized deformation of space is perceived as a point particle the latter being a kind of "corse-grain" description. We show how the Hamiltonian formulation of the metadynamics of the hypershell defines the Newtonian dynamics of the centers of the phase objects (the *quasi-particles*). The particles are not moving *through* the continuum. They are propagating *over* its surface.

Thus, in a Universe governed by the proposed here analogy, the dualism of De Broglie is replaced by a wave-particle syncretism.

A neutral particle is a flexural localized wave (flexon) sustained by the balance between the generating effect of the membrane anti-tension and the restraining effect of the cubic nonlinearity and dispersion. It is a solution of the Generalized Dispersive Wave Equation (GDWE), or the "master equation" of the wave mechanics. A charged particle appears when a *twiston* (see Part I) and a *flexon* form a bound state.

REFERENCES

1. L. Pauling, and J. E. B. Wilson, *Introduction to Quantum Mechanics With Application to Chemistry*, Dover, New York, 1985.
2. C. H. Hinton, *Speculations on the Fourth Dimension. Selected Writings of C. H. Hinton*, Dover, New York, 1980.
3. E. Schrödinger, *Annalen der Physik*, **79**, 743 (1926).
4. H. Neuber, *ZAMM*, **29**, 97–xx,142–xx (1949).
5. M. Dikmen, *Theory of Thin Elastic Shells*, Pitman, Boston, 1982.
6. C. I. Christov, *Annuaire de L'Universite de Sofia*, **89**, 129–140 (1995).
7. C. I. Christov, "Discrete out of continuous: Dynamics of Phase patterns in Continua," in *Continuum Models and Discrete Systems – Proceedings of CMDS8*, edited by K. Markov, World Scientific, Singapore, 1996, pp. 370–394.
8. S. Timoshenko, and S. Woinowsky-Krieger, *Theory of Plates and Shells*, McGraw-Hill, New York, 1959.
9. J. S. Russell, "Report on the Committee on Waves," in *Report of 7th Meeting (1837) of British Assoc. Adv. of Sci., Liverpool*, John Murray, London, 1838, pp. 417–496.
10. J. V. Boussinesq, *Journal de Mathématiques Pures et Appliquées*, **17**, 55–108 (1872).
11. N. J. Zabusky, and M. D. Kruskal, *Phys. Rev. Lett.*, **15**, 57–62 (1965).
12. C. I. Christov, and M. G. Velarde, *J. Bifurcation & Chaos*, **4**, 1095–1112 (1994).

Harmonic Oscillators as Bridges between Theories

Y. S. Kim* and Marilyn E. Noz †

*Department of Physics, University of Maryland College Park, Maryland 20742, U.S.A.
†Department of Radiology, New York University, New York, New York 10016, U.S.A.

Abstract. Other than scattering problems where perturbation theory is applicable, there are basically two ways to solve problems in physics. One is to reduce the problem to harmonic oscillators, and the other is to formulate the problem in terms of two-by-two matrices. If two oscillators are coupled, the problem combines both two-by-two matrices and harmonic oscillators. This method then becomes a powerful research tool to cover many different branches of physics. Indeed, the concept and methodology in one branch of physics can be translated into another through the common mathematical formalism. It is noted that the present form of quantum mechanics is largely a physics of harmonic oscillators. Special relativity is the physics of the Lorentz group which can be represented by the group of by two-by-two matrices commonly called $SL(2,c)$. Thus the coupled harmonic oscillators can therefore play the role of combining quantum mechanics with special relativity. Both Paul A. M. Dirac and Richard P. Feynman were fond of harmonic oscillators, while they used different approaches to physical problems. Both were also keenly interested in making quantum mechanics compatible with special relativity. It is shown that the coupled harmonic oscillators can bridge these two different approaches to physics.

Keywords: Harmonic oscillator, quantum, relativity, partons
PACS: 03.65.Ka, 03.65.Pm, 11.10.-z

INTRODUCTION

Because of its mathematical simplicity, the harmonic oscillator provides soluble models in many branches of physics. It often gives a clear illustration of abstract ideas. In many cases, the problems are reduced to the problem of two coupled oscillators. Soluble models in quantum field theory, such as the Lee model [1] and the Bogoliubov transformation in superconductivity [2], are based on two coupled oscillators. More recently, the coupled oscillators form the mathematical basis for squeezed states in quantum optics [3].

According to our experience, the present form of quantum mechanics is largely a physics of harmonic oscillators. Since the group $SL(2,C)$ forms the universal covering group of the Lorentz group, special relativity is a physics of two-by-two matrices. Therefore, the coupled harmonic oscillator can provide a concrete model for relativistic quantum mechanics.

With this point in mind, Dirac and Feynman used harmonic oscillators to test their physical ideas. In this paper, we first examine Dirac's attempts to combine quantum mechanics with relativity in his own style: to construct mathematically appealing models. We then examine how Feynman approached this problem. He was insisting on his own style. Observe the experimental world, tell the story of the real world, and then write down mathematical formulas as needed.

CP755, ISIS: International Symposium on Interdisciplinary Science
edited by A. Ludu, N.R. Hutchings and D.R. Fry
© 2005 American Institute of Physics 0-7354-0240-X/05/$22.50

In this paper, we use coupled harmonic oscillators to build a bridge between the two different attempts made by Dirac and Feynman.

In section 1, we start with the classical Hamiltonian for two coupled oscillators. It is possible to obtain a explicit solution for the Schrödinger equation in terms of the normal coordinates.

Section 2 examines Dirac's life-long efforts to combine quantum mechanics and special relativity. Starting from Dirac's work, we construct a covariant model of relativistic extended particles by combining Dirac's oscillators with Feynman's phenomenological approach to relativistic quark model. In section 3, it is shown that Feynman's parton model can be interpreted as a limiting case of one Lorentz-covariant bound-state model.

COUPLED OSCILLATORS

Two coupled harmonic oscillators serve many different purposes in physics. It is well known that this oscillator problem can be formulated into a problem of a quadratic equation in two variables. The diagonalization of the quadratic form includes a rotation of the coordinate system. However, the diagonalization process requires additional transformations involving the scales of the coordinate variables [4].

In this paper, we start with a simple problem of two oscillators with equal mass. This contains enough physics for our present purpose. Then the Hamiltonian takes the form

$$H = \frac{1}{2}\left\{\frac{1}{m}p_1^2 + \frac{1}{m}p_2^2 + Ax_1^2 + Ax_2^2 + 2Cx_1x_2\right\}. \tag{1}$$

If we choose coordinate variables

$$y_1 = \frac{1}{\sqrt{2}}(x_1 + x_2), \qquad y_2 = \frac{1}{\sqrt{2}}(x_1 - x_2), \tag{2}$$

the Hamiltonian can be written as

$$H = \frac{1}{2m}\left\{p_1^2 + p_2^2\right\} + \frac{K}{2}\left\{e^{-2\eta}y_1^2 + e^{2\eta}y_2^2\right\}, \tag{3}$$

where

$$K = \sqrt{A^2 - C^2}, \qquad \exp(2\eta) = \sqrt{\frac{A - C}{A + C}}, \tag{4}$$

If y_1 and y_2 are measured in units of $(mK)^{1/4}$, the ground-state wave function of this oscillator system is

$$\psi_\eta(x_1, x_2) = \frac{1}{\sqrt{\pi}}\exp\left\{-\frac{1}{2}(e^{-\eta}y_1^2 + e^{\eta}y_2^2)\right\}, \tag{5}$$

The wave function is separable in the y_1 and y_2 variables. However, for the variables x_1 and x_2, the story is quite different [4]. The key question is how the quantum mechanics in the world of the x_1 variable is affected by the x_2 variable. If the x_2 space is not observed, it corresponds to Feynman's rest of the universe [4].

Let us write the wave function of Eq.(5) in terms of x_1 and x_2, then

$$\psi_\eta(x_1, x_2) = \frac{1}{\sqrt{\pi}} \exp\left\{ -\frac{1}{4} \left[e^{-\eta}(x_1 + x_2)^2 + e^\eta(x_1 - x_2)^2 \right] \right\}. \tag{6}$$

When the system is decoupled with $\eta = 0$, this wave function becomes

$$\psi_0(x_1, x_2) = \frac{1}{\sqrt{\pi}} \exp\left\{ -\frac{1}{2}(x_1^2 + x_2^2) \right\}. \tag{7}$$

The system becomes separable and becomes decoupled.

DIRAC'S HARMONIC OSCILLATORS

Paul A. M. Dirac is known to us through the Dirac equation for spin-1/2 particles. But his main interest was in the foundational problems. First, Dirac was never satisfied with the probabilistic formulation of quantum mechanics. This is still one of the hotly debated subjects in physics. Second, if we tentatively accept the present form of quantum mechanics, Dirac was insisting that it has to be consistent with special relativity. He wrote several important papers on this subject. Let us look at some of his papers on this subject.

TABLE 1. Quantum qechanics and special relativity. There are quantum excitations along the space-like longitudinal direction, but there are no excitations along the time-like direction. The time-energy relation is a c-number uncertainty relation. As for special relativity, Dirac's light-cone system leads to a squeeze transformation illustrated in this table. One way to combine quantum mechanics with special relativity is to combine these two figures.

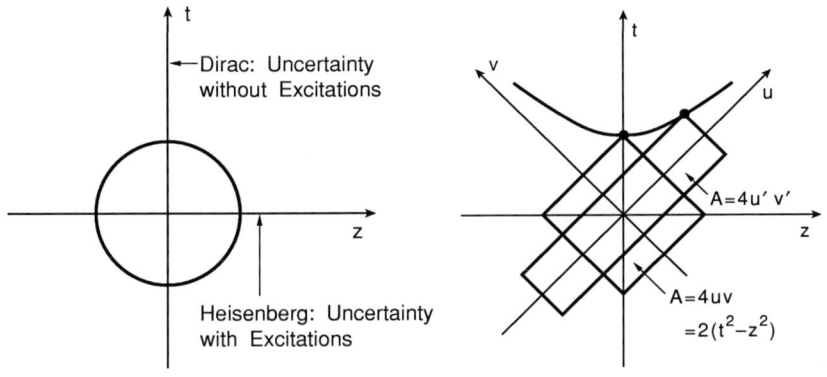

During World War II, Dirac was looking into the possibility of constructing representations of the Lorentz group using harmonic oscillator wave functions [5]. The Lorentz group is the language of special relativity, and the present form of quantum mechanics starts with harmonic oscillators. Presumably, therefore, he was interested in making quantum mechanics Lorentz-covariant by constructing representations of the Lorentz group using harmonic oscillators.

In his 1945 paper [5], Dirac considers the Gaussian form

$$\exp\left\{-\frac{1}{2}\left(z^2+t^2\right)\right\}, \tag{8}$$

where z and t are the longitudinal and time-like variables respectively. This is a strange expression for those who believe in Lorentz invariance. The expression $\left(z^2-t^2\right)$ is invariant, but Dirac's Gaussian form of Eq.(8) is not. Yet, Dirac's expression of Eq.(8) is consistent with his earlier papers on the time-energy uncertainty relation [6]. In those papers, Dirac observed that there is a time-energy uncertainty relation, while there are no excitations along the time axis. He called this the "c-number time-energy uncertainty" relation. When one of us (YSK) was talking with Dirac in 1978, he clearly mentioned this word again. He said further that this is one of the stumbling block in combining quantum mechanics with relativity. This situation is illustrated in Table 1.

In 1949, the Reviews of Modern Physics published a special issue to celebrate Einstein's 70th birthday. This issue contains Dirac paper entitled "Forms of Relativistic Dynamics" [7]. In this paper, he introduced his light-cone coordinate system, in which a Lorentz boost becomes a squeeze transformation. When the system is boosted along the z direction, the transformation takes the form

$$\begin{pmatrix} z' \\ t' \end{pmatrix} = \begin{pmatrix} \cosh(\eta/2) & \sinh(\eta/2) \\ \sinh(\eta/2) & \cosh(\eta/2) \end{pmatrix} \begin{pmatrix} z \\ t \end{pmatrix}. \tag{9}$$

This is not a rotation, and people still feel strange about this form of transformation. In 1949 [7], Dirac introduced his light-cone variables defined as [7]

$$u = (z+t)/\sqrt{2}, \qquad v = (z-t)/\sqrt{2}, \tag{10}$$

the boost transformation of Eq.(9) takes the form

$$u' = e^{\eta/2}u, \qquad v' = e^{-\eta/2}v. \tag{11}$$

The u variable becomes expanded while the v variable becomes contracted, as is illustrated in Table 1. Their product uv remains invariant. In Dirac's picture, the Lorentz boost is a squeeze transformation.

If we combine the two figures in Table 1, we end up with Fig. 1. In mathematical formulae, this transformation changes the Gaussian form of Eq.(8) into

$$\psi_\eta(z,t) = \left(\frac{1}{\pi}\right)^{1/2} \exp\left\{-\frac{1}{2}\left(e^{-\eta}u^2 + e^{\eta}v^2\right)\right\}. \tag{12}$$

Let us go back to section 1 on the coupled oscillators. The above expression is the same as Eq.(5). The x_1 variable now became the longitudinal variable z, and the x_2 variable became the time like variable t.

We can use the coupled harmonic oscillators as the starting point of relativistic quantum mechanics. This allows us to translate the quantum mechanics of two coupled oscillators defined over the space of x_1 and x_2 into the quantum mechanics defined over the space time region of z and t.

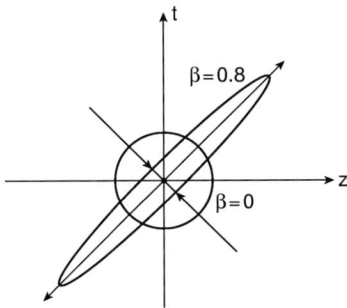

FIGURE 1. Effect of the Lorentz boost on the space-time wave function. The circular space-time distribution in the rest frame becomes Lorentz-squeezed to become an elliptic distribution.

This form becomes Eq.(8) when η becomes zero. The transition from Eq.(8) to Eq.(12) is a squeeze transformation. It is now possible to combine what Dirac observed into a covariant formulation of harmonic oscillator system. First, we can combine his c-number time-energy uncertainty relation described in Table 1 and his light-cone coordinate system given in the same table into a picture of covariant space-time localization illustrated in Fig. 1.

In addition, there are two more homework problems which Dirac left us to solve. First, in defining the t variable for the Gaussian form of Eq.(8), Dirac did not specify the physics of this variable. If it is going to be the calendar time, this form vanishes in the remote past and remote future. We are not dealing with this kind of object in physics. What is then the physics of this time-like t variable?

The Schrödinger quantum mechanics of the hydrogen atom deals with localized probability distribution. Indeed, the localization condition leads to the discrete energy spectrum. Here, the uncertainty relation is stated in terms of the spatial separation between the proton and the electron. If we believe in Lorentz covariance, there must also be a time-separation between the two constituent particles, and an uncertainty relation applicable to this separation variable. Dirac did not say in his papers of 1927 and 1945, but Dirac's "t" variable is applicable to this time-separation variable. This time-separation variable will be discussed in detail in section 3 for the case of relativistic extended particles.

Second, as for the time-energy uncertainty relation. Dirac'c concern was how the c-number time-energy uncertainy relation without excitations can be combined with uncertainties in the position space with excitations. Dira's 1927 paper was written before Wigner's 1939 paper on the internal space-time symmetries of relativistic particles [8].

Both of these questions can be answered in terms of the space-time symmetry of bound states in the Lorentz-covariant regime. In his 1939 paper, Wigner worked out internal space-time symmetries of relativistic particles [8]. He approached the problem by constructing the maximal subgroup of the Lorentz group whose transformations leave the given four-momentum invariant. As a consequence, the internal symmetry of a massive particle is like the three-dimensional rotation group.

If we extend this concept to relativistic bound states, the space-time asymmetry which Dirac observed in 1927 is quite consistent with Einstein's Lorentz covariance. The time variable can be treated separately. Furthermore, it is possible to construct a representations of Wigner's little group for massive particles [9]. As for the time-separation, it is also a variable governing internal space-time symmetry which can be linearly mixed when the system is Lorentz-boosted.

FEYNMAN'S OSCILLATORS

Quantum field theory has been quite successful in terms of Feynman diagrams based on the S-matrix formalism, but is useful only for physical processes where a set of free particles becomes another set of free particles after interaction. Quantum field theory does not address the question of localized probability distributions and their covariance under Lorentz transformations. In order to address this question, Feynman *et al.* suggested harmonic oscillators to tackle the problem [10]. Their idea is indicated in Fig. 2.

Feynman Diagrams

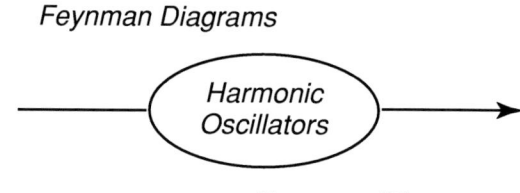

Feynman Diagrams

FIGURE 2. Feynman's roadmap for combining quantum mechanics with special relativity. Feynman diagrams work for running waves, and they provide a satisfactory resolution for scattering states in Einstein's world. For standing waves trapped inside an extended hadron, Feynman suggested harmonic oscillators as the first step.

Before 1964 [11], the hydrogen atom was used for illustrating bound states. These days, we use hadrons which are bound states of quarks. Let us use the simplest hadron consisting of two quarks bound together with an attractive force, and consider their space-time positions x_a and x_b, and use the variables

$$X = (x_a + x_b)/2, \qquad x = (x_a - x_b)/2\sqrt{2}. \qquad (13)$$

The four-vector X specifies where the hadron is located in space and time, while the variable x measures the space-time separation between the quarks. According to Einstein, this space-time separation contains a time-like component which actively participates as in Eq.(9), if the hadron is boosted along the z direction. This boost can be conveniently described by the light-cone variables defined in Eq(10). Does this time-separation variable exist when the hadron is at rest? Yes, according to Einstein. In the present form of quantum mechanics, we pretend not to know anything about this variable. Indeed, this variable belongs to Feynman's rest of the universe.

What do Feynman *et al.* say about this oscillator wave function? In their classic 1971 paper [10], Feynman *et al.* start with the following Lorentz-invariant differential equation.

$$\frac{1}{2}\left\{x_\mu^2 - \frac{\partial^2}{\partial x_\mu^2}\right\}\psi(x) = \lambda\,\psi(x).\tag{14}$$

This partial differential equation has many different solutions depending on the choice of separable variables and boundary conditions. Feynman *et al.* insist on Lorentz-invariant solutions which are not normalizable. On the other hand, if we insist on normalization, the ground-state wave function takes the form of Eq.(8). It is then possible to construct a representation of the Poincaré group from the solutions of the above differential equation [9]. If the system is boosted, the wave function becomes given in Eq.(12).

QUARKS ⟶ PARTONS

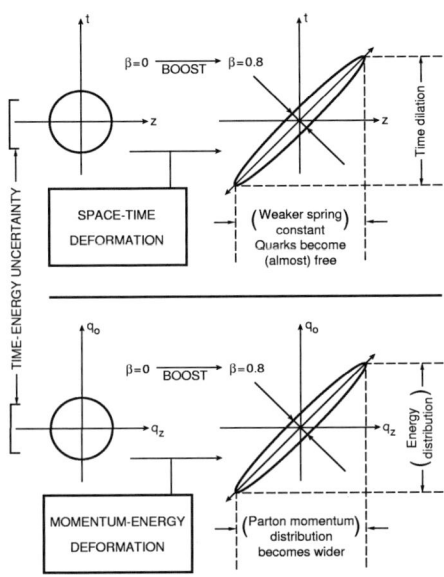

FIGURE 3. Lorentz-squeezed space-time and momentum-energy wave functions. As the hadron's speed approaches that of light, both wave functions become concentrated along their respective positive light-cone axes. These light-cone concentrations lead to Feynman's parton picture.

This wave function becomes Eq.(8) if η becomes zero. The transition from Eq.(8) to Eq.(12) is a squeeze transformation. The wave function of Eq.(8) is distributed within a circular region in the uv plane, and thus in the zt plane. On the other hand, the wave function of Eq.(12) is distributed in an elliptic region with the light-cone axes as the major and minor axes respectively. If η becomes very large, the wave function becomes concentrated along one of the light-cone axes. Indeed, the form given in Eq.(12) is a Lorentz-squeezed wave function. This squeeze mechanism is illustrated in Fig. 1.

There are many different solutions of the Lorentz invariant differential equation of Eq.(14). The solution given in Eq.(12) is not Lorentz invariant but is covariant. It is normalizable in the t variable, as well as in the space-separation variable z. It is indeed possible to construct Wigner's $O(3)$-like little group for massive particles [8], and thus the representation of the Poincaré group [9]. Our next question is whether this formalism has anything to do with the real world.

In 1969, Feynman observed that a fast-moving hadron can be regarded as a collection of many "partons" whose properties appear to be quite different from those of the quarks [12]. For example, the number of quarks inside a static proton is three, while the number of partons in a rapidly moving proton appears to be infinite. The question then is how the proton looking like a bound state of quarks to one observer can appear different to an observer in a different Lorentz frame? Feynman made the following systematic observations.

a. The picture is valid only for hadrons moving with velocity close to that of light.
b. The interaction time between the quarks becomes dilated, and partons behave as free independent particles.
c. The momentum distribution of partons becomes widespread as the hadron moves fast.
d. The number of partons seems to be infinite or much larger than that of quarks.

Because the hadron is believed to be a bound state of two or three quarks, each of the above phenomena appears as a paradox, particularly b) and c) together.

In order to resolve this paradox, let us write down the momentum-energy wave function corresponding to Eq.(12). If we let the quarks have the four-momenta p_a and p_b, it is possible to construct two independent four-momentum variables [10]

$$P = p_a + p_b, \qquad q = \sqrt{2}(p_a - p_b), \tag{15}$$

where P is the total four-momentum. It is thus the hadronic four-momentum.

The variable q measures the four-momentum separation between the quarks. Their light-cone variables are

$$q_u = (q_0 - q_z)/\sqrt{2}, \qquad q_v = (q_0 + q_z)/\sqrt{2}. \tag{16}$$

The resulting momentum-energy wave function is

$$\phi_\eta(q_z, q_0) = \left(\frac{1}{\pi}\right)^{1/2} \exp\left\{-\frac{1}{2}\left(e^\eta q_u^2 + e^{-\eta} q_v^2\right)\right\}. \tag{17}$$

Because we are using here the harmonic oscillator, the mathematical form of the above momentum-energy wave function is identical to that of the space-time wave function. The Lorentz squeeze properties of these wave functions are also the same. This aspect of the squeeze has been exhaustively discussed in the literature [9, 13, 14].

When the hadron is at rest with $\eta = 0$, both wave functions behave like those for the static bound state of quarks. As η increases, the wave functions become continuously squeezed until they become concentrated along their respective positive light-cone axes. Let us look at the z-axis projection of the space-time wave function. Indeed, the width of the quark distribution increases as the hadronic speed approaches that of the speed of light. The position of each quark appears widespread to the observer in the laboratory frame, and the quarks appear like free particles.

The momentum-energy wave function is just like the space-time wave function, as is shown in Fig. 3. The longitudinal momentum distribution becomes wide-spread as the hadronic speed approaches the velocity of light. This is in contradiction with our expectation from non-relativistic quantum mechanics that the width of the momentum distribution is inversely proportional to that of the position wave function. Our expectation is that if the quarks are free, they must have their sharply defined momenta, not a wide-spread distribution.

However, according to our Lorentz-squeezed space-time and momentum-energy wave functions, the space-time width and the momentum-energy width increase in the same direction as the hadron is boosted. This is of course an effect of Lorentz covariance. This indeed is the key to the resolution of the quark-parton paradox [9, 13].

After these qualitative arguments, we are interested in whether Lorentz-boosted bound-state wave functions in the hadronic rest frame could lead to parton distribution functions. It is thus possible to compare the oscillator-based parton distribution with that observed in high-energy laboratories [15].

Feynman's parton picture is one of the most controversial models proposed in the 20th century. The original model is valid only in Lorentz frames where the initial proton moves with infinite momentum. It is gratifying to note that this model can be produced as a limiting case of one covariant model which produces the quark model in the frame where the proton is at rest.

REFERENCES

1. S. S. Schweber, *An Introduction to Relativistic Quantum Field Theory* (Row-Peterson, Elmsford, New York, 1961).
2. A. L. Fetter and J. D. Walecka, *Quantum Theory of Many Particle Systems* (McGraw-Hill, New York, 1971).
3. Y. S. Kim and M. E. Noz, *Phase Space Picture of Quantum Mechanics* (World Scientific, Singapore, 1991).
4. D. Han, Y. S. Kim, Am. J. Phys. **67**, 61 (1999).
5. P. A. M. Dirac, Proc. Roy. Soc. (London) **A183**, 284 (1945).
6. P. A. M. Dirac, Proc. Roy. Soc. (London) **A114**, 234 and 710 (1927).
7. P. A. M. Dirac, Rev. Mod. Phys. **21**, 392 (1949).
8. E. Wigner, Ann. Math. **40**, 149 (1939).
9. Y. S. Kim and M. E. Noz, *Theory and Applications of the Poincaré Group* (Reidel, Dordrecht, 1986).
10. R. P. Feynman, M. Kislinger, and F. Ravndal, Phys. Rev. D **3**, 2706 (1971).
11. M. Gell-Mann, Phys. Lett. **13**, 598 (1964).
12. R. P. Feynman, *The Behavior of Hadron Collisions at Extreme Energies*, in *High Energy Collisions*, Proceedings of the Third International Conference, Stony Brook, New York, edited by C. N. Yang *et al.*, Pages 237-249 (Gordon and Breach, New York, 1969).
13. Y. S. Kim and M. E. Noz, Phys. Rev. D **15**, 335 (1977).
14. Y. S. Kim, Phys. Rev. Lett. **63**, 348 (1989).
15. P. E. Hussar, Phys. Rev. D **23**, 2781 (1981).

Fourier-Galerkin Method for Time Dependent Problems of Interacting Localized Waves

M. A. Christou* and C. I. Christov†

*Dept of Mathematics, University of Louisiana at Monroe
†Dept of Mathematics, University of Louisiana at Lafayette

Abstract. We develop a Fourier-Galerkin spectral technique for computing solutions of type of interacting localized waves. We use a special complete orthonormal system of functions in $L^2(-\infty,\infty)$. The rate of convergence of the coefficients is shown to be exponential. As a featured example we consider the Boussinesq Paradigm Equation (BPE). We obtain results for the head-on and overtaking collisions of two or three solitons. We evaluate also the phase shifts of solitons.

Keywords: Galerkin Spectral Method
PACS: 02.60.Cb, 02.70.Hm

INTRODUCTION

In recent years a number of physical problems have led to boundary value problems in infinite domains. These are the cases when no boundary conditions are specified at given points, but rather the square of the solution (or some other energy norm) is required to be integrable over an infinite domain. Such solutions are said to belong to the $L^2(-\infty,\infty)$ space. A typical example is furnished by the problem for solitary wave (soliton) solutions. There are many difficulties on the way of application of difference or/and finite-element numerical methods to the problems in $L^2(-\infty,\infty)$, especially when some more subtle characteristics are sought. Very often, the finite-domain problem has a solution only for some denumerable set of intervals of specific length. It can even happen that each of the finite-domain approximations has only a trivial solution, while the original problem possesses a nontrivial one, or *vice versa*.

This difficulty can be surmounted if a spectral method with basis system that consists of localized functions is used instead of a difference method. Such expansion will automatically acknowledge the requirement that the solution belongs to $L^2(-\infty,\infty)$ space. Here we make use of a complete orthonormal (CON) system of functions proposed in [1]. These functions are orthogonal without weight and for them expression are available relating the product of two members of the system into series with respect to the system. This property is crucial because it allows one to use a Galerkin type expansion, the latter being much simpler and faster in implementation than the pseudo-spectral algorithms.

BOUSSINESQ PARADIGM EQUATION

Consider the so-called Boussinesq Paradigm equation (BPE),

$$u_{tt} = (u - 3u^2 + \beta_1 u_{tt} - \beta_2 u_{xx})_{xx}, \tag{1}$$

CP755, *ISIS: International Symposium on Interdisciplinary Science*
edited by A. Ludu, N.R. Hutchings and D.R. Fry

which has an analytical *sech* soliton solution in the moving frame [2]

$$u = \frac{1}{2}(c^2 - 1)\mathrm{sech}^2\left(\frac{x - ct}{2}\sqrt{\frac{c^2 - 1}{\beta_1 c^2 - \beta_2}}\right). \tag{2}$$

As we can see from (2), the soliton exists for $c > 1$ (supercritical phase speeds).

Before proceeding further we mention that re-scaling the spatial variable x does not change the nature of the asymptotic boundary value problem in $L^2(-\infty, \infty)$. Upon introducing $z = \zeta x$ we recast (1) to the following

$$u_{tt} = \zeta^2(u - 3u^2 + \beta_1 u_{tt} - \zeta^2 \beta_2 u_{zz})_{zz} \quad \text{with a.b.c} \quad u(t, z) \to 0, \quad z \to \pm\infty. \tag{3}$$

The scaling parameter ζ can be used to optimize the method in the sense that its introduction allows one to bring the typical length scales of the employed system of functions closer to the length of support of the sought localized solution.

We introduce an auxiliary function q and show that when localized solution is sought, eq.(3) is a corollary of the following system

$$u_t = \zeta^2 q_{zz}, \tag{4}$$
$$q_t = u - 3u^2 + \beta_1 u_{tt} - \beta_2 \zeta^2 u_{zz}. \tag{5}$$

Function q is also a localized function but it can assume nonzero values at infinities. It has the shape of a hydraulic jump (or *kink*). Then the asymptotic boundary conditions for the system (4), (5) have the form

$$u, q_z \to 0 \quad \text{for} \quad z \to -\infty, \infty.$$

The initial condition is the superposition of two *sech*-solutions which are situated far enough from each other in order to neglect their intersection in the initial moment. We can find initial condition for the function $q(z, t)$ of each soliton after we integrate the first equation of (5) twice with respect to z in the moving frame. Then

$$q(z, t) = \frac{-3c_1}{\zeta^2}\sqrt{c_1^2 - 1}\,\tanh\left[\left(\frac{z_1}{2\zeta} - c_1 t\right)\sqrt{\frac{c_1^2 - 1}{\beta_1 c_1^2 - \beta_2}}\right]$$
$$+ \frac{-3c_2}{\zeta^2}\sqrt{c_2^2 - 1}\,\tanh\left[\left(\frac{z_2}{2\zeta} - c_2 t\right)\sqrt{\frac{c_2^2 - 1}{\beta_1 c_2^2 - \beta_2}}\right].$$

Now, function $q \notin L^2(-\infty, \infty)$ and it cannot be developed into series with respect to a CON system of functions whose members vanish at infinities. We use another auxiliary function $r(z)$ which "absorbs" the undesired behavior of $q(z, t)$, namely $p(z, t) = q(z, t) + r(z)$, where

$$r(z) = -\left[\frac{3c_1}{\zeta^2}\sqrt{\frac{c_1^2 - 1}{\beta_1 c_1^2 - \beta_2}} + \frac{3c_2}{\zeta^2}\sqrt{\frac{c_2^2 - 1}{\beta_1 c_2^2 - \beta_2}}\right]\tanh(z).$$

Note that the introduction of the new function $p(z,t)$ does not alter the equation (3) because $r(z)$ does not depend on t. In terms of $p(z,t)$, the system (4), (5) reads

$$u_t = \zeta^2 p_{zz} + \zeta^2 r_{zz}, \tag{6}$$

$$p_t = u - 3u^2 + \beta_1 u_{tt} - \zeta^2 \beta_2 u_{zz} \tag{7}$$

This is an essential point for the application of the method.

THE FOURIER-GALERKIN METHOD IN $L^2(-\infty, \infty)$

From the known spectral techniques, we choose the Galerkin method because it has the advantage of simplicity in implementation in comparison with the spectral collocation method or tau-method (see the arguments in [3]). The Galerkin technique requires explicit formulas expressing the products of members of the CON system into series with respect to the system. A system with the desired property has been proposed in 1982 in [1] and since then applied to Boussinesq equation and some other nonlinear-wave problems (see, [4, 5, 6]). The applications were limited to the stationary in the moving frame waves. A completely new application is started by the present authors [7] for the classical Boussinesq equation, and in [8] for the Reguralized Long-Wave equation [9]. Here we forward the technique for the equation with mixed fourth derivative.

The system

$$\rho_n = \frac{1}{\sqrt{\pi}} \frac{(ix-1)^n}{(ix+1)^{n+1}}, \quad n = 0,1,2,\ldots \tag{8}$$

was introduced by Wiener [10] as Fourier transform of the Laguerre functions (functions of parabolic cylinder). Higgins [11] defined it also for negative indices n and proved its completeness and orthogonality. The significance of (8) for nonlinear problems was demonstrated in [1], where the product formula was derived and the two real-valued subsequences of odd functions S_n and even functions C_n were introduced, namely

$$\rho_n \rho_k = \frac{\rho_{n+k} - \rho_{n-k}}{2\sqrt{\pi}}, \quad S_n = \frac{\rho_n + \rho_{-n-1}}{i\sqrt{2}}, \quad C_n = \frac{\rho_n - \rho_{-n-1}}{\sqrt{2}}. \tag{9}$$

One easily shows that the product (e.g., $C_n C_k$) of members of the real-valued sequences is expanded in series with respect to the system as follows (see, [1]):

$$C_n C_k = \frac{1}{2\sqrt{2\pi}} [\, C_{n+k+1} - C_{n+k} - C_{n-k} + C_{n-k-1}] = \sum_{m=1}^{\infty} \beta_{nk,m} C_m \tag{10}$$

$$\beta_{nk,m} = \frac{1}{2\sqrt{2\pi}} \{\delta_{m,n+k} + \delta_{m,|n-k|} - \delta_{m,n+k+1} - \mathrm{sgn}[|n-k| - 0.5]\delta_{m,[|n-k|-0.5]}\}.$$

Before proceeding further we discuss the way to increase the computational effectiveness of the product formulas. If we consider an even function from the $L(-\infty, \infty)$ space, say $U(x) = \sum_{n=0}^{\infty} u_n C_n$, we can show that

$$U^2(x) = \sum_{n=0}^{N}\sum_{m=0}^{N} u_n u_m C_n C_m = \sum_{l=0}^{N}\left[\sum_{n=0}^{N}\sum_{m=0}^{N}\beta_{nm,l}u_n u_m\right]C_l \overset{\text{def}}{=} \frac{1}{2\sqrt{2\pi}}\sum_{l=0}^{N} b_l C_l, \tag{11}$$

$$b_l = \sum_{n=0}^{l-1} u_n u_{l-1-n} - \sum_{n=0}^{l} u_n u_{l-n} - 2\sum_{n=l}^{N} u_n u_{n-l} + 2\sum_{n=l+1}^{N} u_n u_{n-l-1}. \tag{12}$$

For the second derivative of the basis functions one has (see [1])

$$C_n'' = \sum_{m=0}^{\infty} \chi_{m,n} C_m, \qquad S_n'' = \sum_{m=0}^{\infty} \chi_{m,n} S_m,$$

$$\chi_{m,n} = -\tfrac{1}{4}n(n-1)\delta_{m,n-2} + n^2\delta_{m,n-1} - \tfrac{1}{4}(n+1)(n+2)\delta_{m,n+2}$$

$$-\tfrac{1}{4}n^2 + (2n+1)^2 + (n+1)^2\delta_{m,n} + (n+1)^2\delta_{m,n+1},$$

Here χ is a diagonal matrix (pentadiagonal, more specifically), which can be inverted for $O(N\ln N)$ operations. This gives another edge in the computational efficiency of the developed here technique.

An important property of the proposed expansion is its exponential convergence ([7]).

THE TIME-STEPPING ALGORITHM

We develop the sought solution u, p into series with respect to C_n and S_n namely,

$$u^l(z) = \sum_{n=0}^{\infty} a_n^l C_n(z) + b_n^l S_n(z), \quad p^{l+\frac{1}{2}}(z) = \sum_{m=0}^{\infty} d_m^{l+\frac{1}{2}} C_m(z) + e_m^{l+\frac{1}{2}} S_m(z). \tag{13}$$

Now, we insert the spectral expansion (13), into equations (6), (7) and making use the orthogonality of the system we get the discrete equations

$$\frac{a_m^{l+1} - a_m^l}{\tau} = \zeta^2 \sum_{k=0}^{\infty} d_k^{l+\frac{1}{2}} \chi_{m,k}, \qquad \frac{d_m^{l+\frac{1}{2}} - d_m^{l-\frac{1}{2}}}{\tau} = -\frac{\zeta^2}{2}\sum_{k=0}^{\infty}(a_k^{l+1} + a_k^{l-1})\chi_{k,m}$$

$$+\frac{a_m^{l+1} + a_m^{l-1}}{2} + \beta_1 \frac{a^{l+1} - 2a^l + a^{l-1}}{\tau^2} - \sum_{k_1=0}^{\infty}\sum_{k_2=0}^{\infty}(a_{k_1}^l a_{k_2}^l \beta_{k_1 k_2,m} + b_{k_1}^l b_{k_2}^l \alpha_{k_1 k_2,m})$$

$$\frac{b_m^{l+1} - b_m^l}{\tau} = \zeta^2 \sum_{k=0}^{\infty}\left(b_k^{l+\frac{1}{2}} + r_k\right)\chi_{m,k}, \qquad \frac{e_m^{l+\frac{1}{2}} - e_m^{l-\frac{1}{2}}}{\tau} = -\frac{\zeta^2}{2}\sum_{k=0}^{\infty}(b_k^{l+1} + b_k^{l-1})\chi_{k,m}$$

$$+\frac{b_m^{l+1} + b_m^{l-1}}{2} + \beta_1 \frac{b^{l+1} - 2b^l + b^{l-1}}{\tau^2} - \sum_{k_1=0}^{\infty}\sum_{k_2=0}^{\infty}(a_{k_1}^l b_{k_2}^l \gamma_{k_1 k_2,m} + b_{k_1}^l a_{k_2}^l \gamma_{k_1 k_2,m}).$$

where $\alpha_{k_1 k_2 k_3}$ and $\gamma_{k_1 k_2 k_3}$ are similar to $\beta_{k_1 k_2 k_3}$ given in (10). Respectively, r_k are the coefficients of the spectral expansion of function r_{zz}. In the numerical calculations a truncated version of the above system is used in which the infinity is replaced by N.

The initial conditions $\{a_n^0\}$, $\{b_n^0\}$ and $\{a_n^1\}$, $\{b_n^1\}$ for the Fourier coefficients are calculated for $t = 0$ and $t = \tau$ by means of numerical quadrature of the analytic solution formulas after multiplying them by C_n or S_n. In its turn, the initial conditions for the coefficients $\{d_n^{\frac{1}{2}}\}$, $\{e_n^{\frac{1}{2}}\}$ of p are computed via numerical quadrature of its analytic expression after multiplying the latter by C_n or S_n. Note that the initial conditions have to be calculated anew every time when the value ζ of the scaling parameter is changed. Having specified the initial conditions we can begin the time stepping. Let us assume that the variables $\{a_n^{l-1}\}$, $\{b_n^{l-1}\}$, $\{a_n^l\}$, $\{b_n^l\}$, $\{d_n^{l-\frac{1}{2}}\}$, $\{e_n^{l-\frac{1}{2}}\}$ are known. Then our systems for the coefficients give two coupled nine-diagonal algebraic systems for $\{b_n^{l+1}\}$, $\{e_n^{l+\frac{1}{2}}\}$ and for $\{a_n^{l+1}\}$, $\{d_n^{l+\frac{1}{2}}\}$, respectively. After these systems are solved and a time step is completed, the time index l is reset, and the process is repeated.

VALIDATIONS OF THE SCHEME

The optimal ζ is different for different initial configurations of the system of solitons. When the solitons aresituated far from each other, one is faced with a wave configuration which is not tightly localized. Then the optimal value of ζ is smaller. Conversely, if in the initial configuration the solitons are close enough, the value of ζ tends to be larger. In the present work we consider initial configurations of solitons that are well separated (in order not to overlap significantly), but not excessively far from each other (not to loose the localization). After conducting extensive numerical experiments we found for these cases that the optimal value of the scaling parameter is in the vicinity of $\zeta = 0.06$. This is the value for which the convergence was faster and more accurate

As far as the time increment is concerned we found that the calculations are perfectly stable for τ as large as 0.1 even for $c = 2$, which is a very large value from the point of view of weakly-nonlinear approximation.

As already mentioned above, the rate of convergence of the Galerkin series is exponential and our calculations comply with this analytical result. In Figure 1 we present the Galerkin coefficients for the case $c_1 = 1.8$ and $c_2 = -1.4$. As we can see, the convergence remains exponential after 300 time steps even though we used a rather large time increment $\tau = 0.1$.

RESULTS AND DISCUSSION

For the Proper Boussinesq Equation, the faster solitons have smaller amplitudes, while in the case of the BPE (as in the KdV and Reguralized Long-Wave equations), the faster solitons are taller. The amplitude of the wave is getting smaller as $c \to 1, c > 1$. When $c = 1$, then $u(z,t) = 0$,

We computed the evolution of several different initial configurations using $\zeta = 0.06$ and $\tau = 0.1$. For c close to unity the interaction is completely shape preserving and after the interaction the two solitons reappear with their exact shapes and the only sign of inelasticity is the phase shift. For larger c we observe that after the separation of

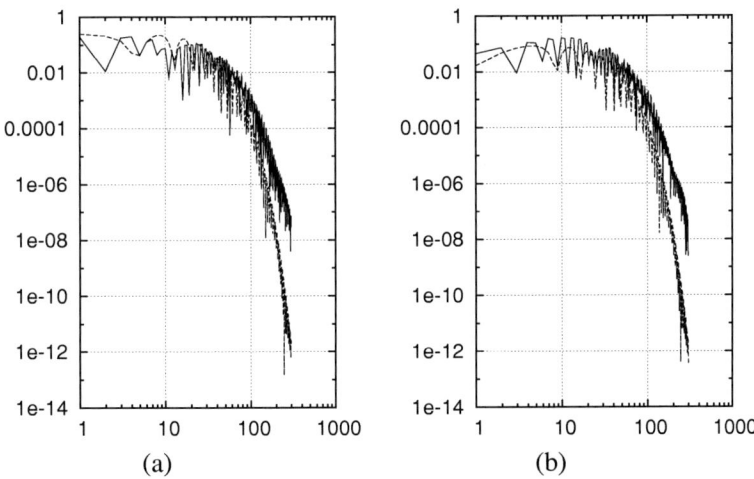

(a)   (b)

FIGURE 1. Exponential decay of the computed coefficients for $\zeta = 0.06$ and $N = 250$: (a) Even coefficients; (b) Odd coefficients.

the solitons a residual signal appears, which was also reported in the work using finite difference scheme [12].

The scheme proposed here conserves the energy of the system up to six decimal digits which made possible the investigation of the interaction for very long times. Note that even the slightest but persistent "leakage" of energy during the calculations would have led to eventual linear dispersion of the solution and disappearance of the permanent (*sech*) shapes. The quantitative agreement with the finite-difference scheme is very good.

The cases of two equal solitons require only even functions in the expansion while the general case needs also the odd functions and reveals better the effectiveness of the method developed here. We present the interaction of two nonequal solitons in Fig. 2.

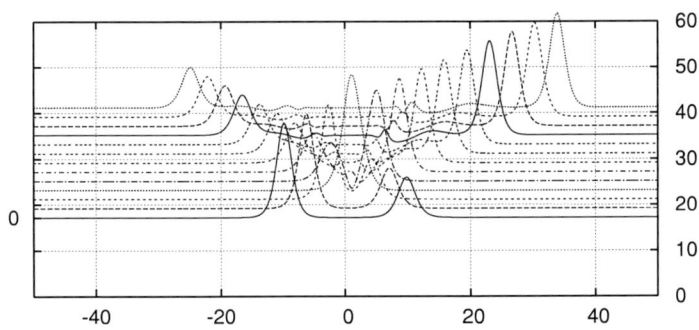

FIGURE 2. Collision of two nonequal solitons for $c_l = 1.8$ and $c_r = -1.4$

An important advance of the numerical technique here in comparison with our previous works is that we succeeded to apply it also to the overtaking collisions of solitons. The specific difficulty here is that the whole pattern moves away from the origin of the coordinate systems which decreases the role of the lower-order terms in the expansion and degrades the practical convergence. This requires larger number of terms. Yet, our calculations show that $N = 250$ is fully adequate number even for the case of tipple interaction of solitons as presented in Fig. 3.

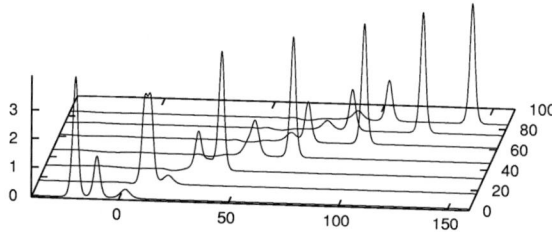

FIGURE 3.   Overtaking of three nonequal solitons for $c_1 = 1.9$, $c_2 = 1.3$ and $c_3 = 1.1$.

In the end we focus our attention on the phase shift, describing the difference between the positions of a soliton with and without interaction after the same number of time steps. The idea of phase shift is illustrated in Fig. 4 for the collision of two solitons with $c_l = 1.5$ and $c_r = -1.1$ and for dimensionless time $t = 30$.

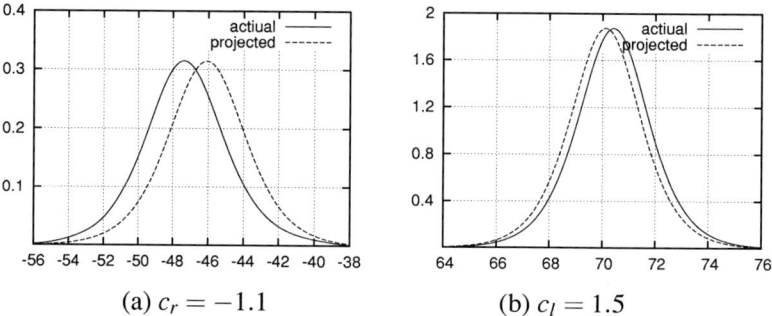

(a) $c_r = -1.1$ (b) $c_l = 1.5$

FIGURE 4.   The phase shifts for a system of solitons. Solid line: actual position of a soliton; dashed line - the projected position for a noninteracting soliton at the same time $t = 30$.

Table 1. Computed phase shifts

c_1	Shift	c_2	Shift
1.4	1.2	-1.8	0.52
1.5	0.84	-1.5	0.84
1.2	1.38	-1.9	0.28
-1.2	1.24	-1.9	0.76

We have conducted extensive numerical experiments and obtained results for phase shifts for a large number of initial configurations of the solitons. A selection of phase shifts for different phase velocities of the solitons are compiled in Table 1. The last case in the table is for overtaking collision in which the taller soliton suffers relatively larger shift than in the head-on collision with the same smaller soliton.

CONCLUSIONS

A complete orthonormal (CON) basis system in $L^2(-\infty, \infty)$ is used to develop a localized solution into Fourier series with Galerkin identification of the coefficients. It is applied for finding the solitary waves for the so-called Boussinesq Paradigm Equation (BPE) which contains two kinds of dispersion. It is a development upon our previous works where only purely spatial or only purely mixed-derivative dispersions were considered.

In the two-soliton case under consideration, the localization of the solution is much less tighter and the number of terms needed for good approximation is larger. The treatment of the problem required very efficient implementation of the Fourier-Galerkin scheme. The numerical experiments confirm the exponential convergence of the method which allows to obtain highly accurate results for the time dependent problem with as few as $N = 40$ terms. This demonstrates the efficiency of the proposed technique and encourages the future use of the CON system.

It is shown that the solitons recover their exact shapes after the collision but experience phase shifts. When the phase speeds are large, the model does not comply with the weakly-nonlinear assumptions and residual signals are observed trailing the main humps.

REFERENCES

1. C. I. Christov, *SIAM J. Appl. Math.*, **42**, 1337–1344 (1982).
2. C. I. Christov, "Conservative Difference Scheme for Boussinesq Model of Surface Waves," in *Proc. ICFD V*, edited by W. K. Morton, and M. J. Baines, Oxford University Press, 1995, pp. 343–349.
3. J. P. Boyd, *Chebyshev and Fourier Spectral Methods*, Dover Press, 2000.
4. C. I. Christov, and K. L. Bekyarov, *SIAM J. Sci. Stat. Comp.*, **11**, 631–647 (1990).
5. C. I. Christov, *Annuaire de l'Univ. Sofia, Livre 2 – Mathématiques Appliqee et Informatique*, **89**, 169–179 (1995).
6. M. Christou, and C. I. Christov, *Journal of Comp. Analysis and Appl.*, **4**, 63–77 (2002).
7. C. I. Christov, and M. A. Christou (2004), to be submitted.
8. M. A. Christou, and C. I. Christov, *Journal of Mathematics and Computers in Simulation* (2004), to appear.
9. T. B. Benjamin, J. L. Bona, and J. J. Mahony, *Phil. Trans. Roy. Soc. London*, **A272**, 47–78 (1972).
10. N. Wiener, *Extrapolation, Interpolation and smoothing of stationary time series*, Technology Press MIT and John Wiley, New York, 1949.
11. J. R. Higgins, *Completeness and basis properties of sets of special functions*, Cambridge University Press, London, 1977.
12. C. I. Christov, and M. G. Velarde, *J. Bifurcation & Chaos*, **4**, 1095–1112 (1994).

Generalized Quasilinearization Method for Reaction Diffusion Equation with Numerical Applications

A. S. Vatsala, Jie Yang

Department of Mathematics, University of Louisiana at Lafayette, Lafayette, LA, 70504

Abstract. We have developed generalized quasilinearization method for reaction diffusion systems in [11], when the forcing functions are the sum of convex and concave functions. In this paper, we present theoretical result of the scalar case of reaction diffusion equation when the forcing function is the sum of a convex and concave function without proof. Further we demonstrate the application of the GQL method to reaction diffusion equation with numerical applications.

Keywords: Generalized Quasilinearization, Quadratic Convergence, Finite Difference Method, Reaction Diffusion Equation
PACS: 00.02.60-x

INTRODUCTION

The method of Quasilinearization developed by Bellman and Kalaba [1, 2] has been extended, generalized and refined so that it includes the situation when the forcing function is the sum of a convex and concave functions. See [6, 7] for details. The method is popular in applications such as quantum mechanics and in physics [4, 5, 8, 9] due to its faster rate of convergence. The method of quasilinearization has been extended to reaction diffusion systems in [11]. The numerical application of theoretical results of [11] are presented in [12]. In this paper we merely state two scalar results of [11]. However, our scalar result is an improvement over the usual GQL method (the generalized quasilinearization method) of [6], since we do not require the forcing functions to be twice differentiable in the nonlinear term, which is needed in the GQL method. Reaction diffusion equations occurs in population models and in chemical reactions. We present two numerical results which demonstrate the application of our theoretical results.

PRELIMINARIES

In this section we present some basic definitions and theorem related to reaction diffusion systems that we need in our main results.

Consider the reaction diffusion system:

$$\mathcal{L}u = F(t,x,u) \ in \ Q_T, \tag{1a}$$

$$\mathcal{B}u = \Phi \ on \ \Gamma_T, \tag{1b}$$

$$u(0,x) = u_0(x) \ in \ \overline{\Omega}, \tag{1c}$$

CP755, *ISIS: International Symposium on Interdisciplinary Science*
edited by A. Ludu, N.R. Hutchings and D.R. Fry
© 2005 American Institute of Physics 0-7354-0240-X/05/$22.50

where Ω is a smooth bounded domain in $\mathbb{R}^m$, $Q_T = (0,T) \times \Omega$ for some $T > 0$. Further, $\Gamma_T = (0,T) \times \partial\Omega$, for any $T < \infty$, $u = (u_1, u_2, u_3, \cdots, u_N) \in \mathbb{R}^N$, $F = (F_1, F_2, F_3, \cdots, F_N)^T \in C^{\frac{\alpha}{2},\alpha}[[0,T] \times \overline{\Omega} \times \mathbb{R}^N, \mathbb{R}^N]$, and $\Phi \in C^{\frac{1+\alpha}{2},1+\alpha}[\Gamma_T, \mathbb{R}^N]$, $u_0(x) \in C^{2+\alpha}[\overline{\Omega}, \mathbb{R}^N]$. $\mathscr{L} = (\mathscr{L}_1, \mathscr{L}_2, \mathscr{L}_3, \cdots, \mathscr{L}_N)^T \in \mathbb{R}^N$, and ν be a unit outward normal vector field on Γ_T. Here $\mathscr{L}_k = \frac{\partial}{\partial t} - L_k$, where $a_{ij}^k, b_i^k \in C^{\frac{\alpha}{2},\alpha}[\overline{Q}_T, R]$,

$$L_k u = \sum_{i,j=1}^{m} a_{ij}^k(t,x) \frac{\partial^2 u}{\partial x_i \partial x_j} + \sum_{i=1}^{m} b_i^k(t,x) \frac{\partial u}{\partial x_i},$$

and $\mathscr{B} = (\mathscr{B}_1, \mathscr{B}_2, \mathscr{B}_3, \cdots, \mathscr{B}_N)^T \in \mathbb{R}^N$, $\mathscr{B}$ is the boundary operator given by $\mathscr{B}_i u_i = p_i(t,x) u_i + q_i(t,x) \frac{\partial u_i}{\partial \nu}$ where $\frac{\partial u_i}{\partial \nu}$ denotes the outward normal derivative of u and $\nu(t,x)$ the unit outward normal vector field on $\partial\Omega$ for $(t,x) \in \Gamma_T$, and $p,q \in C^{\frac{1+\alpha}{2},1+\alpha}[\Gamma_T, \mathbb{R}_+^N]$.

Also, we list the following assumptions for convenience.

(A_1) (i) For each $k = 1, 2, \cdots, N$, L_k is uniformly elliptic operator on $\overline{\Omega}$, and the co-efficients of L_k belong to $C^{\frac{\alpha}{2},\alpha}[\overline{Q}_T, R]$, that is for each $i,j \in \{1, 2, \cdots, m\}$ and $k = 1, 2, \cdots, N$, $a_{ij}, b_j \in C^{\frac{\alpha}{2},\alpha}[\overline{Q}_T, R]$;

(ii) $p_k, q_k \in C^{\frac{1+\alpha}{2},1+\alpha}[\Gamma_T, R]$, $p_k > 0$, $q_k \geq 0$ on Γ_T;

(iii) $\partial\Omega \in C^{2+\alpha}[\mathbb{R}, \mathbb{R}]$;

(iv) $\Phi \in C^{\frac{1+\alpha}{2},1+\alpha}[\Gamma_T, \mathbb{R}^N]$, $u_0(x) \in C^{2+\alpha}[\overline{\Omega}, \mathbb{R}^N]$;

(v) The IBVP (1a) – (1c) satisfies the compatibility condition of order $\frac{\alpha+1}{2}$.

(vi) $F \in C^{\frac{\alpha}{2},\alpha}[\overline{Q}_T \times \mathbb{R}^N, \mathbb{R}^N]$, that is each F_k is Hölder continuous in t and (x,u) with exponent $\frac{\alpha}{2}$, α respectively;

In order to state our next known result, the Lax Equivalence Theorem which we use in our numerical applications, we consider the linear diffusion equation of the form

$$\frac{\partial u}{\partial t} = Lu + f, \quad x \in U, t \geq 0 \tag{2}$$

where $U \subset \mathbb{R}^s$, $u = u(x,t)$, $f = f(x,t)$ and L is a linear elliptic operator with a general full discretion form

$$u_{\Delta x}^{n+1} = A_{\Delta x} u_{\Delta x}^n + k_{\Delta x}^n, \quad n = 0, 1, \cdots, \tag{3}$$

Theorem 1. *Lax Equivalence Theorem*

Provided that the linear evolutionary PDE (2) is well posed, the fully discretized numerical method (3) is convergent if only if it is stable and of order $p \geq 1$.

MAIN RESULTS, NUMERICAL APPROACH

In this section, we recall the special case of result relative to (1a)–(1c) where $N = 1$, the scalar case. Although this special case has been established in [6], our result has been developed under weaker assumptions. The assumption that f_{uu} or g_{uu} to exist is not really required in our result. We rewrite the special case when $f, g \in C^{\frac{\alpha}{2},\alpha}[\overline{Q}_T \times \mathbb{R}, \mathbb{R}]$ as follows:

$$\mathscr{L}(u) = f(t,x,u) + g(t,x,u) \quad \text{in } Q_T, \tag{4a}$$
$$\mathscr{B}(u) = \Phi \quad \text{on } \Gamma_T, \tag{4b}$$
$$u(0,x) = u_0(x) \quad \text{in } \overline{\Omega}, \tag{4c}$$

In this section, we merely state two theoretical results of [11] for $N = 1$ without proof. We present numerical examples which verify our theoretical results of Generalized quasilinearization method for (4a)–(4c). The following is the first theoretical result.

Theorem 2. *Assume that* $(i) - (v)$ *of* (A_1) *holds, Further assume that*

(A_2) $v_0, w_0 \in C^{1,2}[\overline{Q}_T \times \mathbb{R}, \mathbb{R}]$ *be lower and upper solution of (4a)–(4c) of the form*

$$\mathscr{L}v_0 \leq f(t,x,w_0) + g(t,x,v_0) \quad \text{in } Q_T$$
$$\mathscr{B}v_0 \leq \Phi \quad \text{on } \Gamma_T \tag{5}$$
$$v_0(0,x) \leq u_0(x) \quad \text{in } \overline{\Omega}$$

and

$$\mathscr{L}w_0 \geq f(t,x,v_0) + g(t,x,w_0) \quad \text{in } Q_T,$$
$$\mathscr{B}w_0 \geq \Phi \quad \text{on } \Gamma_T \tag{6}$$
$$w_0(0,x) \geq u_0(x) \quad \text{in } \overline{\Omega}$$

such that $v_0(t,x) \leq w_0(t,x)$ *on* $\overline{Q}_T$;

(A_3) $f, g, f_u, g_u \in C^{\frac{\alpha}{2},\alpha}[\overline{Q}_T \times \mathbb{R}, \mathbb{R}]$ *such that* f_u *is nondecreasing in* u *and* $f_u \leq 0$, g_u *is nonincreasing in* u *on* $\Lambda = [u : v_0(t,x) \leq u(t,x) \leq w_0(t,x) \text{ for } (t,x) \in Q_T]$, *for each* $(t,x) \in Q_T$ *we have*

$$f_u(t,x,u) - f_u(t,x,v) \leq m(u-v) \quad \text{for } u \geq v;$$
$$g_u(t,x,v) - g_u(t,x,u) \leq k(u-v) \quad \text{for } u \geq v;$$

where constants $m > 0, k > 0$.
Then there exist monotone sequences $\{v_n(t,x)\}, \{w_n(t,x)\}$ *which converges uniformly to the unique solution* $u(t,x)$ *of (4a)–(4c) on* $\overline{Q}_T$ *and the convergence is quadratic.*

Theorem 3. *Assume that the assumptions* (i)–(v) *in* (A_1) *hold and*

(A_2) $v_0, w_0 \in C^{1,2}[\overline{Q}_T \times \mathbb{R}, \mathbb{R}]$ *are lower and upper solution of (4a)–(4c) of the form*

$$\mathscr{L}v_0 \leq f(t,x,v_0) + g(t,x,w_0) \quad \text{in } Q_T,$$
$$\mathscr{B}v_0 \leq \Phi \quad \text{on } \Gamma_T, \tag{7}$$
$$v_0(0,x) \leq u_0(x) \quad \text{on } \overline{\Omega},$$

and

$$\mathscr{L}w_0 \geq f(t,x,w_0) + g(t,x,v_0) \quad \text{in } Q_T,$$
$$\mathscr{B}w_0 \geq \Phi \quad \text{on } \Gamma_T, \tag{8}$$
$$w_0(0,x) \geq u_0(x) \quad \text{on } \overline{\Omega},$$

such that $v_0(t,x) \leq w_0(t,x)$ *on* $\overline{Q}_T$;

(A_3) $f, g, f_u, g_u \in C^{\frac{\alpha}{2}, \alpha}[\overline{Q}_T \times \mathbb{R}, \mathbb{R}]$ *such that* f_u *is nondecreasing in u, g_u is nonincreasing in u and $g_u \leq 0$ on* $\Lambda = [u : v_0(t,x) \leq u(t,x) \leq w_0(t,x)$ *for* $(t,x) \in Q_T]$. *For each* $(t,x) \in Q_T$, *we have*

$$f_u(t,x,u) - f_u(t,x,v) \leq m(u-v) \ for \ u \geq v,$$

$$g_u(t,x,v) - g_u(t,x,u) \leq k(u-v) \ for \ u \geq v,$$

where constants $m, k > 0$.

Then there exists monotone sequences $\{v_n(t,x)\}, \{w_n(t,x)\}$ *which converges uniformly and monotonically to the unique solution $u(t,x)$ of (4a)–(4c) on $\overline{Q}_T$ and the convergence is quadratic.*

We have presented numerical examples related to the theoretical results of [11] in [12]. In this paper, we present two numerical results relative to Theorem 2 and 3 respectively. Although the examples we have presented f_{uu} and g_{uu} exist and $f_{uu} > 0$, $g_{uu} < 0$, it is to be noted that in our results, we do not require that. In fact, according to our theoretical results, we can develop numerical methods even when f_{uu} and g_{uu} does not exist.

Our first example is related to Theorem 2.

Example 1. *Let*

$$u_t - \frac{1}{16}\frac{\partial^2 u}{\partial x^2} = f + g \quad 0 < x < 1, \ 0 < t < 1,$$
$$u(0,t) = u(1,t) = 0, \quad 0 < t \leq 1, \qquad\qquad (9)$$
$$u(x,0) = \sin(\pi x) + x(1-x) \quad 0 \leq x \leq 1,$$

where $f = \frac{1}{2}u^2 - 2u$ *and* $g = -3u^2 + 3u + 4$ *in (9).*

It is easy to see that $v_0 \equiv 0, w_0 \equiv 2$ are coupled lower and upper solutions of the type (5), (6). One can easily verify that (9) satisfies all the hypothesis of Theorem 2. Also observe that the forcing functions $f + g$ is not increasing or decreasing in u for each t, x on Q_T. By Theorem 2, we construct the sequences $\{v_n(t,x)\}, \{w_n(t,x)\}$ which are the unique solutions of the following linear systems and converge to the unique solution of (9) quadratically.

$$\mathscr{L}v_{n+1} = f(t,x,w_n) + f_u(t,x,w_n)(w_{n+1} - w_n) + g(t,x,v_n) + g_u(t,x,w_n)(v_{n+1} - v_n)$$
$$\mathscr{B}v_{n+1} = \Phi,$$
$$v_{n+1} = u_0(x). \qquad\qquad (10)$$

$$\mathscr{L}w_{n+1} = f(t,x,v_n) + f_u(t,x,w_n)(v_{n+1} - v_n) + g(t,x,w_n) + g_u(t,x,w_n)(w_{n+1} - w_n)$$
$$\mathscr{B}w_{n+1} = \Phi,$$
$$w_{n+1} = u_0(x). \qquad\qquad (11)$$

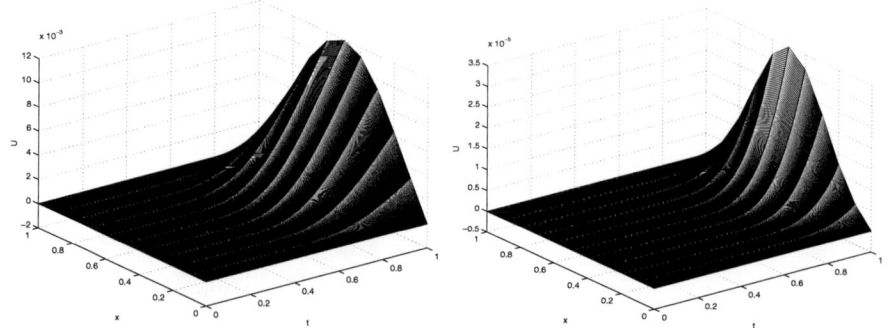

FIGURE 1. The distance between upper solution w and lower solution v in iterations $5, 6$ respectively

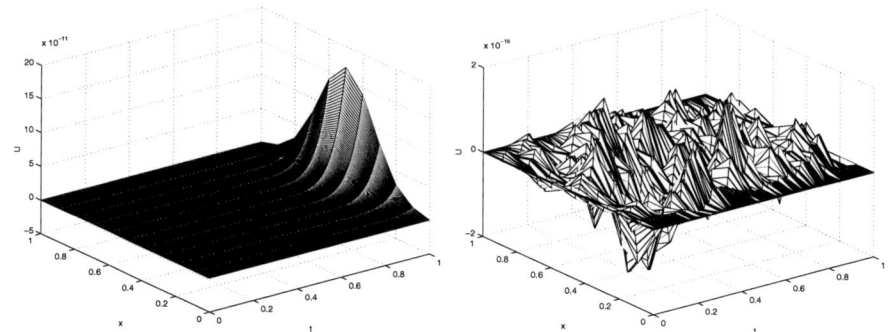

FIGURE 2. The distance between upper solution w and lower solution v in iterations $7, 8$ respectively

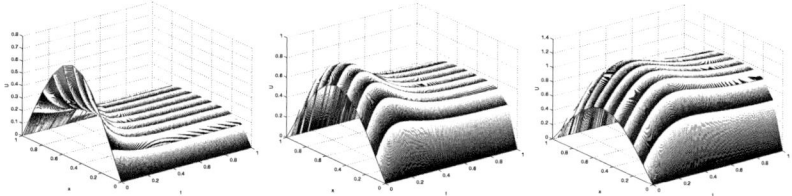

FIGURE 3. The lower solution v in the first 3 iteration

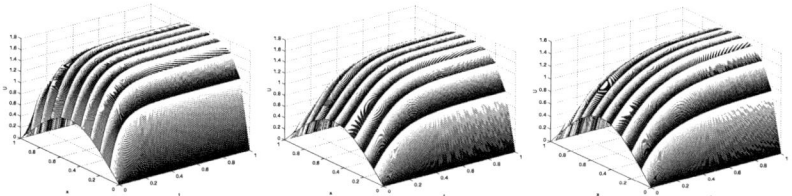

FIGURE 4. The upper solution w in the first 3 iteration

We use explicit method to approximate (10) and (11) as in [3] by choosing $M = 500, N = 10$ and $\Delta t = \frac{1}{M}, \Delta x = \frac{1}{N}$. The numerical results are shown in Figure 1 to Figure 4. After 8 iterations, the l_∞ norm for the distance between upper and lower solutions reaches 10^{-15}. In order to prove the finite difference scheme is accurate of order $(2, 1)$ for system (10), (11), see [10] for theoretical details. The discussion about the stability for parabolic equations can be found in [3, 10]. Since we have $\frac{\Delta t}{(\Delta x)^2} \leq 0.5$ in our example here, using Lax Equivalent Theorem, the numerical scheme for example 1 is stable and convergent.

The next scalar example satisfies all the hypothesis of Theorem 3.

Example 2. *Consider*

$$u_t - \frac{\partial^2 u}{\partial x^2} = f + g \quad 0 < x < 2, \ 0 < t,$$

$$u(0,t) = u(2,t) = 0, \quad 0 < t, \tag{12}$$

$$u(x,0) = \sin(\frac{\pi}{2}x), \quad 0 \leq x \leq 2,$$

where $f = \frac{5}{2}u^2 - 6u$ and $g = \cos u$ in (12).

It is easy to see that $v_0 \equiv 0, w_0 \equiv 1.5$ are coupled lower and upper solutions that satisfy (7) and (8). Using Theorem 3 for (12), we construct the sequences $\{v_n(t,x)\}, \{w_n(t,x)\}$ which are the unique solutions of the following linear systems:

$$\mathcal{L}v_{n+1} = f(t,x,v_n) + f_u(t,x,v_n)(v_{n+1} - v_n) + g(t,x,w_n) + g_u(t,x,v_n)(w_{n+1} - w_n),$$
$$\mathcal{B}v_{n+1} = \Phi, \quad v_{n+1}(0,x) = u_0(x),$$

$$\tag{13}$$

$$\mathcal{L}w_{n+1} = f(t,x,w_n) + f_u(t,x,v_n)(w_{n+1} - w_n) + g(t,x,v_n) + g_u(t,x,v_n)(v_{n+1} - v_n),$$
$$\mathcal{B}w_{n+1} = \Phi, \quad w_{n+1}(0,x) = u_0(x).$$

$$\tag{14}$$

For each iteration in the generalized quasilinearization method, we approximate the solution of the linear IBVPs (13) (14) by explicit method. Realizing that the explicit method is stable and converges when $\frac{\Delta t}{(\Delta x)^2} \leq 0.5$, we take $M = 400, N = 10, \Delta t = \frac{1}{M}, \Delta x = \frac{2}{N}$. One can develop numerical results for example 2 which are similar to example 1.

CONCLUSION

Our numerical results confirm the theoretical results of Theorem 2 and 3. The generalized quasilinearization method linearizes a nonlinear problem which make it easier to approximate the solutions numerically. The rapid convergence of GQL method can be visualized in the numerical results.

REFERENCES

1. R., Bellman, *Methods of Nonlinear Analysis*, Vol. II, Academic Press, New York, 1973.
2. R., Bellman, and R., Kalaba, *Quasilinearization and Nonlinear Boundary Value Problems*, American Elsevier, New York, 1965.
3. A., Iserles, *A First Course in the Numerical Analysis of Differential Equations*, Cambridge University Press, United Kingdom, 1996.
4. R. Krivec, V.B. Mandelzweig, Numerical Investigation of Quasilinearization Method in Quantum Mechanics, Computer Physics Communications, 138 (2001), 69-79.
5. R. Krivec, V.B. Mandelzweig, Quasilinearization Approach to Quantum Mechanics, Physics, 0112024, to be appear.
6. V. Laksmikantham and S. Köksal, *Monotone Flows and Rapid Convergence for Nonlinear Partial Differential Equations*, V7, Taylor and Francis, 2003.
7. V. Lakshmikantham, A.S. Vatsala, *Generalized Quasilinearization for Nonlinear problems*, Kluwer Academic Publishers, 1998.
8. V.B. Mandelzweig, Quasilinearization Method and its Verification on Exactly Solvable Models in Quantum Mechanics, Journal of Mathematical Physics, Vol.40(12), 6266-6291.
9. V.B. Mandelzweig, F. Tabakin, Quasilinearization Approach to Nonlinear Problems in Physics with Application to Nonlinear ODEs, Computer Physics Communications, 141 (2001), 268-281.
10. J.W. Thomas, *Numerical Partial Differential Equations Finite Difference Methods*, Springer , 1995.
11. A.S. Vatsala, J. Yang, Generalized Quasilinearization Method for Reaction Diffusion Systems, Nonlinear Differential Equation, Theory, Methods and Applications, Vol 8, No.2, pp1-24, 2004.
12. A.S. Vatsala, J. Yang, Numerical Investigation of Generalized Quasilinearization Method for Reaction Diffusion Systems, Computers and Mathematics with Applications, 2004, to appear.

2D Solitary Waves of Boussinesq Equation

Jayanta Choudhury* and Christo I. Christov*

Dept of Mathematics, University of Louisiana at Lafayette

Abstract. In this paper, the 2D stationary-propagating localized solutions of Boussinesq's equation are investigated numerically. An algorithm for treating the bifurcation and finding a nontrivial solution is created. The scheme is validated employing different grid sizes and different size of the box that contains the solution. The results obtained show that there is pseudo-Lorentzian elongation of the scale of the solitons but it is only in the direction transverse to the propagation velocity. In longitudinal direction the scales are slightly contracted, so kind of "relative" contraction takes place. Results are shown graphically and discussed.

Keywords: Boussinesq, Soliton,2D, Numerical
PACS: 02.60.Cb, 02.70.Bf

INTRODUCTION

Localized waves that propagate without change of shape at long distance and/or for long times are called solitary waves. They were first observed by Russell [1, 2] who named the phenomenon "The Great" or "The Permanent" wave. Since then, permanent waves have been found in many other fields of modern physics, such as metal lattices (phonon propagation), optical fibers, Bose–Einstein Condensate etc.

Boussinesq [3, 4] proposed an equation to model the permanent wave, considering shallow water layer with much smaller thickness compared to the horizontal length scale of the motion. He showed that the balance between the nonlinearity and dispersion maintains the shape of the permanent wave. Various Boussinesq equations were proposed during the years (see, e.g. [5, 6] for the literature).

Zabuski and Kruskal [7] showed numerically that two solitaty waves can interact without losing their identities and coined the name "soliton" to delineate the particle-like behavior of these waves (see [8] for the story). Several analytical techniques have been developed for solving soliton problems of different nonlinear dispersive equations, such as Bäcklund transformation, Inverse Scattering Method, and Hirota Bilinear Method (see [9] for extensive review on the subject). The limitations of analytical techniques necessitate the development of different numerical techniques and an extensive literature is available on this subject. However, the predominant part of the known results are concerned with the 1D cases. In this short note we investigate numerically the shapes of the stationary solitons in two spatial dimensions.

We focus here on the so-called "good" (or "proper") Boussinesq equation (PBE):

$$u_{tt} = \Delta(\gamma^2 u + \alpha u^2 - \beta \Delta u) \tag{1}$$

where $\beta > 0$ is the dispersion parameter, α is the amplitude parameter, and γ is the characteristic speed of the small disturbances (linear waves).

CP755, *ISIS: International Symposium on Interdisciplinary Science*
edited by A. Ludu, N.R. Hutchings and D.R. Fry
© 2005 American Institute of Physics 0-7354-0240-X/05/$22.50

In 1D, PBE possesses *sech* solution

$$u(x,t) = -\frac{3}{2}\frac{\gamma^2 - c^2}{\alpha}\operatorname{sech}^2\left(\frac{x - ct}{2}\sqrt{\frac{\gamma^2 - c^2}{\beta}}\right) \qquad (2)$$

where c is the phase speed of the localized wave. This solution demonstrates that the balance between the nonlinearity and dispersion can maintain the shape of the localized wave, making the latter permanent. For $\alpha > 0$, the solitons are depressions. When one considers PBE equation as a generic class, however, one can assume for convenience that $\alpha < 0$ and have bell-like shapes for the solitons.

The most salient features of the *sech* soliton (2) are that

1. it exists for subcritical phase speeds $|c| < |\gamma|$;
2. if c increases, the amplitude decreases;
3. This is due to the scale factor $\sqrt{(\gamma^2 - c^2)/\beta}$, the shape of solution spreads for increased c.

In 2D, there is no analytical solution for the above Generalized Dispersive Wave Equation (GDWE). Hence devising robust difference scheme is important.

BIFURCATION PROBLEM FOR THE 2D SHAPE

When investigating numerically the dynamics of solitary waves (see, [5]), the analytical *sech* solution (2) is taken as an initial condition. In 2D there is no analytical solution and the first task to be surmounted is to find the shape of the stationary propagating soliton. To this effect we consider the moving frame $\xi = x - c_1 t$ and $\eta = y - c_2 t$, where c_1, c_2 are the components of the velocity of the center of soliton. For the shape function in the moving frame $v(\xi, \eta)$, one gets the following stationary equation

$$0 = -(v_{\xi\xi}c_1^2 + 2c_1c_2v_{\xi\eta} + c_2^2v_{\eta\eta}) + \gamma^2(v_{\xi\xi} + v_{\eta\eta}) + \alpha[v_{\xi\xi}^2 + v_{\eta\eta}^2]$$
$$- \beta(v_{\xi\xi\xi\xi} + 2v_{\xi\xi\eta\eta} + v_{\eta\eta\eta\eta}). \qquad (3)$$

If localized solutions are sought, the following boundary conditions are imposed

$$v(-L_1, \eta) = v(L_1, \eta) = 0 = v_x(-L_1, \eta) = v_x(L_1, \eta) = 0,$$
$$v(\xi, -L_2) = v(\xi, L_2) = v_y(\xi, -L_2) = v_y(\xi, L_2) = 0, \qquad (4)$$

where L_1 and L_2 are called "actual infinities" and they define the size $[2L_1 \times 2L_2]$ of the rectangular region to which the infinite domain is reduced.

The localized solution exists alongside with the trivial solution and hence we are faced with a bifurcation problem. Obtaining a numerical solution requires avoiding the trivial solution. In the present work we propose to fix the value of the function in one point, say in the middle point, namely $v(0,0) = \theta$ and to introduce a new function $v = \theta u$. The equation for new function u is the same as for function v with the only differences that the coefficient of the nonlinear term is now $\alpha\theta$ and that we have the additional constraint

$$u(0,0) = 1. \qquad (5)$$

The condition (5) makes the problem overposed and in order to get a well posed problem we need some flexibility. The latter comes through the fact that θ is not known and should be adapted so that the original equation is also satisfied at the point $\xi = 0, \eta = 0$. Following this idea we use the original equation in the middle point as definitive equation for θ, namely

$$\theta = \left. \frac{\Delta(\gamma^2 u - \beta \Delta u) - (u_{\xi\xi} c_1^2 + 2c_1 c_2 u_{\xi\eta} + c_2^2 u_{\eta\eta})}{\alpha \Delta u^2} \right|_{\xi=0,\eta=0}. \tag{6}$$

Eq.(3) written for u, together with (5) and (6) form a well posed problem. Because of nonlinearity, this boundary-value problem has to be solved iteratively. One way to do this is to add artifical time derivative u_t in the governing equation and to start from arbitrary non-zero initial condition and timestep the solution until convergence is reached in a sense that two consecutive time steps do not differ significantly.

Without fear of confusion we will use once again the independent variable x, y in the moving frame *in lieu* of ξ, η.

DIFFERENCE SCHEME

We consider rectangular region $L_1 = L_2$ and specify the same number of grid intervals N in both directions. The continuous function u is replaced by the grid function

$$u_{ij}^n = u(ih, jh; nk), \qquad i = 0, 1, \ldots, N, j = 0, 1, \ldots, N,$$

where h is the spacing in both x- and y-directions and $k = \Delta t$ is time increment

We use an explicit difference scheme which is stable only when $\tau \leq \frac{1}{2}h^4$. It makes it inefficient for a bigger numerical investigation, but the purpose of this work is to investigate the practical properties of the approximation and to prepare a solution that can be used in further works as benchmark. Also, the main purpose is to find the actual shape of the wave and to discuss its physical relevance.

The iterative finite difference scheme reads

$$\frac{u_{i,j}^{n+1} - u_{i,j}^n}{k} = (\gamma^2 - c_1^2) \frac{u_{i+1,j}^n - 2u_{i,j}^n + u_{i-1,j}^n}{h^2} + (\gamma^2 - c_2^2) \frac{u_{i,j+1}^n - 2u_{i,j}^n + u_{i,j-1}^n}{h^2}$$
$$- 2c_1 c_2 \frac{u_{i+1,j+1}^n - u_{i+1,j-1}^n - u_{i-1,j+1}^n + u_{i-1,j-1}^n}{4h^2} + \alpha \left(\frac{(u_{i+1,j}^n)^2 - 2(u_{i,j}^n)^2 + (u_{i-1,j}^n)^2}{h^2} \right.$$
$$+ \frac{(u_{i,j+1}^n)^2 - 2(u_{i,j}^n)^2 + (u_{i,j-1}^n)^2}{h^2} \right) - \beta \left(\frac{u_{i+2,j}^n - 4u_{i+1,j}^n + 6u_{i,j}^n - 4u_{i-1,j}^n + u_{i-2,j}^n}{h^4} \right.$$
$$+ 2 \frac{u_{i+1,j+1}^n - 2u_{i,j+1}^n + u_{i-1,j+1}^n - 2(u_{i+1,j}^n - 2u_{i,j}^n + u_{i-1,j}^n)}{h^2 h^2}$$
$$+ \frac{u_{i+1,j-1}^n - 2u_{i,j-1}^n + u_{i-1,j-1}^n}{h^2 h^2} + \left. \frac{u_{i,j+2}^n - 4u_{i,j+1}^n + 6u_{i,j}^n - 4u_{i,j-1}^n + u_{i,j-2}^n}{h^4} \right). \tag{7}$$

A difference version of the equation for θ is coupled to (7).

The grid b. c. are $u_{0j} = u_{1j} = u_{N-1j} = u_{Nj} = 0$, $u_{i0} = u_{i1} = u_{iN-1} = u_{iN} = 0$.

VALIDATION OF THE SCHEME

Since the iterations with respect to the artificial time are conducted until convergence, the term approximating the time derivative disappears for large number of iterations. Then the accuracy of the computations is affected by two main factors: magnitudes of spatial increment h and the cut-off value L of the region (the "actual infinity"). Our numerical experiments show that the difference of the calculated shapes for two different boxes with $L = 25$ and $L = 50$, differ only by 0.003 which means that size $2L = 100$ is fully enough to provide ample space for the solution to decay properly for large x, y.

In order to assess the consistency of the scheme we have performed calculations for the same value $2L = 100$ on grids with different N: 100, 200, and 400. Respectively, h is 1, 0.5, or 0.025. The norm of the difference between the solutions on different grids can be defined as $e(N) = \max_{i,j} |\hat{u}_{ij} - \tilde{u}_{ij}|$, where $\hat{u}$ and $\tilde{u}$ are the functions on the respective grids. The truncation error of the scheme in our case is $e(N) \simeq O(h^2)$. For a secon-order scheme, reducing the spacing twice should reduce the error four times. Indeed, we have observed that $e(100) = 0.035$ and $e(200) = 0.008$ which are approximately in the ratio 4:1.

The absolute value of error $e(200) = 0.008$ gives the estimate $e \sim 0.03h^2$. This ensures one that the truncation error for $N = 400$ will be of order of 0.002 which is small enough to claim that a grid with 400x400 points is reliable for the problems under consideration. All results reported in what follows are obtained with grid 400×400 points.

RESULTS

In Fig. 1, the comparison of the shapes of a soliton at rest, and a soliton propagating in x-direction with phase speed $c_2 = 0.6$, is shown. The lower pannesl present the lines of sonstant elevation/depression and the negative isolines are taken much denser, because the depressions are much smaller than the elevations. In both cases there is slight depression around the main hump which is the main difference between the 2D soliton and the *sech* soliton in the 1D case.

However, when the soliton is moving, the depression is deeper in front and behind the soliton. A better quantitative description of this effect is presented in Fig. 2 where the x- and y- cross sections are presented. The upper row of graphs show the function itself, the lower row shows the absolute value of the function. The downward spikes in the lower row are in the places where the function changes it sign.

Now it is seen that the depression is present even for $c = 0$, but in the x cross section (right panels) the first change of sign is approximately in the same place, just the depth is increasing with c. In the y cross-section the passage through zero elevation takes much farther from the center which means that the soliton is elongated in y-direction. Thus the elongation of the soliton takes place in the transverse direction which is completely non-intuitive if one is guided by the 1D results where the elongation takes place in longitudinal direction.

In the end we present the cross section of the soliton shape. We chose the value 0.1 for the contour line to be compared, because the zero line spans much bigger domain and is not convenient for presentation. The left panel in Fig. 3 presents the contour for

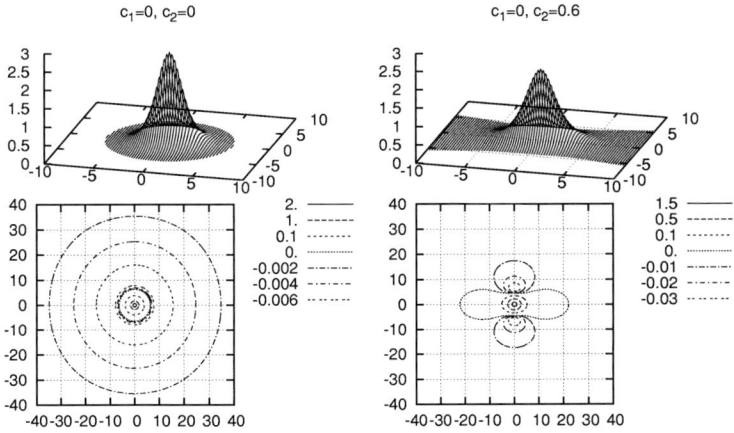

FIGURE 1. The shapes of the 2D Boussinesq soliton. Left: $c_1 = 0, c_2 = 0$. Right: $c_1 = 0.6, c_2 = 0$.

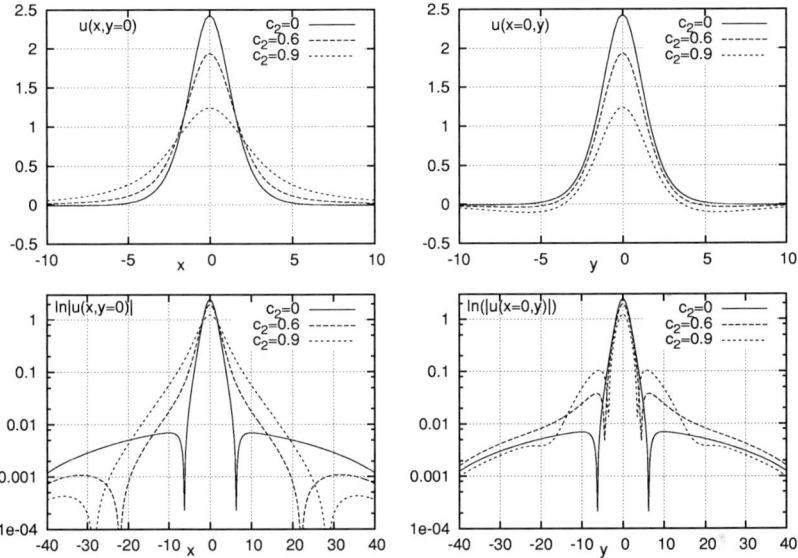

FIGURE 2. The comparison between the cross-sections of the soliton's shape. Left: $y = o$. Right: $x = 0$. Upper panels show the shape, the lower panels show the logarithm of the absolute value.

$c_1 = 0$ and different c_2. The shape becomes elliptical in a similar fashion as in Lorentz contraction. It is seen that the transverse scale is slightly contracted. The conclusion is that the 2D Boussinesq quasi-particle undergo relative pseudo-Lorentz contraction in the direction of motion.

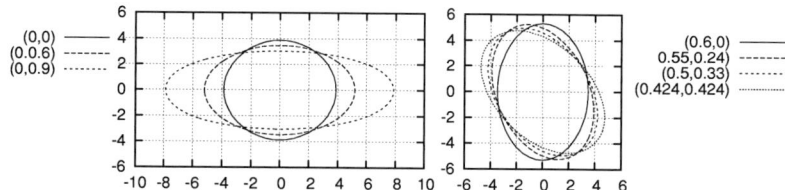

FIGURE 3. Left: Contour line 0.1 for motion in y-direction with phase speeds $c = 0, 0.6, 0.9$. Left: Contour line 0.1 for different c_1 and c_2, but for the same magnitude $c = \sqrt{c_1^2 + c_2^2} = 0.6$ of the phase speed.

The right panels of the Fig. 3 depicts the orientation of the elliptic cross section for different combination of the components of the phase velocity, but when the amplitude of the latter is kept constant $c = \sqrt{c_1^2 + c_2^2} = 0.6$. It is seen that within the error of approximation we have the same ellipse, with its major and minor axis aligned exactly with the direction of the phase velocity.

CONCLUSION

In the present paper the shape of the 2D Boussinesq soliton is calculated numerically. This is a result not known in the literature and shows two main physical differences in comparison with the 1D case: there is pseudo-Lorentzian elongation in the transverse direction, and there are depressions in front and in the rear of the soliton. This shows that one has to proceed with caution when comparing the qualitative features of quasi-particles in 2D to the well studied 1D cases.

The solution is obtained on high density grid 400×400 and can be used in future works as initial condition for numerical experiments for collision of Boussinesq solitons.

REFERENCES

1. J. S. Russell, "Report on the Committee on Waves," in *Report of 7th Meeting (1837) of British Assoc. Adv. of Sci., Liverpool*, John Murray, London, 1838, pp. 417–496.
2. J. S. Russell, "On Waves," in *Report of 14th Meeting (1844) of the British Association for the Advancement of Science*, York, 1845, pp. 311–390.
3. J. V. Boussinesq, *Comp. Rend. Hebd. des Seances de l'Acad. des Sci.*, **73**, 256–260 (1871).
4. J. V. Boussinesq, *Journal de Mathématiques Pures et Appliquées*, **17**, 55–108 (1872).
5. C. I. Christov, and M. G. Velarde, *J. Bifurcation & Chaos*, **4**, 1095–1112 (1994).
6. C. I. Christov, *Wave Motion*, **34**, 161–174 (2001).
7. N. Zabusky, and M. D. Kruskal, *Physics Review Letters*, **15**, 240 (1965).
8. M. Kruskal, "The birth of the soliton," in *Nonlinear Evolution Equations Solvable by the Spectral Transform*, edited by F. Calogero, Pitman, London, 1978, pp. 1–8.
9. A. C. Newell, *Slitons in Mathematics and Physics*, SIAM, New York, 1986.

Nonlinear modeling of 3-d flagellar dynamics

Andrei Ludu[*] and Nathan Hutchings[†]

[*]*Northwestern State University, Dept. Chemistry and Physics, Nathitoches, Louisiana, USA*
[†]*Northwestern State University, Dept. Biology, Nathitoches, Louisiana, USA*

Abstract. We develop exact equations formulas for the sliding between arbitrary filament pairs in a bundle subjected to a general 3-d deformation. We introduce a classification of deformations, and we study particular examples for bending, twisting, helical and toroidal shapes. We prove that simultaneous combination of twisting and bending can produce a drastically drop in the sliding. These results can relate the geometry of shapes and motions with the distribution of molecular motors activity.

Keywords: Flagellum, nonlinear dynamics, bending, twisting, deformation
PACS: 87.16.Ka, 05.65.+b, 87.10.+e, 87.17.-d, 02.40.Hw

1. INTRODUCTION

In the present work we focus on the deformations and dynamics of the flagellum, in particular the axoneme-like structures, a conserved structure found in all flagella and cilia [1]. The present approaches [1-7]provide accurate predictions for polymers, and for equilibrium configurations, but predict less about viscously dominating motion where the solvent viscosity produces the friction forces responsible for propulsion [6]. Depending on the dynamical mechanism in each model, different nonlinear features are predicted to describe the self-organizational behavior of such complex systems. For example, in the transverse force model [4] a dynamical feed-back between the geometry and the dynamics is proposed. The local sliding between doublets generates local bending, and when the axoneme is bent, some transversal forces develop across the axoneme that change the 9-folded symmetry of the doublets, which in turn changes the inter-doublet spacing. This change of separation of the doublets controls the activation of motors and creates a positive feedback to bending, which provides spontaneous beats and oscillations. In the present paper we develop mathematical formalisms based on differential geometry of curves, on the theory of motion of curves, and on functional analysis approaches to obtain a new and general formula for sliding between two arbitrary filaments in a bundle of parallel filaments subjected to a 3-d deformation. This allow us to bring a direct connection between any observed shape of a flagellum and the distribution of activity of the motors: from experiments and movies of motion we identify and interpolate shapes and we obtain the distribution of motors activity through the slide formulas. Some interesting nonlinear effects are discussed and compared to the present literature. The results in this article can be used both in numerical or analytical further modeling, and in experimental studies, in order to predict distributions of forces and torques starting from experimentally observed shapes. This formalism can be also used to analyze the dynamics of the flagellum and to better understand the cellular

CP755, *ISIS: International Symposium on Interdisciplinary Science*
edited by A. Ludu, N.R. Hutchings and D.R. Fry
© 2005 American Institute of Physics 0-7354-0240-X/05/$22.50

propulsion systems using flagellum or hair-like systems and bending/twisting traveling or stationary deformations.

2. BUNDLE DEFORMATIONS

The axoneme of flagella and cilia is a rod-like elastic structure that moves, bends, twists and oscillates as a result of the internally generated forces and torques created by molecular motor proteins (dynein). The geometry and dynamics of the structure are dependent on the flagellum's interaction with the viscous environment. The axoneme consists of a cylindrical structure of 9 parallel pairs (doublets) of filaments (microtubules). In most, but not all organisms, the axoneme contains one central pair of filaments which is connected to the other doublets by protein spokes [1,3-5]. Cohesiveness and elasticity between microtubules is provided by structural proteins, and the internal forces and torques necessary for movement are generated by a distribution of ATP-dependent molecular motors located in-between the adjacent microtubule doublets. The molecular motors form arms between the outer doublet microtubules, resulting in a local relative sliding displacement between adjacent microtubules [3,4,6]. In addition to a relatively parallel shifting along the longitudinal axis of the axoneme (providing microtubules are free to move), the boundary conditions imposed on the doublets creates bending in different directions and twists around the lateral axis of symmetry. Bending is the result of holding the doublets together during a shift (quasi-elastic constrains). Bending is possible if the local sliding is dominant for a given pair, or for two opposite pairs acting in the same direction. The deformation is related to the local distribution of axial forces and the normal torques along the flagellum. Local twist is possible if all of the pairs sliding in the same direction are producing a torque oriented along the local flagellum axis, which consequently produces a longitudinal compression/extension. Unfortunately, local sliding and twist cannot be simply attributed to a one-to-one correspondence with the local action of motors, since bending can occur in some passive sections of the filament due to the mechanical action of other motor-active areas elsewhere in the flagellum. For example, if the longitudinal displacement of the filaments is restricted at both ends of flagellum, the total sliding integrated along the flagellum should be zero. So due to the mechanical constrain of the total system, regions with positive bending (considered by convention as positive) should be compensated with regions of negative bending, even if some of them are not motor active. Very recent experiments performed with micron calibrated glass needles show if a bend is forced into an inactive flagellum, the other half bends in opposite direction trying to keep the total integral curvature zero. In this situation these results suggest that in this situations the attachments between doublets within the axoneme do not allow too much relative sliding at the ends. Interestingly, the deformations in some inactive sectors can induce the activation of the motors, which creates a cooperative feed-back process, resulting in stabilizing certain type of waves or adapting movement to the environment.

Although many flagellated cells have additional structures within the flagellum, in the current model we study the 3-dimensional deformations of a simpler axoneme-like structure consisting of a flexible bundle of pairs of filaments. In this configuration, the filament pairs are constrained to stay "parallel" in order to preserve the structural unity, and because they are connected through quasi-rigid molecular arms. By *parallel* we mean that the two filaments in a pair, and any two pairs are separated by almost constant distances along their common normal directions in any of their points. We describe the filaments in a bundle by a finite family of parameterized regular curves $\vec{r}_k(\alpha)$, where $k = 1, 2, \ldots, 2N$ labels the filaments in the bundle and N is the number of pairs (9 outer pairs plus 1 central pair, for example in the axoneme case). Two different filaments k and j are "parallel" if their equations fulfil the condition

$$\vec{r}_k(\alpha) = \vec{r}_j(\alpha) + d_{kj}\vec{n}(\alpha) + e_{kj}\vec{b}(\alpha), \tag{1}$$

where d_{kj}, e_{kj} are constant parameters determining the separation between the pairs j and k in the normal plane. The versors $\vec{n}(\alpha)$ and $\vec{b}(\alpha)$ are the normal and binormal directions to the central filament at α. Such two "parallel" curves are called Bertrand curves and their curvature and torsion should be linearly dependent in any point along the filament. With the exception of plane curves, the only curve that accepts a Bertrand mate is the circular helix, which also accepts infinitely many Betrand mates. Since we have more than two pairs in a bundle, the Bertrand criterium applies, and consequently we can define two basic types of deformations: (i) Bending, where the curves are coaxial parallel circles (degenerated helices), and (ii) Coherent torsion, where the pairs have a helical shape. Of course, the connections between the pairs are neither constant nor rigid, but in the following calculations their variation will be neglected compared to the global deformation of the whole bundle, so we can use this classification of deformation as a good approximation.

Consequently, the flagellar deformations can also be described by two basic types: bending by curvature changes, and twisting by torsion changes. Of course we can also have simultaneously a combination of bending and twisting [8]. In the following we describe the formulation of bending and twisting. We mention that our approach for the geometry of bundles of filaments can also be applied to several other systems from biology like DNA topology, or the intracellular cytoskeleton (both the radial structure and the conical structure).

We consider a number of identical filaments distributed around a central filament which intersects the origin O. The un-deformed bundle, $z < 0$, is parallel to the Oz axis, Fig.1. the relative sliding of filaments, we refer to the *terminal normal plane* of the filament curves, intersecting each such curve at its end point (the normal plane of a curve is the plane perpendicular to the tangent to the curve) [3]. We choose this reference plane since the flagellar and ciliar beats are predominantly restricted to the direction perpendicular to the central pair of the axoneme. The relative sliding between any two filaments placed on the circumference, see Fig. 2, called tangent sliding δ_{tan}, is the difference between their radial slidings:

$$\delta_{tan}(\phi) = 2Lr\sin\left(\frac{\delta\phi}{2}\right)\frac{\sin(\frac{\delta\phi}{2} + \phi - \phi_0)}{R_0 - r\cos(\phi + \delta\phi - \phi_0)}. \tag{2}$$

Eq.(2) is the bending equation, and it represents the relative sliding δ_{tan} between any two filaments in the bundle (of angular position ϕ and angular separation $\delta\phi$) for a bending (constant radius R_0) of arbitrary orientation ϕ_0, and subtend bundle arclength L. Eq.(2)

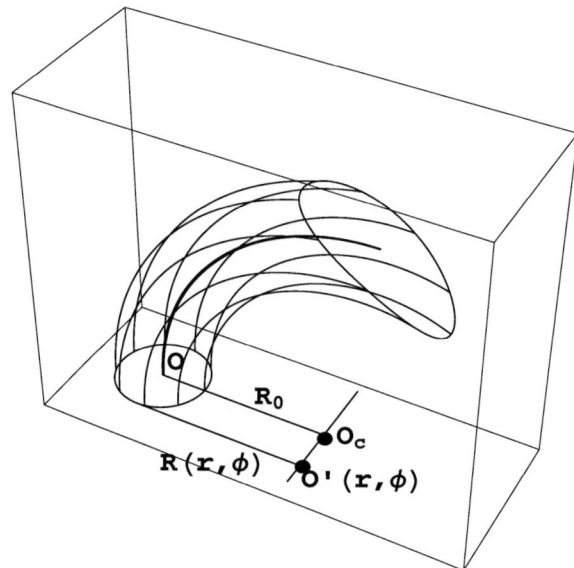

FIGURE 1. Sliding induced by bending of nine equal length filaments distributed on a base circle around a central filament. The center of rotation O_c is along Ox axis ($\phi_0 = 0$) and the bending is in the xOz plane. The origin of the filament bundle is rigidly kept in the xOy plane. The ends of all 9+1 filaments are lying on a terminal surface described by the locus of points of equal arclength on all filaments.

approaches the 2-dimensional bending formula in the limit $r \ll R_0$ and $\delta\phi \to 0$. We prove this by choosing $\phi - \phi_0 = \pi/2$ without any loss of generality, namely

$$\delta_{tan} \to 2Lr\frac{\delta\phi}{2}\frac{\sin(\frac{\delta\phi}{2} + \frac{\pi}{2})}{\frac{1}{k}} \simeq kr\delta\phi \ ds \to a\int kds,$$

where $a = r\delta\phi$ is the linear separation between filaments, and $k = 1/R$ is the curvature of the arclength L.

If we take for $\delta\phi$ the angular separation between two adjacent pairs, we notice that the tangent sliding has maximum value at angles

$$\phi_{\text{max sliding}} = \phi_0 - \frac{\delta\phi}{2} \pm arccos\left[\frac{r}{R_0}cos\left(\frac{\delta\phi}{2}\right)\right], \tag{3}$$

which is neither in the bending plane nor orthogonal to it. For example, in the case of the *T. brucei* axoneme we have measured $\delta\phi = 1.4^0$ [3,4] and from Eqs.(2,3) the maximum sliding occurs at about 76° to the right and left of the bending plane, Fig. 5, which is very close to the separation between 3 consecutive pairs (2×40^0). This result agrees with the experimental evidence of the maximum change in doublet spacing occurring at doublets 3 and 8, in the geometric clutch model [4]. In a full 3D approach for curvature and twist based on the Euler equations of momentum conservation [4] this effect is explained by the action of the quasi-elastic bridges between outer doublets.

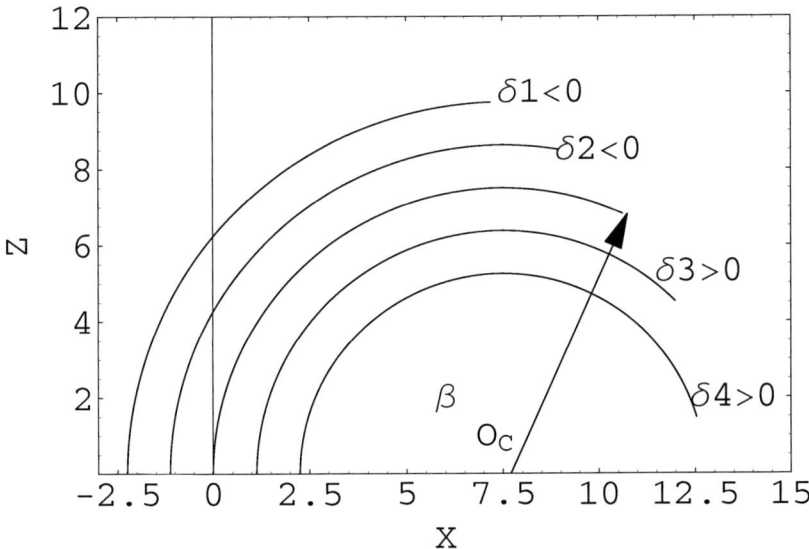

FIGURE 2. A section in the xOz bending plane of Fig. 1. The parameter β is the bending angle around the bending center O_c, made by the final position vector of the central filament with the Ox axis (the arrow). Different relative positive and negative displacements with respect to the central filament are denotcd with $\delta 1$, etc. Because the terminal length of the filaments 1 and 2 is shorter than the central filament we define their relative slide as negative.

The obtained maxima of the sliding in Eq.(3) are responsible for the Principal and Reverse forces producing dynein bridges in the Lindemann geometric clutch model [4]. Since the first term in the RHS of Eq.(3) is arbitrary, and the second term is negligible, the angular deviation of the most stressed (shifted) pair with respect to the bending plane is given by $\phi_{\text{max sliding}} - \phi_0 \simeq rcos[\frac{r}{R_0}cos(\frac{\delta\phi}{2})]$. For a given angular separation $\delta\phi$ between pairs in the axoneme, and for a given structural radius of the axoneme r this angle of maximum stress depends only on the radius of curvature of the bending, R_0. The maximum internally driven bending of such a flagellum would be given by

$$R_{0\ max} = r\frac{cos\left(\frac{\delta\phi}{2}\right)}{cos\left(\phi - \phi_0 + \frac{\delta\phi}{2}\right)}. \tag{4}$$

In the case of *T. brucei*, the maximum bend obtained is $r/R_0 \simeq 0.75$ which gives a minimum radius of curvature of $R_0 \simeq 1.8\mu m$ which is in good agreement with experimental measurements of curvature [3,4,6]. In the following, we can relate the radius of bending

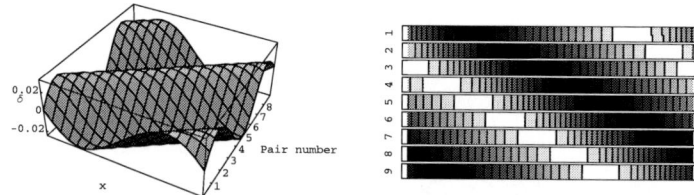

FIGURE 3. Distribution of tangent sliding along a right helix bundle. Left: sliding versus arclength x and number of the pair. Right: sliding versus x for each pair in contour plot. If the helix rotates the waves and patterns are traveling along x. The diagonal oscillating patterns of sliding can be related with the activity of motors, showing "metachronism" type of self-organization.

(radius of curvature) to the maximum sliding $\delta_{tan\ max}$ of the most active pair

$$R_0 = \frac{r\sqrt{\delta_{tan\ max}^2 + 4L sin\left(\frac{\delta\phi}{2}\right)^2 (L - \delta_{tan\ max})}}{\delta_{tan\ max}}, \tag{5}$$

where $r, \delta\phi$, are provided by the structural model for the axoneme. Since $\delta_{tan\ max} << L$, Eq.(4) can be very well approximated with

$$R_0 \simeq \frac{2rL}{\delta_{tan\ max}} sin\left(\frac{\delta\phi}{2}\right) \simeq \frac{rL\delta\phi}{\delta_{tan\ max}}. \tag{6}$$

Eq.(5) provides a simple intuitive connection the between bending radius and the maximum tangent sliding in the appropriate pair. For a typical flagellum we obtain $\delta_{tan\ max}/L = r\delta\phi/R = .03$, which means that a maximum bend of $R_0 = 5\mu m$ can be obtained if the active pairs slide with as little as 3% of their total length.

We conclude this section by presenting a direct application of the bending equations. We consider three traditional flagellar shapes, namely the helix (left and right), planar waves, and rotation around a fixed point of a rigid flagellum (conical shape) [8]. For each such shape we calculate the bending (tangent) deformation in all pairs of the filament bundle. In real 3-d cases, the bending should be coupled with twisting, a situation described in the following sections. In Fig. 3, we present the distribution of the tangent sliding along a helical bundle (right and left chirality, respectively). In time, when the helix rotates, the waves presented in the left frame are traveling along the x-direction. In the right frame we show the same result in a contour plot, for all 9 pairs. The diagonal oscillating patterns, which represent the amplitude of sliding, could be related in a physical model with the activity of motors, showing a very well self-organized synchronism. These patterns are in good agreement with recent results obtained in complex numerical computation of a flagellar model, namely the pair "metachronism" [5]. From analysis of Eq.(5) in different situations we noticed that the dependence of sliding versus separation of pairs $\delta\phi$ is not very relevant, while the sliding depends asymptotically on the radius of bending, i.e. after a certain limit $(R > 15 \div 20r)$ the increasing in R is not any more relevant.

3. UNIFORM TWISTING

Long flexible structures like flagella and cilia are subject to twisting deformations in addition to bending as evidenced by both planar and helical bending patterns within the same organism under different conditions [4]. Internally driven bending is produced by a distribution of local superficial torques oriented asymmetrically with respect to the normal of the bundle envelope. The torques have the same absolute direction in diametrally opposite points of the circumference (if at a certain point the torque is parallel to the normal, then at the diametrally opposite point torque is antiparallel to the normal). In contrast to bending, twisting can only result from torque having the same orientation with respect to the normal. Moreover, bending deformation may be the action of any number of pairs, while twisting deformation requires simultaneous action of two (or more) pairs, since all pairs have to slide coherently in the same direction. In this section we consider the envelope of the filament bundle being a cylindrical surface where parallel pairs of filaments are symmetrically distributed at equal separation angles $\delta\phi$. The shape of the bundle is determined by the action of the internal shear forces between adjacent filaments in each pair. In the Hines-Blum 3D sliding filament model, the twist is produced by fixed links located asymmetrically. The resulting twist is a second-order effect proportional to the square of the flagellum radius. We consider a circular bundle of radius r consisting of N equidistant parallel pairs of length L. For each pair, the two parallel filaments (aa' and bb') are separated by an angle $\delta\phi$. In a twisting deformation we assume that the upper ends a' and b' of the filaments are rotated with the angle ϕ, while their basis remain fixed, and consequently the height of the bundle decreases from L to h. In order to calculate the relative shift between the two filaments in this twisted pair, we note that the filaments take the shape of two parallel cylindrical helices of radius r and pitch b, related to the height of the twisted bundle by $h = b\phi$. In a twist, all pairs are sliding in the same direction, like in a collective coherent deformation. If we unfold the cylindric bundle envelope into a rectangle of base $r\phi$ and height h, the twisted filaments became inclined parallel lines, all making an angle $\alpha = tan^{-1}b/r$ with horizontal basis (the unfolded base circle). By using the formula for the length of a helix $L = \sqrt{r^2 + b^2}\phi$ we obtain the twisting slide in the form

$$\delta_{twist} = \frac{ar\phi}{L} = \frac{r^2\delta\phi}{\sqrt{r^2+b^2}}, \tag{7}$$

where $a = r\phi$ is the linear separation between filaments in a pair. Indeed, this twisting slide is in the second order with respect to filament separation, and hence Eq.(6) is in agreement with the Hines-Blum model [3]. From the energetics perspective, the deformation is larger for bending than for twisting for a given amount of work (sliding). That is twisting can act like a lower gear, while bending can act like a higher gear in terms of self-propulsion effectiveness. External mechanical stress of the environment can also induce twist. An important fact about twisting is the amount of slide between doublets is equal along the axis of the bundle (the helix), and it does not increase as in the case of the bending deformation.

The compression of the total length of the bundle for a given twist is

$$\delta L = L - \sqrt{L^2 - r^2 \phi^2}, \tag{8}$$

and the relative compression can be approximated with

$$\frac{\delta L}{L} \simeq \frac{r^2 \phi^2}{2L^2} = \frac{rk}{2}, \tag{9}$$

which not only means it is constant along the bundle, but it is a sort of average between the radius and the radius of curvature of the helix k. Eq.(8) is in itself an interesting geometric result. For example, for a bundle of radius $r = 1\mu m$, length $L = 10\mu m$, twisted with a complete turn $\phi = 2\pi$, with pairs separated at $\delta\phi = 5^o$ the shift between the adjacent filaments is approximatelly $100nm$ and the relative length compression is $\delta L/L = 1\%$. Using the bending equation Eq.(2), we can test the twisting slide formula Eq.(6). If we consider that all filaments are initially parallel and perpendicular to a base circle, a uniform helix shape is produced by an infinitesimal bending at the base circle. The rest the filaments remain parallel and do not suffer any additional slide. Thus, an initial infinitesimal bend should provide the same effect as a twist. Indeed, if we consider bending slide Eq.(2) with $R = a, L = acos(\alpha)$ we have

$$\delta_{bend}\Big|_{inf} = L \frac{a}{R_0} = acos\left(\frac{\pi}{2} - \alpha\right) = \delta_{twist}. \tag{10}$$

We conclude this section with a remark. In this model, we consider that all filaments (hence all pairs) remain parallel between them, and are contained in the cylindric surface. This is a simplifying hypothesis, since in some real situations the filaments begin to taper, splay, or intertwine and loose the parallelism in long rectilinear segments placed between shorter segments with large twisting or bending. This situation is considered in the following section by providing a formula for twisting of a pair around a variable geometry cylinder (variable radius and shape).

4. GENERALIZED 3-DIMENSIONAL DEFORMATIONS

We introduce in this section a general sliding formula between two "close" enough filaments, for any shape of the filament bundle. This approach can accommodate the case of simultaneous bending and twisting in 3-dimensions. The term "close" will be defined in the following. The shape of the bundle is defined as the side surface Σ of a finite cylinder with variable radius and variable shape. Such a bundle can be constructed starting from a finite open smooth curve of equation $\vec{r}_C(s)$ and tangent $\vec{t}_C(s)$, i.e. the central filament. We can build Σ as a smooth family of circles labeled by s (the arclength parameter). All circles have finite radius described by a smooth function $r = r(s)$, centered in $\vec{r}_c(s)$, and are inclosed in the normal plane to $\vec{r}_C(s)$. In other words, the surface of the bundle is a smooth 2-d differentiable manifold which is locally the

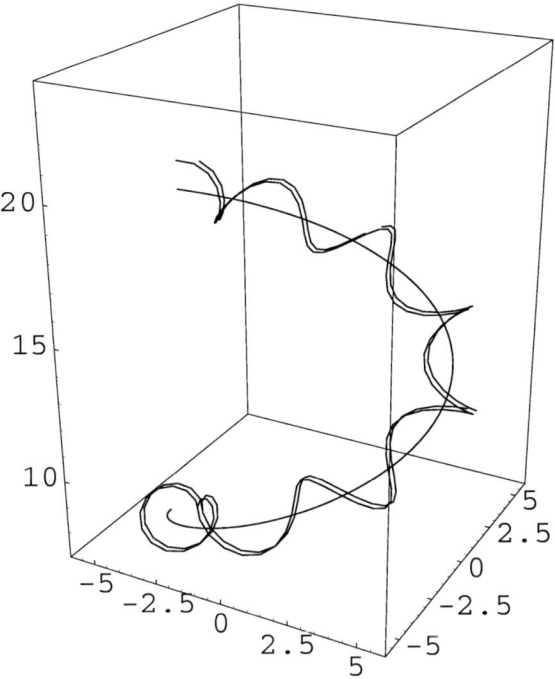

FIGURE 4. A pair of filaments winding around a helical-like shape bundle. The total relative sliding induced by the helical path can be seen at the end of the pair. This shift is less than the shift for a similar bending without torsion.

cross product between the unit segment and a circle. The outer filaments are $2N$ non-intersecting smooth curves γ_i, $i = 1, 2, \ldots, 2N$ lying on the bundle surface Σ. They are described by equations $\vec{r}_i(s_i)$, and unit tangent vectors $\vec{t}_i(s_i)$, where s_i is the arclength parameter along each such curve. In addition, we assume that the outer filaments are "parallel," that is no outer filament can return towards the initial point. This constrain is fulfilled if the outer filament equations satisfy the inequality $\vec{t}_c(s) \cdot \vec{t}_i(s) > 0$ for any s. We consider the initial ends of the circularly uniformly distributed $2N$ filaments are kept rigid and perpendicular on their base circle during deformation (zero initial sliding initial condition). Let us consider two "close" filaments in a general deformation of the bundle, namely two curves $\gamma_{1,2}$ of equations $\vec{r}_{1,2}(s_{1,2})$, and unit tangent vectors $\vec{t}_1(s_1), \vec{t}_2(s_2)$ Fig. 4. In the initial plane ($s_1 = s_2 = 0$) the two curves are separated by a linear distance a. We are interested in the relative "sliding" $\delta(L)$ between these two curves at a certain length $s = L$ along one of them, say γ_1. Now we can introduce the hypothesis of "closeness" as requesting that $a, \delta(L) << L$.

Under the above two hypotheses (parallelism and closeness of two filaments in each pair) we construct a geometric definition of the "sliding." We consider the intersection between the normal plane to γ_1 (defined by $\vec{n}_1(L)$ and $\vec{b}_1(L)$) and γ_2, which occurs at some $s_2 = s^* \neq L$. The existence of such a point is provided by the parallelism hypothesis: the normal plane to γ_1 intersects Σ in a circle, and γ_2 has to intersect this circle because it never returns backwards. The "sliding" is defined as $\delta(L) = L - s^*$. The intersection between the normal plane and γ_2 is described by the equation $\vec{n}_1(L)C_1 + \vec{b}_1(L)C_2 + \vec{r}_1(L) = \vec{r}_2(s^*)$, where $C_{1,2}$ are local orthogonal coordinates in the normal plane. Under the hypothesis of small sliding and small separation of pairs compared to the length of the bundle we can expand $\vec{r}_2(s^*)$ in a Taylor series around $s_2 = L$, and we have $\vec{n}_1(L)C_1 + \vec{b}_1(L)C_2 + \vec{r}_1(L) - \vec{r}_2(L) = -\vec{t}_2(L)\delta$. If we multiply this equation with $\vec{t}_1(L)$, and we approximate $\vec{t}_1(L) \simeq \vec{t}_2(L)$ we can write the infinitesimal sliding in the form

$$\delta(L) = \Omega|_{s,a}(\vec{r}_2 - \vec{r}_1)_{s=L} \cdot \vec{t}_1(L), \tag{11}$$

where Ω is the antisymmetric part of the second order differential operator with respect to s, a, $\Omega|_{s,a} = (1 + d_s + d_a + d_{s,a}^2)|_{s=a=0}$. We have to consider the account order two because twisting produces slidings of one order of magnitude smaller than bending. Also, infinitesimal sliding should be an odd function of s, a, so we neglect the symmetric part of the second order differentials, i.e. $d_{s,s}$. Technically, Ω is provided by the second-order antisymmetric terms in the Taylor expansion of Eq.(10) with respect to s and a around $(0,0)$. The total slide is obtained by integration of Eq.(10) along γ_1. From a geometrical perspective, we consider the 2-dimensional manifold of coordinates s, a generated by the translation of the curve γ_1 through different values of the separation parameter a into γ_2. In this manifold, Ω is the second order prolongation of the infinitesimal translation generator in s and a [9], and the local infinitesimal slide is the local action of this infinitesimal generator on the $(\vec{r}_1 - \vec{r}_2) \cdot \vec{t}$ function.

We test Eq.(10) in some particular cases investigated above: bending and twisting. In the case of twisting, we choose the two parallel helices of radius r, and pitch b provided by the equations

$$\vec{r}_1(s) = \left(r\cos\frac{s}{r}, r\sin\frac{s}{r}, \frac{bs}{r} \right)$$

$$\vec{r}_2(s) = \left(r\cos\frac{s-a}{r}, r\sin\frac{s-a}{r}, \frac{bs}{r} \right), \tag{12}$$

where $\frac{a}{r} = \delta\phi_0$ is the angular separation parameter. We have

$$\vec{t}_1(s) = \frac{1}{\sqrt{r^2 + b^2}} \left(-r\sin\frac{s}{r}, r\cos\frac{s}{r}, b \right),$$

and consequently

$$(\vec{r}_2 - \vec{r}_1)_{s=L} \cdot \vec{t}_1(L) = \frac{r^2 \sin\frac{a}{r}}{\sqrt{b^2 + r^2}},$$

which provides

$$\delta_{bend} = \frac{ar}{\sqrt{r^2 + b^2}} = \frac{\phi_0 r^2}{\sqrt{r^2 + b^2}}, \tag{13}$$

which is identical with Eq.(6) for twisting. Eq.(10) works equally good for bending. For example, for the two filaments

$$\vec{r}_1(s) = \left(R_0 cos\frac{s}{R_0}, R_0 sin\frac{s}{R_0} \right)$$

$$\vec{r}_2(s) = \left((R_0 - a)cos\frac{s}{R_0 - a}, (R_0 - a)sin\frac{s}{R_0 - a} \right), \tag{14}$$

Eq.(10) generates the sliding

$$(\vec{r}_2 - \vec{r}_1)_{s=L} \cdot \vec{t}_1(L) = (R_0 - a)sin\frac{aL}{R_0(a - R_0)} \rightarrow -\frac{aL}{R_0},$$

which is exactly the bending slide expression for the linear separation parameter $a = \delta R$ and $\phi - \phi_0 \pi/2$, $\delta\phi_o \simeq sin(\delta\phi_0)$, like in Eq.(2).

5. DYNAMICS OF DEFORMATIONS OF FILAMENT BUNDLES

The dynamics of deformations of filament bundles is purely local through the action of motors, yet is influenced by geometric global constrains like the total integral curvature, total torsion, length, area etc. For example clamped end boundary condition in flagellar dynamics restricts the total curvature along the flagellum to zero [8]. Examples of such dynamics include axonemal mechanics dynamics of free boundary incompressible fluids, conservation of length and topological parameters in polymer dynamics [9], etc. One of the most important features of these types of dynamics is their connection with nonlinear dynamical systems and conserving quantities [9]. In the following we present how the deformation of the filament bundle can be modeled through the analysis of the differential properties of a family of smooth curves, and how the dynamics of the curvature of these curves can be related with the integrable (nonlinear) "modified Korteweg-de Vries" system (mKdV) [8].

We describe the dynamics of the filament bundle as a one-parameter family of curves $\vec{r}(s_\beta, \beta)$, where s_β is the arclength parameter along the β curve in the family, and β labels the family. There are two possible interpretation of β parameter in the dynamics of a filament bundle. On one hand β can describe the deformation of a certain filament during the deformation of the whole bundle. In this sense β can be interpreted as time. On the other hand, β can describe the parallel transport of an already deformed filament into its neighbor pair filament, also deformed. In this case time is frozen, and we assume a certain fixed geometry for the bundle.

6. THREE-DIMENSIONAL DEFORMATIONS

The situation is similar for the deformation of filaments in 3-d, except for a few additional calculations. The variation of position of a point on a filament is described now by a 3-d formula

$$\vec{dr} = \vec{t}ds + (\vec{t}\delta + \vec{n}\Delta + \vec{b}\Lambda)d\beta, \tag{15}$$

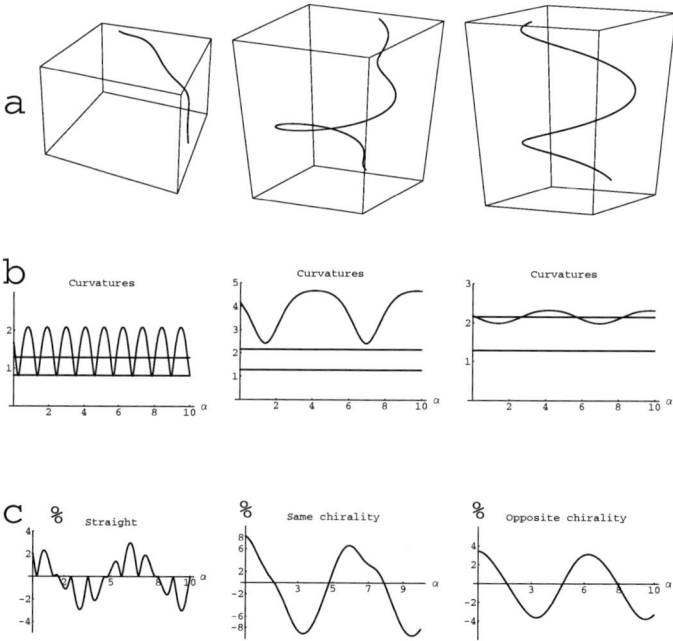

FIGURE 5. Aspect of bundle, curvatures and sliding for helical pair twisting around a helical bundle plotted versus the α angle for a realistic *T.brucei* case. (a). Left parameters for a straight relaxed flagellum $R = 10\mu m$, $r = 0.5\mu m$, $B = \frac{30}{2\pi}\mu m$, $b = \frac{12}{2\pi}\mu m$. Center: parameters for an over-twisted flagellum with pairs and bundle twisted with the **same chirality** $R = 6\mu m$, $r = 0.5\mu m$, $B = \frac{6}{2\pi}\mu m$, $b = \frac{12}{2\pi}\mu m$. Right: parameters for the same over-twisted flagellum, but having pairs and bundle twisted with **opposite chirality** $R = 6\mu m$, $r = 0.5\mu m$, $B = \frac{12}{2\pi}\mu m$, $b = \frac{12}{2\pi}\mu m$; (b) Graphical representation of the three curvatures (arbitrary units). Always the curvature of the bundle is larger than the curvature of the cylindric helix of radius r. (c) Relative slide in percentage, versus α angle. The twisted case (central) with same chirality has the sliding double compared to the other cases.

where in addition to the terms explained in Eq.(1) we have one more degree of freedom along the binormal direction, and hence one parameter, Λ. The equations for the tangent variation are more complicated, and in addition to curvature we have the torsion $\tau(s,\beta)$ function involved in equations, too, but the metrics is described by the same equation. By using a similar approach as in the 2-d case we obtain the relation $\partial\delta/\partial s = k\Lambda.$, and we can write

$$\delta(s,\beta) = \delta(0,\beta) + \int_0^s k\vec{dr}\cdot\vec{n}ds. \tag{16}$$

In the following we calculate the relative separation vector $\vec{dr}$ between the ends of the two filaments (denoted i, j) in the pair, $\vec{dr}(s) = \vec{r}_i(s) - \vec{r}_j(s) = \vec{r}_{ij}(s)$, $\vec{r}_{ij}(0) = \vec{r}_{ij0}$, $|\vec{r}_{ij0}| = a$, where a is the initial linear separation between these 2 filaments. In Figs. (5,6) we present a numerical estimation of Eq.(15) for helical pairs twisted and bent around a helical deformed bundle. The decomposition of $d\vec{r}$ along the local versors of the Serret-Frenet frame is not sufficient, because it does not contain information about the position of a specific filament. Moreover, if the filament has some straight line segment, the normal and binormal are not defined along those segments. Consequently we need to establish an "internal" local frame in addition to the Serret-Frenet frame. During the displacement ds of the j pair the vector $\vec{r}_i$ performs an infinitesimal rotation of angle $d\phi = kds$ around the binormal (local bending), and a rotation of angle $d\psi = \tau ds$ around the tangent (local twist), so its variation reads

$$d\vec{r}_i = (\vec{b} \times \vec{r}_i)d\phi + (\vec{t} \times \vec{r}_i)d\psi,$$

and the differential equation governing $\vec{r}_i$ vector is

$$\frac{d\vec{r}_i}{ds} = \vec{\omega} \times \vec{r}_i, \quad \vec{\omega} = k\vec{b} + \tau\vec{t}. \tag{17}$$

with initial conditions $\vec{r}_i(0) = \vec{r}_{i0}$, namely the structural position of the filament i in the initial cross section of the un-deformed bundle (the base circle). Eq.(16) conserves the norm of vector $\vec{r}_i$, and consequently is a rotation which actually follows the motion of the normal unit vector, since we have $d\vec{n}/ds = \vec{\omega} \times \vec{n}$, with the exception that the origin of the vector $\vec{r}$ also translates along the central filament with the arclength s. Eq.(16) is integrable and its solution has the form

$$\vec{r}_i(s) = e^{\int_0^s \hat{\omega} ds'}\vec{r}_{i0}. \tag{18}$$

where the exponential is defined in the operatorial sense (exponential matrix), and $\hat{\omega}$ is the dual tensor of the vector $\vec{\omega}$, that is $\hat{\omega}_{nm} = \varepsilon_{nmp}\vec{\omega}_p$ with $n, m, p = 1, ...3$ and ε_{nmp} being the Levi-Civitta tensor.

We illustrate these equations with the following example. For a helix shape of curvature $k = \frac{R}{R^2+b^2}$ and torsion $\tau = \frac{b}{R^2+b^2}$ we have $\vec{\omega} = (0, 0, g^{-\frac{1}{2}})$ so the only nonzero elements in the matrix are $\hat{\omega}_{12} = -\hat{\omega}_{21} = -g^{-\frac{1}{2}}$, and the exponential of this matrix is a rotation matrix around the Oz axis. For a bundle of nine pairs $i = 1, 2, \ldots, 9$ the solution Eq.(17) reads

$$\vec{r}_i(s) = \left(-R\cos\frac{2\pi(i-1)}{9}\cos\frac{s}{\sqrt{R^2+b^2}} \mp bR\frac{\sin\frac{2\pi(i-1)}{9}}{\sqrt{R^2+b^2}}\sin\frac{s}{\sqrt{R^2+b^2}}, \right.$$

$$-R\cos\frac{2\pi(i-1)}{9}\sin\frac{s}{\sqrt{R^2+b^2}} \pm bR\frac{\sin\frac{2\pi(i-1)}{9}}{\sqrt{R^2+b^2}}\cos\frac{s}{\sqrt{R^2+b^2}},$$

$$\left. \frac{bR}{\sqrt{R^2+b^2}}\sin\frac{2\pi(i-1)}{9} \right). \tag{19}$$

Eq.(18) represents the position vector of the filament i with its origin on the central axis at arclength s measured from the origin. The most general sliding formula for a pair labeled i and for a filament bundle described by normal $\vec{n}$ and curvature k reads

$$\delta_i(s) = \vec{r}_{i0} \cdot \int_0^s \vec{n}(s')e^{\int_0^{s'} \hat{\omega} ds''}k(s')ds'. \tag{20}$$

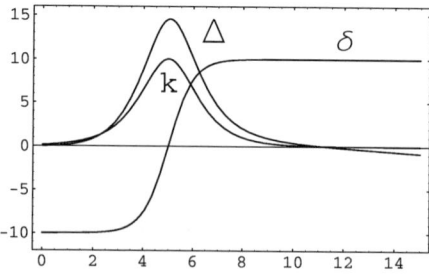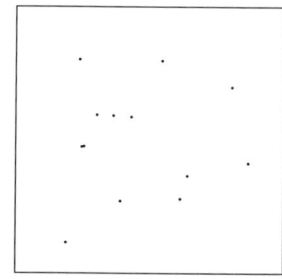

FIGURE 6. A kink-like sliding traveling wave along the bundle produces soliton profiles in separation and curvature (left) and a certain symmetric pattern of 2-d beats in the bundle (right)

We can use Eqs.(16,17,19) to calculate and predict the relative sliding for different pairs in different configuration of simultaneous bending and twisting.

Sliding analysis provides local and global information about the dynamics and the geometric constrains of the bundle. For example, if the total slide at the end of a given pair is zero, it results that the global integral curvature of this curve is zero. Also, sliding interacts with the dynamics in a very direct way. For example, certain boundary conditions at the end of the pairs (like clamped ends) introduce restrictions in the class of admissible shapes. An interesting situation (which has not been mentioned in literature) occurs in the mixed bending and twisting case. For pure bending the relative sliding of a pair in a bundle is constantly increasing versus the length of the bent segment (also proportional to the bending angle). If we first twist the un-deformed (rectilinear) bundle around its symmetry axis, and then we bend it exactly in the same configuration as before, the resulting relative sliding of the same pair is much smaller. The larger the twisting the smaller the sliding. This situation is presented in Fig. 7.

The explanation is simple, the twisted pair follows alternatively segments inside the bending (positive sliding) and outside the bending (negative sliding) and the total final slide is compensated almost to zero. The sliding at the end of the pair is actually produced only by the last unfinished turn of twist around the bundle axis. This result was verified with a powerful numerical code and same behavior was confirmed. This fact may give a hint about the induced twist in flagellum around the body of a trypanosome-like cell, or about the propagation of helical flagellar beats. Also, the combination between bending and twisting could work like a molecular gear shift: for the same bending deformation, more twist reduces the amount of sliding.

7. CONCLUSIONS

We describe the dynamics of axoneme-like filament bundles from the perspective of the relative sliding between filament pairs, in an arbitrary shape bundle, in order to understand and explain mechanisms of self-deformation. Such filament bundles are interesting geometric objects placed, from the geometry point of view, in between differential curves theory and surface theory. This study reports several new expressions

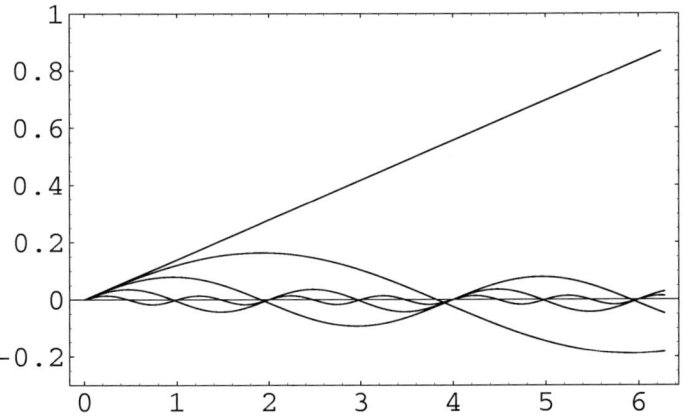

FIGURE 7. Sliding in a twisted and then bent bundle plotted versus the arc (in rad) of the bent region. The straight line is for bending only. The oscillating functions represent the sliding for the same bending angle and arclength, but the bundle was initially twisted uniformly along its length with ϕ angle. Smaller slidings (and more oscillations) are produced by larger twisting angle, and the number of periods of oscillation is given by $\phi/2\pi$.

for sliding. For pure bending or twisting, for bending coupled with twisting (especially in the case of helical shapes), a new geometric definition for sliding, and the most general sliding expressions for any configuration. We analyze the two basic types of bundle deformations, bending and twisting. The general sliding formula for uniform ordinary bending, for any pair in a bundle, is obtained and compared to previous obtained 2-d bending models. The distribution of sliding, and maximum sliding among pairs in general is obtained in general, and for some examples like planar waves, rigid rotation and helical motion. We obtain the formula for uniform twisting, and for twisting around a deformed shape, and we illustrate these formulas with numerical examples from helical, and trypanosome-like shapes. We also obtain through a functional differential formalism a general integral formula for the sliding produced by any shape 3-d. This result can provide the sliding distribution in any segment of any pair, for a bundle of an arbitrary geometry, stationary or in motion. The coupling of twisting and bending, can drastically reduce the relative sliding. Based on the theory of motion of curves we obtain a differential equation which connects the curvature, the tangent sliding, and the normal shift of the pairs. Based on this equation, we can predict the distribution of the sliding (and consequently of motor activity) for any given flagellar shape and dynamics. Several applications of the obtained sliding expressions are in very good agreement with published geometrical models, and with some dynamical models [3,5]. The results are also discussed form the global geometrical invariants point of view, like total integral curvature and length, and are compared to the Hamiltonian models of similar systems [6], as well as with nonlinear effects observed or predicted in literature. The equations obtained for the distribution of sliding along and among the pairs can be used in further modeling, theoretical approaches and/or experimental studies in order to calculate force and torque distributions, and–starting form experimental observed shapes–to calculate and predict energy transfer between motors and pairs, and in general to analyze the dynamics of the flagellum. Some of the conclusions could help the understanding of the cellular self-propulsion using flagellum or hair-like systems, and can be related to the bending and twisting waves that can generate swimming.

REFERENCES

1. B. Alberts, D. Bray, J. Lewis, M. Raff, K. Roberts and J. D. Watson, *The Molecular Biology of the Cell* (Garland, New York, 1994).
2. M. Murase, *The Dynamics of Cell Motility* (Wiley, New York, 1992); C. B. Lindemann and K. S. Kanous, Cell. Mot. Cytoskel. **31**, 1 (1995).
3. M. Hines and J. J. Blum, Biophys. J. **23**, 41 (1978); **25**, 421 (1979); Quart. Rev. Biophys. **12** 103 (1979); M. Hines and J. J. Blum, Biophys. J., **46** 559 (1984).
4. C. B. Lindemann, Cell Motility and the Cytoskeleton **29**, 141 (1994).
5. C. J. Brokaw, Cell. Mot. Cytoskel. **42**, 134 (1999); C. J. Brokaw, *Proc. Natl. Acad. Sci. USA* **72** (1975) 3102; *Biophys. J.* **48** (1985) 633
6. K. Sekimoto, N. Mori, K. Tawada, and Y. Y. Toyoshima, *Phys. Rev. Lett.*, **75** (1995) 172-175.
7. L. Bourdieu, T. Duke, M. B. Elowitz, D. A. Winkelmann, S. Leibler, and A. Libehaber, *Phys. Rev. Lett.*, **75** (1995) 176-179.
8. D. Fry, N. Hutchings, and A. Ludu, preprint LANL arXiv:physics/0309026, 2003; N. Hutchings, and A. Ludu "A Model of African Trypanosome Cell Motility and Quantitative Description of Flagellar Dynamics", submitted 2004; A. Ludu and N. Hutchings, submitted.
9. K. Nakayama, H. Segur, and M. Wadati, *Phys. Rev. Lett.*, **69** (1992) 2603; H. Hasimoto, *J. Fluid Mech.* **51** (1972) 477.

ACKNOWLEDGMENTS

We gratefully acknowledges the support of the Louisiana Board of Regents under grant LEQSF $(2004 - 2007) - RD - A25$ and are grateful for the IDEAS program (*www.scitech.nsula.edu/IDEAS*) support. A.L. gratefully acknowledges the support of the National Science Foundation under grant 0140274

Symmetry Breaking in a Model for Nodal Cilia

Charles J. Brokaw

Division of Biology, California Institute of Technology
Pasadena, CA 91125, USA

Abstract. Nodal cilia are very short cilia found in the embryonic node on the ventral surface of early mammalian embryos. They create a right to left fluid flow that is responsible for determining the normal asymmetry of the internal organs of the mammalian body. To do this, the distal end of the cilium must circle in a counterclockwise sense. Computer simulations with 3-dimensional models of flagella allow examination of 3-dimensional movements such as those of nodal cilia. 3-dimensional circling motions of short cilia can be achieved with velocity controlled models, in which dynein activity is regulated by sliding velocity. If dyneins on one outer doublet are controlled by the sliding velocity experienced by that doublet, the system is symmetric, and the 3-dimensional models can show either clockwise or counterclockwise circling. My computer simulations have examined two possible symmetry breaking mechanisms: 1) dyneins on doublet N are regulated by a mixture of the sliding velocities experienced by doublets N and N+1 (numbered in a clockwise direction, looking from the base). or 2) symmetry is broken by an off-axis force that produces a right-handed twist of the axoneme, consistent with observations that some dyneins can rotate their substrate microtubules in a clockwise direction.

INTRODUCTION: THE BIOLOGY OF NODAL CILIA

About one in ten thousand humans have the condition known as situs inversus, in which the asymmetry of the internal body organs is reversed [reviewed in 1]. This condition has been known for many years, and is found in other mammals besides humans. It results from a failure of a normal developmental mechanism that solves a fundamental problem in developmental biology: How is the symmetry of the fertilized egg broken, to generate, in almost all cases, a consistent asymmetry of the internal body organs?

By itself, situs inversus usually causes no health problems, and may be undetected for years. However, situs inversus is sometimes associated with significant medical problems. This association, now recognized as "Kartagener-Afzelius syndrome" [2], has been the key to understanding situs inversus and the development of normal body asymmetry. Neither Kartagener nor Afzelius set out to study situs inversus. Manes Kartagener was an MD working in Zurich, Switzerland. He was apparently the first to study the inheritance of the combination of bronchial and sinus infections and situs inversus, and he recognized that situs inversus was not found in all of the family members who inherited the high incidence of respiratory infections. Bjorn Afzelius is a Swedish electron microscopist. Beginning in the 1950s, he was one of the earliest electron microscopists to study the microscopic anatomy of cilia and flagella. In the early 1970s, he was asked to look at the flagella of non-motile spermatozoa from some infertile men. It turned out that some of these men not only had defective sperm

CP755, *ISIS: International Symposium on Interdisciplinary Science*
edited by A. Ludu, N.R. Hutchings and D.R. Fry
© 2005 American Institute of Physics 0-7354-0240-X/05/$22.50

flagella, but also the symptoms recognized by Kartagener. Afzelius made the connection with defective cilia in the respiratory tract, and confirmed this with electron microscopy. Afzelius speculated that cilia in the early embryo might produce a rotational movement responsible for normal asymmetry. It followed from this that if these embryonic cilia were non-motile, the asymmetry might be determined randomly, explaining a 50% incidence of situs inversus in these individuals with defective cilia.

Confirmation of Afzelius's speculation has come only in the last few years, from observations on mouse embryos. A mouse mutant with situs inversus was known since 1959 [3], and Supp et al. [4,5] reported that it resulted from a mutation in a dynein. Dyneins are the motor enzymes responsible for the movement of cilia and flagella. The laboratory of Nobutaka Hirokawa, at the University of Tokyo, was studying a kinesin, a member of the other group of motor enzymes that move along microtubules. This kinesin is essential for the assembly of cilia, and a mutant mouse defective in this kinesin has serious developmental defects. Using mutant and normal mice, Hirokawa's group was able to show that cilia in the embryonic node on the ventral surface of the mouse embryo are motile, and produce a leftward-directed fluid flow that is essential for development of proper asymmetry [6,7,8]. This flow is produced by a circling motion of these nodal cilia, which is clockwise when looking at the surface of the embryo. However, in the cilia and flagella field, it is conventional to look at cilia from base to tip, and the circling movement of nodal cilia is therefore counterclockwise in this context.

A leftward-directed fluid flow could be produced by counterclockwise circling of nodal cilia if they are, on average, tilted towards the posterior end of the embryo (which has already been determined by the critical time for left-right asymmetry appearance). When tilted posteriorly, the cilium will be moving from left to right when it is closest to the surface of the epithelium, and generation of rightward-directed fluid flow will be retarded by the surface. The cilium will be moving from right to left when it is farthest from the surface, where generation of leftward-directed flow is most effective [9]. However, the fluid movement in the nodal depression is unlikely to be this simple [10]. The sequence of steps by which a leftward-directed flow induces the development of normal asymmetry is still under active investigation, and contains multiple events that could fail and cause situs inversus that is unrelated to ciliary motility. However, the focus of the remainder of this article will be ciliary motility; specifically, what is the initial symmetry-breaking factor that causes the circling movement of nodal cilia to be counterclockwise?

Nodal cilia are very small, and difficult to see by ordinary light microscopy, so no detailed description of their bending is available. Scanning electron microscopy shows that they are only 2 or 3 μm long [11]. Transmission electron microscopy of cross-sections of nodal cilia shows the absence of the central pair of microtubules found in most motile cilia and flagella [8]. In most cases, the central pair of microtubules is required for planar bending, and three-dimensional or helical movements are found when the central pair is absent.

The cytoskeleton of cilia and flagella is known as an axoneme. The structure of the axoneme, as revealed by electron microscopy of transverse sections, is asymmetric. The dynein motor enzymes are located in the "arms" located on the A tubule of each of the 9 outer doublets (Fig. 1).

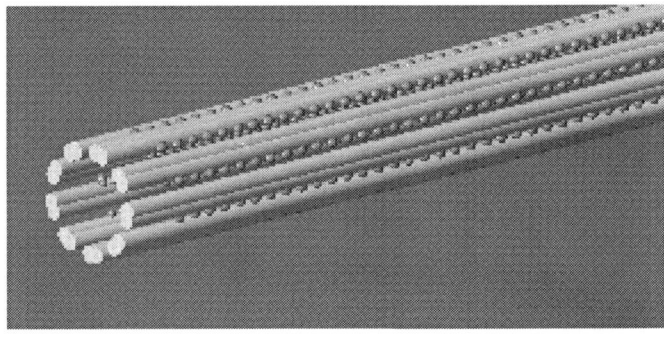

FIGURE 1. Basic elements of an axoneme, the common cytoskeleton of cilia and flagella. Nine outer microtubular doublets are arranged to form the surface of a cylinder. Dynein motor enzymes, represented here as spheres, are lined up in two rows on the A-tubule of each doublet, and produce sliding movements by interacting with the B-tubule of the adjacent doublet.

These arms are attached stably to the A tubule of the doublet, and reach around in a clockwise direction to interact with the B tubule of the adjacent doublet. This interaction produces sliding displacements between adjacent doublets [12]. In spite of the universal asymmetry of this structure, flagella from some organisms produce bending patterns that require clockwise progression of activity around the circumference of the axoneme, and flagella from other organisms require counterclockwise progression of activity. So asymmetry of function is not explained in a simple manner by asymmetry of structure. My approach to this problem is to construct a computer model of a nodal cilium and ask what features need to be incorporated into the model to obtain consistent counterclockwise circling movements.

COMPUTER MODELLING OF THREE-DIMENSIONAL MOVEMENT OF FLAGELLA AND CILIA

Three-dimensional modelling of flagella has been described in detail [13] and modified for modelling short flagella and cilia such as nodal cilia [14]. The models start by arranging the 9 outer doublets uniformly on the surface of a cylinder, as in Fig. 1, and then the doublets are just represented by parallel lines on the surface. There must be important connections and forces that are involved in maintaining this arrangement, but these are not considered in my modelling. The only distortion that is allowed is sliding (shear) between adjacent doublets; when this shear is not uniform along the length, the model must bend. Sliding between doublets is required to be 0 at the basal end of the model.

When an axoneme is bent, it acts as an elastic structure, and tries to return to a straight configuration. In other words, it has an elastic bending resistance. There are reasonably good quantitative estimates of this resistance, which have led to use of a value of 2×10^8 pN nm^2 for modelling flagella such as sea urchin sperm flagella [15]. Flagella probably also have elastic shear resistances, and they operate in a viscous environment that resists the bending movements. These resistances are important, but they are not essential components of a computer model [16].

For numerical analysis of axonemal bending, the length of the axoneme is divided up into straight segments of equal length, with bending occuring at the joints between these segments. The rate of bending at each joint is obtained by solving a set of equations that balances the active and resistive bending moments, and these rates of bending can then be used to calculate the shape of the model as a function of time [13, 17].

Bending of the axoneme results from shear forces generated by dynein motor enzymes that are distributed uniformly along the length of each outer doublet (Fig. 1). Although no specific information about the spacing and number of dyneins in nodal cilia is available, information from other flagella has been extrapolated for this modelling. A typical flagellum appears to contain at least 9 different dyneins, with some functional differences between these dyneins. There is only very limited knowledge about how these functional differences might be significant for flagellar function [18], and for the present work they are ignored. If all of the dyneins are always producing active shear forces, they will antagonize each other, and no bending will result. So the usual assumption has been that there must be some form of local control mechanism that turns dynein activity on and off. There is no direct experimental evidence that dyneins can be turned on and off -- it has just been considered to be necessary to get a cilium or flagellum to produce bending. Most computer modelling of flagella and cilia has utilized feedback from bending to control dynein activity; these models can be referred to as "curvature controlled models" [13, 17, 19, 20].

Compared with understanding of the other motor enyzmes, myosin and kinesin, knowledge of dynein function is very limited. Dynein may operate in the manner suggested by Huxley [21] for myosin, by attaching to its substrate microtubule while it releases strain by executing a power stroke [22], and then recovering strain energy while detached from the substrate and hydrolyzing ATP [reviewed in 23]. However, dynein is significantly different and more complex that the other motor enzymes [24]. At present, we do not understand this complexity enough to introduce it into computer simulations, and the models used for dynein are basically just myosin models dressed up to look like dynein. For modelling cilia and flagella, two approaches are possible. One approach computes the force by stochastically modelling the chemical kinetics of each individual motor enzyme [13, 25]. The other approach simply calculates the active shear force generated on each doublet for each segment of the model from a simple mathematical model for a force generator. Since the stochastic modelling is computationally more demanding and requires more estimates of unknown parameters, both approaches are useful.

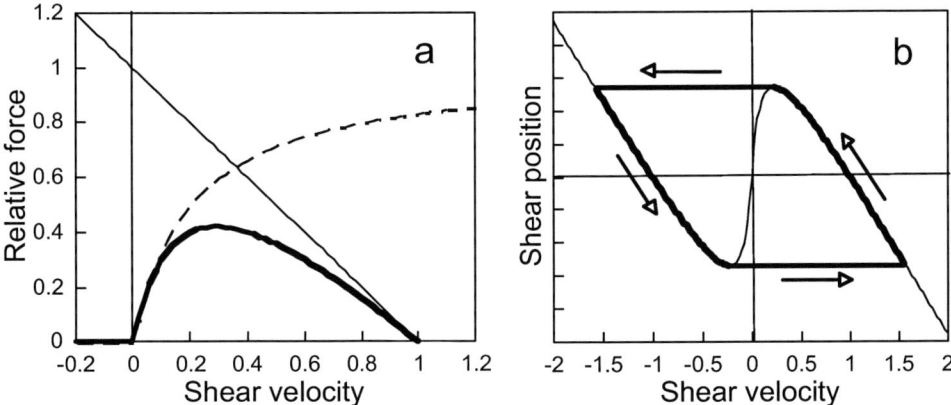

FIGURE 2. Steady-state behavior of a simple model for shear force generation. (a) The thin straight line is the force given by Equ. 1 and 2. The dashed line is the control function given by Equ. 3. The heavy solid line is the combination of the force function and the control function. (b) Two force generators, oriented in opposite directions, combined with a linear elastic shear resistance. With linear elastic resistance, force and shear displacement are proportional, giving a plot that is a phase diagram, showing limit cycle oscillation.

A MODEL FOR NODAL CILIA USING A MATHEMATICAL MODEL FOR FORCE GENERATION AND CONTROL BY SLIDING VELOCITY

This model has been presented in [14]. It starts with a simple mathematical model for a force generator with the property that active shear force decreases with sliding velocity. The model is designed to resemble actual motor enzymes, which produce work by releasing strain. At 0 sliding velocity, active shear force per unit length, $f = F$. F represents the maximum isometric force that can be generated by the currently active population of motors. In response to a sudden shear, the force acts like an elastic element with an elastic constant $= E_{SCB} F$. When $f \neq F$, the force recovers towards F with a first-order rate constant, k_1. The model is described by a simple differential equation for f as a function of sliding velocity $\dot{\alpha}$ [16]:

$$\frac{df}{dt} = -E_{SCB} F \dot{\alpha} + k_1 (F - f). \tag{1}$$

The steady-state solution (constant sliding velocity) is

$$f = F(1 - E_{SCB} \dot{\alpha}/k_1), \tag{2}$$

giving a linear decrease in force with sliding velocity, with a "maximum" sliding velocity at $\dot{\alpha} = k_1/E_{SCB}$. To create an self-oscillatory system, this force generator is combined with a control function that controls F by sliding velocity:

$$\frac{dF}{dt} = k_2\left(\frac{F_0}{1+k_3/\dot{\alpha}} - F\right) \qquad \text{for } \dot{\alpha} > 0$$

$$\frac{dF}{dt} = -k_2 F \qquad \text{for } \dot{\alpha} \leq 0.$$

(3)

where F_0 is the maximum isometric force that can be produced when all of the motors are fully activated, and k_3 is a constant equal to the sliding velocity at which $F = 0.5$ F_0. The time delay process constant k_2 determines how rapidly F follows changes in $\dot{\alpha}$. Figure 2a illustrates the steady state force given by Equ. (1) with constant F, and the control function given by Equ. (3) with very large k_2. Under steady state conditions, the combination of Equations (1) and (3) gives the heavy curve shown in Fig. 2a with the force rising from 0 at $\dot{\alpha} \leq 0$ to a maximum and then decreasing to 0 at a velocity close to the maximum sliding velocity $= k_1 l/E_{SCB}$.

The steady state behavior shown in Fig. 2 is not used for modelling, and is only a rough guide to understanding the behavior of this system. Instead, Equ. (1) is integrated to obtain f at the end of a time interval that is assumed to be short enough so that $\dot{\alpha}$ and F on the right hand side of Equ. (1) can be assumed to be constant during the time interval [16]. Two of these force generators operating in opposite directions, in combination with a linear elastic shear resistance, create an oscillator, described by the phase diagram in Fig. 2b [26]. The interpretation of these specifications is that when dyneins on some doublets are producing sliding in one direction, these dyneins remain active as long as the sliding velocity is greater than the velocity at the peak force. If there is an elastic resistance, the force will gradually increase, the velocity will slow down, and once it falls below the velocity for peak force, the active dyneins are shut off, the velocity reverses, and the antagonistic set of dyneins becomes active.

Numerical integration of this system is feasible by separation of time scales, with the linear elastic terms resulting from E_{SCB} and elastic load treated implicitly, and other terms treated explicitly [16]. Non-linear elastic shear resistance, appropriate for elastic interdoublet linkages, is calculated as proposed by Hines and Blum [19]:

$$m_S = E_S\sigma\left(1 - \frac{1}{\sqrt{1+0.75\sigma^2}}\right),$$

(4)

where m_S represents shear moment per unit length and σ represents shear, in angular units (rad).

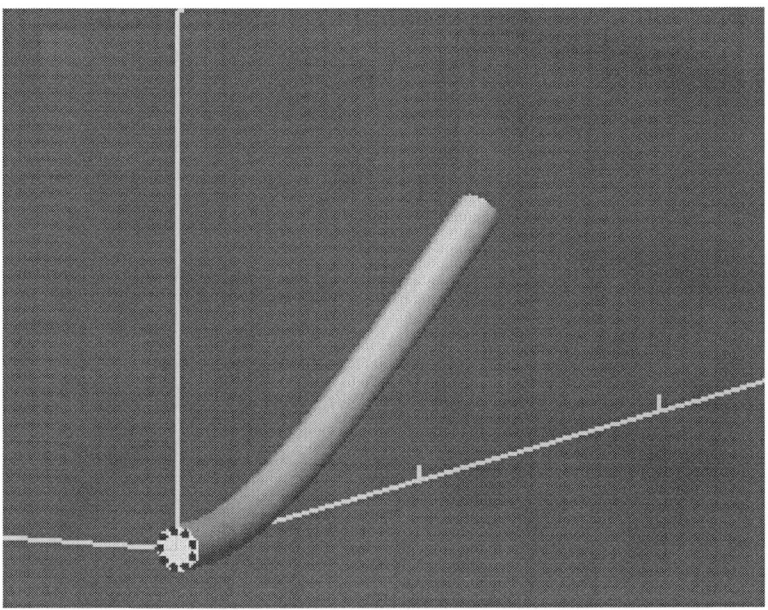

FIGURE 3. One position of a model nodal cilium that is clamped at the basal end and generating a counterclockwise circling motion. Ticks on the Z axis are at 2 μm intervals, and the vertical, Y axis is 2 μm tall. Length of the model is 2.6 μm, circling frequency approx. 10/sec. Parameters used for this computation are the same as for the model in Fig. 2d of [14], except for $E_B = 1.5 \times 10^8$ pN nm^2 and the elastic constant for non-linear shear elasticity is $E_S = 44$ pN (Equ. 4). Computed with 13 length segments and 0.1 ms time steps.

A model of a nodal cilium, with this force generator distributed along each doublet and a large value of k_2, attempts to oscillate with planar bending, but the bending plane wanders around apparently at random. A stable three-dimensional circling movement is obtained with lower values of k_2 (Fig. 3). To produce this circling movement, force generation must propagate from doublet to doublet around the circumference of the model. No mechanism for circumferential propagation has been programmed into the model. The circumferential propagation is "self-organized" by the mechanical interactions imposed by requiring the force generators on each doublet to reside on the surface of a cylinder [14]. Similar self-organization was observed previously with curvature controlled models for flagella, and termed "doublet metachronism", to emphasize its similarity to metachronal coordination of independent cilia on a surface [13].

What determines the direction of circling by this model? In an earlier study of three-dimensional bending of flagella in which active shear force was controlled by curvature, it was found that the direction of circling could be controlled by the direction of sensitivity to curvature [13]. The same device can be used with the current model, by suggesting that the controlling sliding velocity for dyneins on a particular doublet is measured in a direction that is slightly different than the direction of the sliding produced by those dyneins. While this device is successful [14], it requires, in effect, that dyneins on doublet N are influenced by the sliding velocity on doublet

N+1. A more attractive idea is that dyneins act to rotate or twist their substrate microtubules, resulting in twist of the axoneme. Several examples of clockwise (viewed from base to tip) microtubule rotation by dyneins during in vitro motility assays support this idea [27, 28].

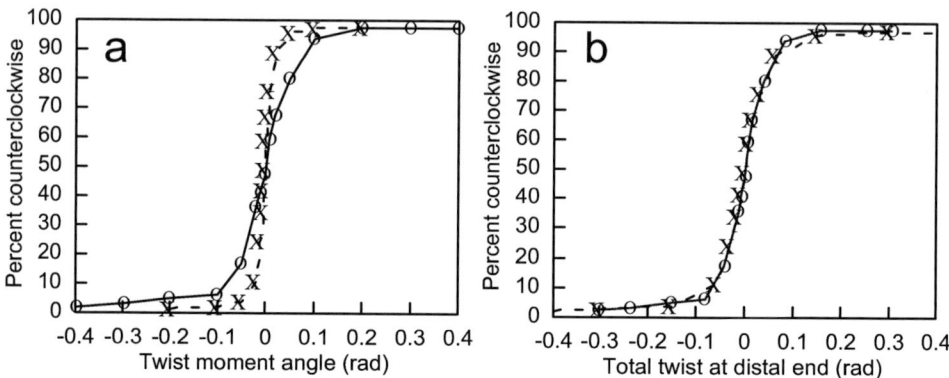

FIGURE 4. Results of computations with a nodal cilium model similar to the model shown in Fig. 3, except that the elastic shear resistance was linear, with $E_S = 4$ pN. The percentage of counterclockwise circling was determined from 400 computations for each point shown. Points on the solid line, from [14], were obtained with the standard twist resistance, and points on the dashed line were obtained with 25% of that twist resistance.

The effect of off-axis force, which produces a moment that twists the axoneme, is shown in Fig. 4. An off-axis force in the direction that would cause clockwise rotation produces clockwise twist of the axoneme, and results in predominantly counterclockwise circling. Results in Fig. 4 are shown for two series of computations, using two different values of axonemal twist resistance. In Fig. 4a, the results are plotted against the angle of the off-axis force, and in Fig. 4b, the results are plotted against the resulting twist measured at the end of the ciliary model. In the latter case, the curves are superimposed, indicating that it is the actual twist that regulates the circling direction of this model. To obtain this result, it was necessary for the model to include a significant amount of elastic shear resistance, as well as the elastic bend resistance [14].

Once a circling direction is established, the symmetry-breaking twisting moment can be removed or reversed, without any change in circling direction [14]. Computations designed to examine the initial determination of circling direction reveal interesting complexity in the behavior of this dynamic system. An example with one particular set of parameters is shown in Fig. 5. This model uses the non-linear elastic shear resistance of Equ. (4), which is designed to replicate the behavior of elastic linkages between the outer doublets. Maintaining the twist for a time equivalent to half the circling time, or 0.05 sec, is sufficient to obtain results equivalent to maintaining twist throughout the computation, as in Fig. 4. These results are shown along the top edge of the plot in Fig. 5. Along the bottom edge, with no twist, and along the vertical line for 0 twist moment, the results are random, with approximately 50% counterclockwise circling. When the twist is maintained for 0.2 to 0.25 sec, the response to the higher values of twist is reversed. This can probably be explained if the amount of sliding at times less than 0.25 sec is not sufficient for the non-linear elastic shear resistance to become important. Under this condition, the model then behaves in the same manner as models without elastic shear resistance, with a reversed and weaker effect of twist on circling direction [14]. With intermediate values of twist moment angle and cutoff time, there are other regions of the plot that show unexpected and unexplained behavior. Whether these complexities of behavior have any relevance to the real biological systems remains to be established.

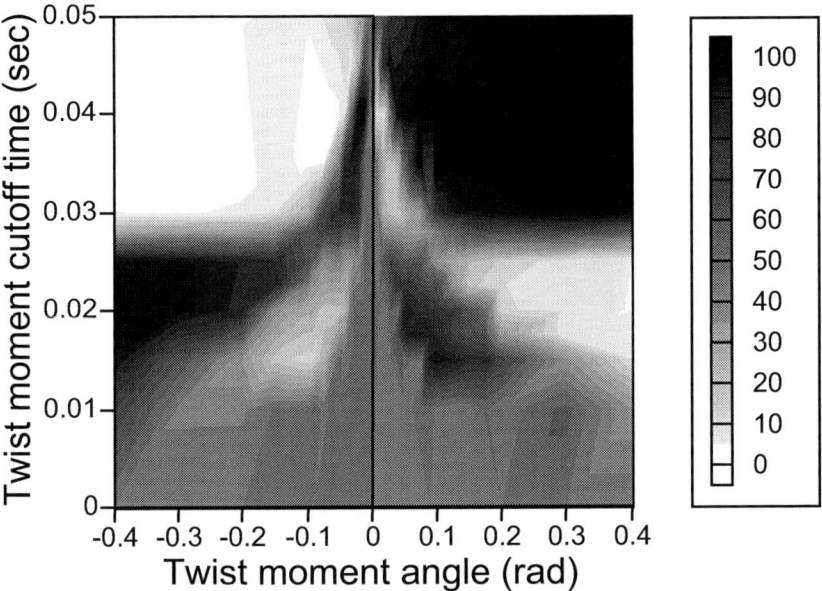

FIGURE 5. Results of computations with the nodal cilium model shown in Fig. 3. The intensity scale represents the percentage of counterclockwise circling, determined from 400 computations at each of 229 pairs of values of twist moment angle and twist moment cutoff time.

EXTENSION TO MODELS WITH STOCHASTIC DYNEINS

A nodal cilium probably contains of the order of 1000 to 3000 dynein motors. If each of these dyneins behaves as an independent force generator, stochastic fluctuations in active shear force will be significant, and the behavior of the model might be expected to be very different from the deterministic model discussed in the previous section. Using stochastic computation of the individual dyneins, the present models can produce circling movements, but the direction of circling is not stable unless additional averaging processes are introduced. These models are not yet adequate for satisfactory examination of symmetry breaking in nodal cilia.

Models that more successfully replicate the movements of real nodal cilia may require consideration of models for dynein motor function that do not assume that each dynein operates independently of its neighbors. Alternatively, oscillation may result from some form of control mechanism that is much less sensitive to stochastic fluctuations in force and sliding caused by independent operation of individual dyneins. Models in which oscillation results from regulation of dynein activity by sliding velocity are designed to have an instability that leads to oscillation. This may be a good design for a deterministic system, but not for a stochastic system. Despite intensive investigation for more than 50 years, the mechanism for oscillation of cilia and flagella remains a mystery.

REFERENCES

1. McManus, C., *Right Hand, Left Hand*, Cambridge: Harvard University Press, 2002.
2. Berdon, W. E., and Willi, U., *Pediatr. Radiol.* **34**, 38-42 (2004).
3. Hummel, K. P., and Chapman, D. B., *J. Hered.* **50**, 9-13 (1959).
4. Supp, D. M., Witte, D. P., Potter, S. S., and Brueckner, M., *Nature* **389**, 963-966 (1997).
5. Supp, D. M.,Brueckner, M., Kuehn, M. R., Witte, D. P., Lowe, L. A., McGrath, J., Corrales, J., and Potter, S. S. *Development* **126**, 5495-5504 (1999).
6. Nonaka, S., Tanaka, Y., Okada, Y., Takeda, S., Harada, A., Kanai, Y., Kido, M., and Hirokawa, N., *Cell* **95**, 829-837 (1998).
7. Okada, Y., Nonaka, S., Tanaka, Y., Sijoh, Y., Hamada, H., and Hirokawa, N., *Molecular Cell* **4**, 459-468 (1999).
8. Nonaka, S., Shiratori, H., Saijoh, Y., and Hamada, H., *Nature* **418**, 96-99 (2002).
9. Blake, J., and Sleigh, M. A., *Biol. Revs.* **49**, 85-125 (1974).
10. Cartwright, J. H. E., Piro, O., and Tuval, I., *Proc. Natl. Acad. Sci.. USA* **101**, 7234-7239 (2004).
11. Bonnafe, E., Touka, M., AitLounis, A., Baas, D., Barras, E., Ucla, C., Moreau, A., Flamant, F., Dubruille, R., Couble, P., Collignon, J., Durand, B., and Reith, W., *Mol. Cell. Biol.* **24**, 4417-4427 (2004).
12. Summers, K. E., and Gibbons, I. R., *Proc. Natl. Acad. Sci. USA*, **68**, 3092-3096 (1971).
13. Brokaw, C. J., *Cell Motil. Cytoskel.* **53**, 103-124 (2002).
14. Brokaw, C. J. *Cell Motil. Cytoskel.* **60**, 35-47 (2005).
15. Omoto, C. K., and Brokaw, C. J., *J. Cell Sci.* **58**, 385-409 (1982).
16. Brokaw, C. J., *Biophys. J.* **48**, 633-642 (1985).
17. Brokaw, C. J., *Biophys. J.* **12**, 564-586 (1972).
18. Brokaw, C. J., *Cell Mot. Cytoskel.* **28**, 199-204 (1994).
19. Hines, M., Blum, J. J., *Biophys. J.* **23**, 41-57 (1978).
20. Lindemann, C. B., *Cell Mot. Cytoskel.* **29**, 141-154 (1994).
21. Huxley, A. F., *Prog. Biophys.* **7**, 255-318 (1957).
22. Burgess, S. A., Walker, M. L., Sakakibara, H., Knight, P. J., and Oiwa, K., *Nature.* **421**, 715-718 (2003).
23. Brokaw, C. J., *Biophys. J.* **73**, 938-951 (1997).
24. Kon, T., Nishiura, M., Ohkura, R., Toyoshima, Y. Y., and Sutoh, K., *Biochemistry.* **43**, 11266-11274 (2004).
25. Brokaw, C. J., *Biophys. J.* **16**, 1013-1027 (1976).
26. Brokaw, C. J., *Proc. Natl. Acad. Sci.. USA* **72**, 3102-3106 (1975).
27. Vale, R. D., and Toyoshima, Y. Y., *Cell.* **52**, 459-469 (1988)
28. Kagami, O., and Kamiya, R., *J. Cell Sci.* **103**, 653-664 (1992).

What Organizes the Molecular Ballet that Promotes the Movement of the Axoneme in Such a Way that its Molecular Machinery Seems to be a Whole?

Christian Cibert

Groupe de "Morphométrie et de Modélisation Cellulaire", Département de Biologie du Développement, Institut Jacques Monod, CNRS, Universités Paris 6, Paris 7, Tour 43, 2 place Jussieu, F-75251 Paris CEDEX 05. E-mail: cibert@ijm.jussieu.fr

Abstract. The axonemal machinery constitutes a highly organized structure whose mechanisms seem to be very simple but whose regulation remains unknown. This apparent simplicity is reinforced by the fact that many models are able to perfectly mimic the axonemal wave trains that propagate along cilia and flagella. However nobody knows what are the actual mechanisms that coordinate the molecular ballet that exist during the beat. Here we present some theoretical elements that show that if the radial spokes are one of the main elements that promote axonemal regulation, they must be involved in a complex mechanism that makes the axoneme a discrete structure whose regulation could depend on local entropy that promotes the emergence of new molecular properties.

INTRODUCTION

The axoneme is the active axial structure of the flagella and the cilia of eucaryotic cells. It is basically constituted by a cylinder of nine outer doublets of microtubule that bear the dynein arms (engines [1, 2]), the nexin links (probably elastic molecules [3]) and the radial spokes that link each outer doublet to the central apparatus [4] (**Figure 1**A). Two antagonistic forces are involved in the propagation of the wave train along these organelles, first the creation of the shear of the outer doublets, and second the balance of this force due to the elastic properties of the axoneme [5] and the fluidic characteristics of the swimming medium [6].

This architecture and its basic mechanical properties are easy to manipulate, and because the axonemal machinery could be reduced to a ensemble of simple equations, the axoneme appears as an "ethereal structure", as if these basic mechanisms were not supported by molecular hardware. However, the axoneme is a molecular edifice and the question is the following. What organizes the molecular ballet that promotes the movement of the axoneme, in such way that the molecular machinery seems to be a whole?

The characteristics of the dynein arms illustrate perfectly the incidence of the internal molecular cross talk that promotes the integration of a system. They are constituted by only one chain of amino acids that defines different sub-domains whose inter-

CP755, *ISIS: International Symposium on Interdisciplinary Science*
edited by A. Ludu, N.R. Hutchings and D.R. Fry

actions generate a cyclic conformational change and consequently a force that allows the sliding of the outer doublets [1, 2]. The dynein arms could be considered as the paradigm of integration, because it is possible to establish a link between their ultra-structural organization, their biochemical characteristics and their biological function even if these three aspects are not deduced one from the other. This defines "integration" and explains what is "emergence". Emergence promotes the effective properties of a molecular edifice that are not the simple sum of the properties of their components [7, 8]. The same reasoning could be used about metabolons [9], *i.e.* many molecular complexes involved in different metabolic pathways [8, 10] and *a fortiori* about the axoneme.

Because of the limitations in the spatial and temporal resolution of microscopic techniques, it is now impossible to observe in real time what happens within the axoneme. Consequently, considering observations of fixed cilia and flagella, only the most obvious mechanisms are theoretically proposed. Moreover, in spite of the excellent fitting between calculated and real wave trains that propagate along cilia and flagella, the computation of these movement depends on hypotheses proposed about the real functions of the different axonemal elements. Yet nobody knows what these mechanisms are effectively. Thus, instead of considering a possible molecular mechanism and modeling a wave train, I have chosen to consider a wave train and modeled the possible behavior of the molecular events that occur along the axoneme during the propagation of a theoretical wave train.

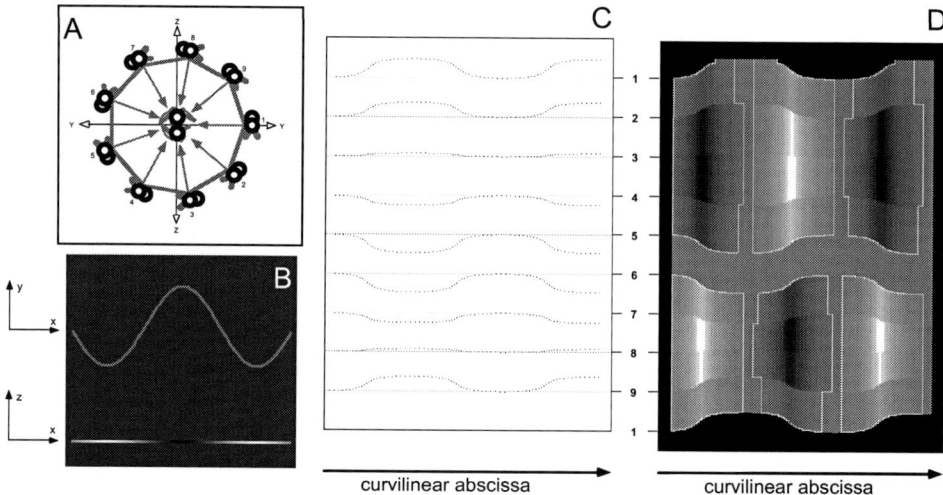

Figure 1: A, is the schematic transversal cross-section of the axoneme, whose orientation is defined by the two systems of reference indicated in B. The central shears of the outer doublets are indicated in C, and the first derivative of the "distances" between these curves defines the "tangential shear" of the outer doublets (D).

The superposition of the central shears calculated during the tipward displacement of a theoretical sine shaped wave train along the model shows that there is a point where the shear is always nil (*P0 points*) indicated by the arrow (**Figure 2**A).

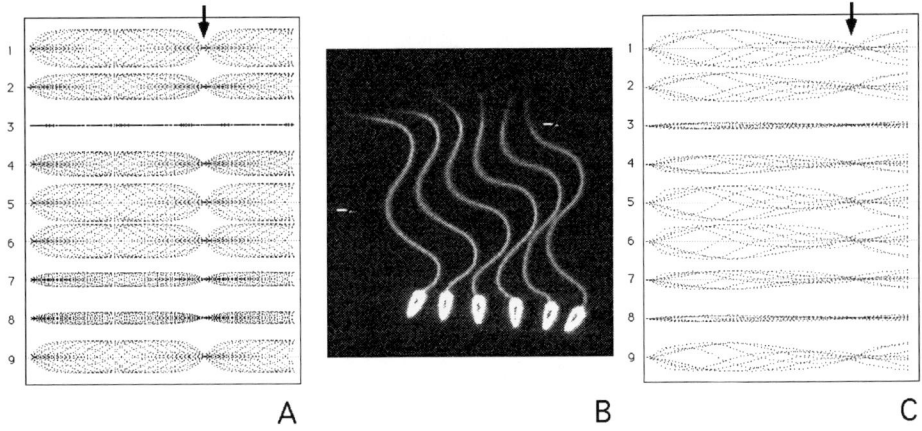

Figure 2 : A, superposition of the central shears of the outer doublets calculated from the theoretical sine shaped wave train. B, traces of a real sea urchin spermatozoon. C, superposition of the central shears of the outer doublets calculated from the traces of the sea urchin spermatozoon. The arrows indicate the location of the P0 points.

TWO CLASSES OF RESET POINTS

Consider a simple sine shaped wave train that propagates along a flagellum [11] (this is evidently a limit case that does not represent all the possible waveforms du to the axonemal activity) (**Figure 1**B). In this scheme, the beat plane and the neutral surface correspond to the (x,y) and the (x,z) plane of the axoneme, respectively (**Figure 1**A). Along the model, the shear between each of the outer doublets and the central apparatus (*central shear*) is modeled as a curve (**Figure 1**C). Because of the sine shaped wave train considered here, the amplitudes of the central shear of the doublets 3 and 8 are the lowest because these doublets are located in the neutral surface. On the other hand, the amplitudes of the central shears of doublets 1 and 9 and the pair 5-6 are the highest because these doublets are located in the bending plane.

If one represents the first derivative of the variation of the distance between two consecutive curves as a gray level image, one schematizes the activity of the dynein arms along a wave train (*tangential shear*), where the white and the black areas show the highest activities of the axonemal cylinder (**Figure 1**D). As expected, the doublets 3 and 8 that do not slide along the central apparatus, are involved in the largest tangential shear, and doublets 1, 9 and pair 5-6 located in the bend plane, involved in the highest central shear, are involved in the lowest tangential shear.

We calculated the central shears within a real sea urchin spermatozoon, whose traces are recorded using a stroboscopic illuminator (**Figure 2**B), and the superposition of the diagrams that we obtained (**Figure 2C**) is the same than the one that we calculated from a theoretical sine shaped wave train (**Figure 2A**); specifically, we observed the same P0 point.

These observations have two consequences. First, it could be postulated that the stable nil shear point P0 and the basal anchor define a sort of module along the axoneme necessarily involved in the definition of the principles that drive the regulation of the axonemal machinery. These stable P0 points are the first reset points. Second, along a given module the central and the tangential shears that involve a given doublet are inversely proportional [11].

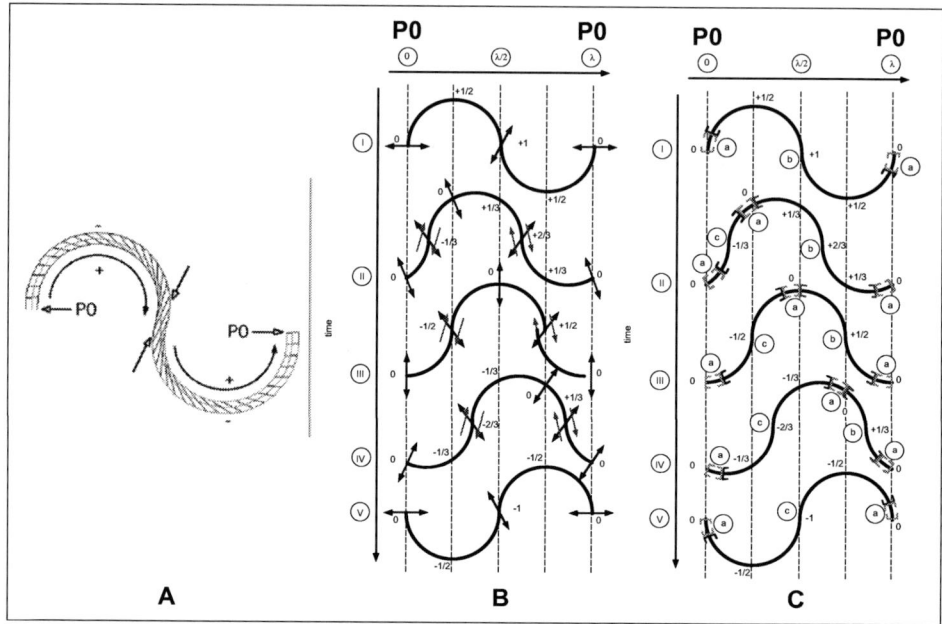

Figure 3 : A, central shear calculated when two "complete" waves involve a module. B, central shear that corresponds to the displacement of the wave train along a module. C, behavior of the radial spokes relative to the displacement of the wave train along a module. Circled a, b and c defined the segments of the axoneme (delineated by two square brackets) where the radial spokes have a homogeneous behavior (see text); the same code is used for the other figures.

According to admitted principles, along two consecutive waves of opposite polarity, the two opposite sides of the neutral surface of the axoneme must be alternatively active as indicated by the black arrows and the + signs in **Figure 3A**. But, considering a wave train that propagates along a module, the polarity of the central shear is the same whatever the sign of the curvature (**Figure 3B**), and a nil shear point propagates along the module (**Figure 3B**). These motile nil shear points are the second "reset points". Consequently, the balance between the activities of the two opposite sides of the neutral surface depends on the displacements of these motile "reset points" along a module defined by the first series of stable P0 points [12].

THE SPECIFIC FUNCTION OF THE RADIAL SPOKES

It is established that the radial spokes and the central apparatus play a regulative function of the axonemal movement whose mechanism remains however to be precisely defined [13].

We modeled the behavior of the radial spokes during the creation of a bent segment and conclude that it is not uniform along the segment. In **Figure 4**, along the segments "a" (circled a (see **Figure 3C**)) the radial spokes follow the bend movement only. Along the segments "c" (circled c) the radial spokes must jump to adjust their length and their tilt along the curved segment [12]. During the reverse phase it is impossible for the radial spokes to passively follow the reverse movement. Were they to do so, their conformation would be the one indicated in the trace number 10 in **Figure 5A**, but this was never observed. During the reverse phase the radial spokes are necessarily involved in a ratchet mechanism that keeps them from having this inverted polarity. In **Figure 5B**, along the segments "a" the radial spokes become perpendicular to the axis of the central apparatus. Along the segments "b" (circled b) the radial spokes maintain this conformation until the axoneme has a straight conformation. These propositions are compatible with the observations of Warner [14].

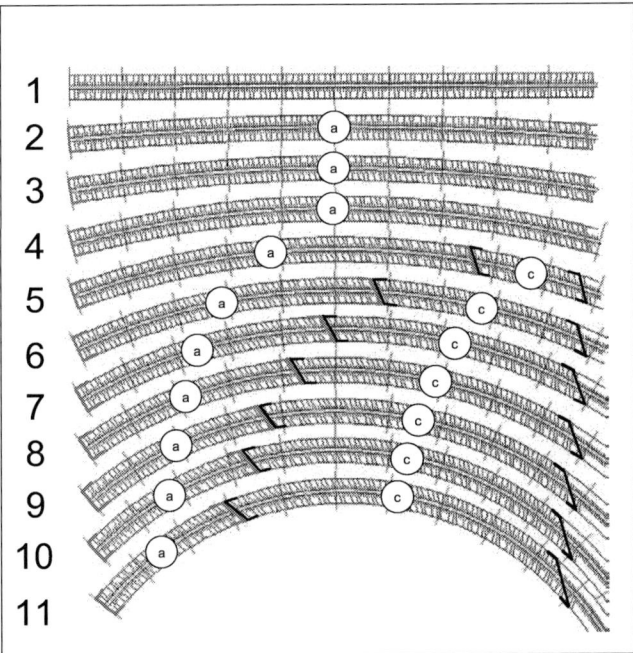

Figure 4 : behavior of the radial spoke during relative to the bending of an axonemal segment.

It is possible to establish a direct relationship between the propagation of the wave train and the local behavior of the radial spokes (**Figure 3C**), (i) along the "a" segments the radial spokes follow the bending, (ii) along the "b" segments the ratchet mechanism must occur, and (iii) along the "c" segments the radial spoke must jump. This means that the behavior of the radial spokes does not depend on the curvature only but on the partition of a module defined by two consecutive P0 points by the motile nil shear point. On the other hand, because the radial spokes are able to recognize a nil shear point, they could be involved in the regulation mechanism of the axonemal machinery in a cooperative manner with the central apparatus [12, 13].

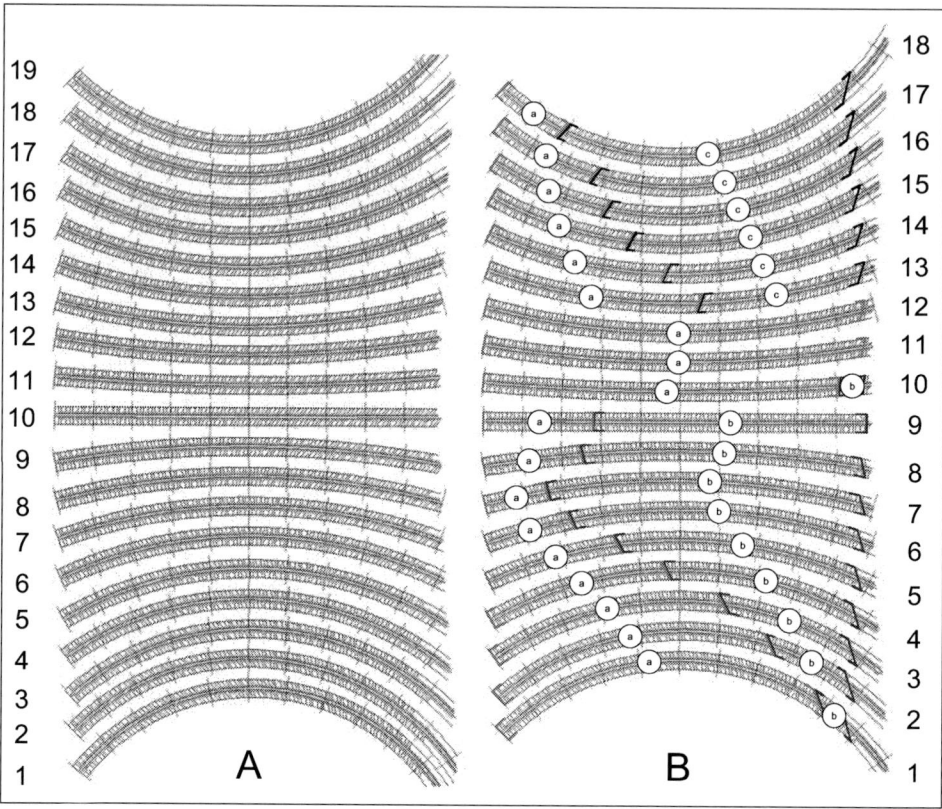

Figure 5 : two reverse phases of the bending of an axonemal segment.

THE "DEVIATED BENDING" OF THE OUTER DOUBLETS

Most of the time, the doublets are considered as the passive supports of the active structures of the axoneme. We used the finite element calculations and considered that the doublets should be compared to beams. Then, we demonstrated that because of their geometry only and consequently because of the difference that exists between the inertia along their major and minor axes, they are involved in "deviated bending" during the increase in their curvature.

This means that when a moment constrains the end of a segment of a doublet, this segment does not bend in the direction defined by the moment, but in a direction that depends on its inertia as indicated in **Figure 6**.

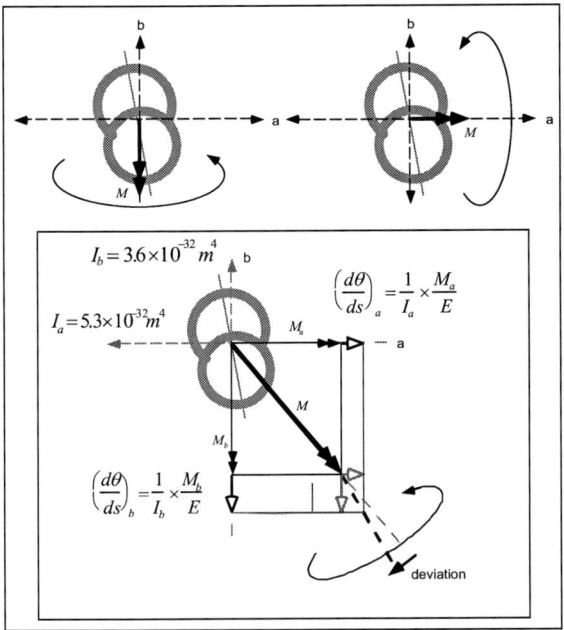

Figure 6 : Definition of the deviated-bending. A cross-section of an axonemal external doublet is schematized, and its two axes of inertia are designed as *a* and *b*. If a moment (double headed arrow) is parallel to either *a* or *b*, the doublet bends around the direction of the moment as indicated in the two upper schemes. Because the two inertias of a doublet are different, when a moment is parallel to neither *a* nor *b* (as shown in the lower scheme), the doublet bends around a direction that depends on the projection of the moment on *a* and *b*, and on the vectorial addition of the individual values of dθ/ds (open headed arrows) calculated on the two axes *a* and *b* as a function of E (the Young modulus of the doublet) and I (the inertia). The broad dashed line indicates the orientation of the real axis of rotation of the doublet under the constraint due to the moment M.

Figure 7 shows the deformation of the axonemal cylinder due to the "deviated bending" when the bending moment increases and when the outer doublets are free [15]. It is noteworthy that the central pair is always located on the outer side of the curved segment, because of the values of its inertia (not shown).

However, because the cylindrical shape of the axoneme is preserved during the bend, the "deviated bending" must induce the compression and the extension of the radial spokes, and the rotation of the doublets around the axonemal cylinder or, at least, the compression and the dilation of the inter-doublet intervals (i.e. the dynein arms themselves). As clearly seen in **Figure 8**, three triplets of doublets [(1, 4, 7), (2, 5, 8) and (3, 6, 9)] are involved in these movements. When the axoneme bends in the +Y direction, the distance between the doublets # 3, 4, 5 and 6, and 7 and 8 increases.

When the axoneme bends in the −Y direction, the distance between the doublets # 2 and 3 and 8 and 9 increases. It is noteworthy that the rotation of the doublets induces a change in the relationship between the dynein arms and the surface of tubule B in terms of distance and orientation. If the orientations of the dynein arms must be preserved during the beat, it is necessary to assume that the link between the doublet and the radial spoke rotates.

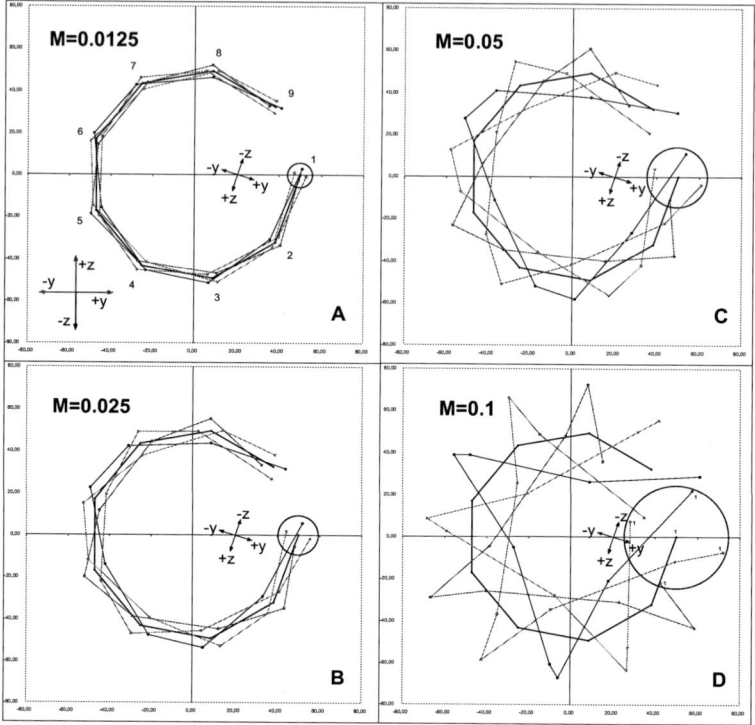

Figure 7: deviated bending of the axonemal cylinder when the bending moment increases. The directions of the constraints are indicated on left lower corner of A. To interprete the graphs, the various locations of the doublet #1 are included in a circle, and a reference mark defines the constraints that defined each of them.

The finite element approach allowed us also to show that the microtubules themselves could be involved in oscillations that could modify the entropy of the axonemal machinery. When diffusing molecules impact the microtubules, the tubulin dimers located in the protofilaments are involved in oscillations depending on the organization of the microtubules and their degree of freedom. These movements whose amplitude is about 6nm and whose frequency is high (ten to hundred times higher than the Brownian motion of free tubulin) are certainly involved in the regulation of the interaction between a microtubule and its ligand [16-18].

Within the axoneme these movements could be involved in the regulation of the interactions between the dynein arms and the surface of tubule B, modifying the local entropy of the axonemal system, even if this hypothesis remains to be explored.

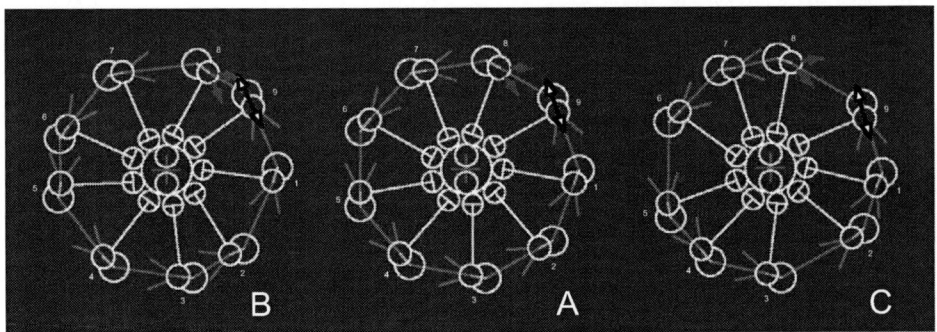

Figure 8 : bending in –Y (B) and +Y (C) directions of the axoneme.

THE AXONEME HAS A DISCRETE ORGANIZATION

To interpret these data, it is possible to propose that the axoneme would be constituted by a series of equivalent "active units" whose conformations are either "positive" or "negative", and whose ensemble of conformations depends on their location along the module delineated by two consecutive stable nil shear points. Each "active unit" associates all the structures that constitute an axonemal slice when the axoneme is straight [12].

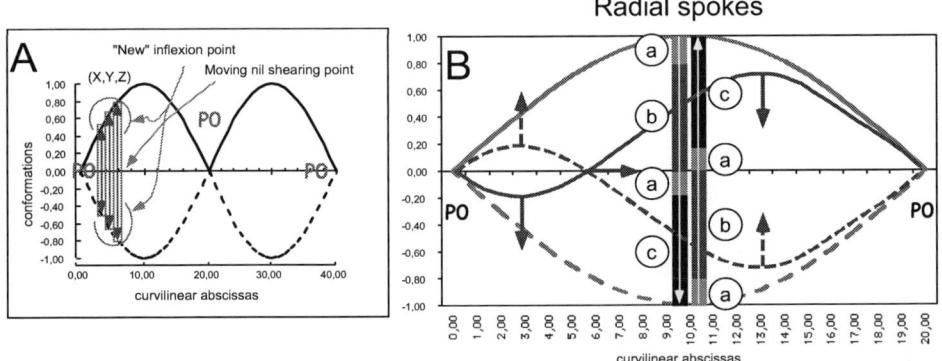

Figure 9 : definition of the active units along a module (A) and behavior of the radial spokes during the wave displacement along a module (B).

In **Figure 9**A, the bold and the dashed lines show the two extreme conformations of each of the "active units"; they describe the axonemal state when the module is involved in only two complete waves of opposite polarities. When the basal anchor creates a new wave, a new inflection point is formed and all the active units must change their conformation.

The speed of these changes must be coordinated between them in such way that a wave train is created along the module. **Figure 9B** presents the cyclic behavior of the radial spokes during the cyclical changes of conformation of the "active units". The moving nil shear point propagates tipward from the left to the right of the diagram along the axoneme. The crossing of the axis of the abscissas corresponds to the change of the activities between the two opposite sides of the neutral surface, because it corresponds to the change in the behavior of the radial spoke that begins a new cycle. It is noteworthy that the segments "b" and "c" decrease in length when they are closer to the stable P0 points that delineate the module.

ENTROPY, INFORMATION AND INTEGRATION

The term "entropy" was used without definition, and includes the notion of global organization. However, considering a series of events, entropy is calculated based on the probability of occurrence of different events and could be used to define either the information associated with this series or the integration of this series [7, 19].

Consider two ensembles of events X and Y (the same reasoning could be used for a larger number of series):

$$X = \{x_i\} | i = \{0,...,m\} \tag{1}$$

$$Y = \{y_j\} | j = \{0,...,n\} \tag{2}$$

where the elements x_i and y_j exist by themselves (e.g.: two series of letters, two series of molecular conformations of two molecular complexes). Then, it is possible to define the individual entropy of the occurrence of the series of events $H(X)$ and $H(Y)$ as:

$$H(X) = -\sum_i p(x_i) \log_2 (p(x_i)) \tag{3}$$

$$H(Y) = -\sum_i p(y_j) \log_2 (p(y_j)) \tag{4}$$

where $p(x_i)$ and $p(y_j)$ are the probabilities of occurrence of the events x_i and y_j. If the two ensembles are considered together, it is possible to calculate the conditional probability that quantifies the simultaneous occurrence of two events of these ensembles as:

$$p(x_i | y_j) = \frac{p(x_i \cap y_j)}{p(y_j)} \tag{5}$$

and to deduce from this value the conditional entropy associated with this probability of occurrence:

$$H(X | Y) = -\sum_i \sum_j p(x_i | y_j) \log_2 (p(x_i | y_j)) \tag{6}$$

Or, considering the joint probability of occurrence of the two events:

$$p(x_i, y_j) = p(x_i) p(y_i | x_i) = p(y_j) p(x_i | y_j) \tag{7}$$

126

the joint entropy could be defined as:

$$H(X,Y) = -\sum_i \sum_j p(x_i, y_j) \log_2 \left(p(x_i, y_j) \right) \tag{8}$$

Shannon defines the information $I(X{:}Y)$ by the following equation:

$$I(X:Y) = H(X) - H(X \mid Y) = H(Y) - H(Y \mid X) \tag{9}$$

or by this other equation:

$$I(X:Y) = H(X) + H(Y) - H(X,Y) \tag{10}$$

that are equivalent and link the probability of occurrences of each of the individual events and the occurrences of each of the couples of events.

For example, consider the paradigm of the alphabet. When X and Y are two Arabian alphabet, it is possible to calculate in a given text the probability of all the couples of letters. In English for example, it is clear that the letters "w" and "z" have a certain probability of occurrence in a text, but that the occurrence of the couples "wz" or "zw" are characterized by a very low probability of occurrence.

In this example, the information is always positive or nil, first because the discriminating term is the conditional or the joint entropy that characterizes the correlation of the occurrence of the two events considered, and second because the entropy of the occurrence of the two letters is a given value that depends only of the text. Then:

$$H(X) + H(Y) \geq H(X,Y) \tag{11}$$

and this equation is always respected.

Now, consider two ensembles of events:

$$\tilde{X} = \{\tilde{x}_i\} \mid i = \{0,...,m\} \tag{12}$$

$$\tilde{Y} = \{\tilde{y}_j\} \mid j = \{0,...,n\} \tag{13}$$

where some subsets of elements do not exist by themselves. Considering for example the two alphabets, this could occur when (in a particular language) "w" and "z" exist in a given text only under certain conditions and specifically when they are associated to construct the specific phoneme "wz".

In this condition and as in the previous case, the individual entropy associated with the occurrences of "w" and "z" and the joint entropy associated with the occurrence of the couple "wz" can be calculated by the following equations that resemble the equations 3, 4 and 8:

$$\left\langle H(\tilde{X}) \right\rangle = -\sum_i p(\tilde{x}_i) \log_2 \left(p(\tilde{x}_i) \right) \tag{14}$$

$$\left\langle H(\tilde{Y}) \right\rangle = -\sum_i p(\tilde{y}_j) \log_2 \left(p(\tilde{y}_j) \right) \tag{15}$$

$$\left\langle H(\tilde{X}, \tilde{Y}) \right\rangle = -\sum_i \sum_j p(\tilde{x}_i, \tilde{y}_j) \log_2 \left(p(\tilde{x}_i, \tilde{y}_j) \right) \tag{16}$$

where the brackets indicate the conditional existence of the letters of the two alphabets.

The integration (in place of information) of the system constituted by the two alphabets could then be calculated by the following equations:

$$\langle I(\tilde{X}:\tilde{Y})\rangle = \langle H(\tilde{X})\rangle + \langle H(\tilde{Y})\rangle - \langle H(\tilde{X},\tilde{Y})\rangle \tag{17}$$

Integration could be either nil or negative and characterizes the cooperativity that exists between "w" and "z" because of the existence of the phoneme "wz".

These notions could be used to characterize the integration of the molecular complexes such as the dynein arms and the discrete axonemal machinery where the conformations of the successive "active units" are mechanically linked.

CONCLUSION

It seems clear that the basis of the axonemal integration depends on the existence of the outer doublets that insure the longitudinal architectural link between the "active units". However, this structural link alone is unable to support all the aspects of the regulation of the axonemal machinery. Specifically, it explains neither why the radial spokes are necessarily involved in a ratchet mechanism as proposed because of electron microscopy observations, nor the physiological significance of the moving nil shear points that propagate along the axoneme we suggest is involved in the behavior of the radial spokes, nor the metachronism that necessarily involves the dynein arms along a bent segment. One important notion implicitly associated with the notion of integration is that it depends only on the probability of occurrence of the conformations of two (or more than two) associated complexes. This means that the regulation of a system does not depend on a necessary exchange of ligand between different molecular complexes as in a transduction cascade involved in hormonal regulation. It is noteworthy that this general mechanism of transduction is involved in the initiation and regulation of the wave shape that depend on the maturation of certain axonemal elements [20]. Only the interactions of the molecular complexes whose conformations vary could be involved in the rapid entropic regulation of the axonemal machinery; this is one of the basis of emergence that characterize the properties of very organized molecular complexes.

ACKNOWLEDGEMENTS

We thank the CNRS and Institut Jacques Monod for providing financial support, and A.-L. Haenni for English language corrections.

BIBLIOGRAPHY

1. Burgess, S. and P. Knight, *Is the dynein motor a winch?* Curr. Opin. Struct. Biol., 2004. **14**: p. 138-46.
2. Burgess, S., et al., *Dynein struture and power stroke.* Nature, 2003. **421**: p. 715-8.
3. Cibert, C., *Elastic extension and jump of the flagellar nexin links: a theoretical mechanical cycle.* Cell Motil. Cytoskel., 2001. **49**: p. 161-75.

4.   Gibbons, I., *Cilia and flagella of eukaryotes.* J. Cell Biol., 1981. **91**(3): p. 107s-24s.

5.   Brokaw, C.J., *Direct measurements of sliding between outer doublet microtubules in swimming sperm flagella.* Science, 1989. **243**: p. 1593-1596.

6.   Brokaw, C., *Microtubule sliding, bend formation, and bend propagation parameters of Ciona sperm flagella altered by viscous load.* Cell Motil. Cystoskel, 1996. **33**: p. 6-21.

7.   Ricard, J., *What do we mean by biological complexity?* C R Biol, 2003. **326**(2): p. 133-40.

8.   Cibert, C., *A cascade of biochemical events creates emergence.* Biol Cell, 2004. **96**(9): p. 677-9.

9.   Srere, P.A., *The metabolon.* Trends Biochem Sci, 1985. **10**: p. 109-10.

10.   Goldbeter, A. and G. Dupont, *Allosteric regulation, cooperativity, and biochemical oscillations.* Biophys. Chem., 1990. **37**: p. 341-53.

11.   Cibert, C., *Axonemal activity relative to the 2D/3D-waveform conversion of the flagellum.* Cell Motility and the Cytoskeleton, 2002. **51**: p. 89-111.

12.   Cibert, C., *Entropy and information in flagellar axoneme cybernetics : a radial spokes integrative function.* Cell motility and the Cytoskeleton, 2003. **54**: p. 296-316.

13.   Mitchell, D.R. and M. Nakatsugawa, *Bend propagation drives central pair rotation in Chlamydomonas reinhardtii flagella.* J Cell Biol, 2004. **166**(5): p. 709-15.

14.   Warner, F., *Cross-bridge mechanism in ciliary motility: the sliding-bending conversion*, in *Cell Motility (Part C)*, R. Goldman, T. Pollard, and J. Rosenbaum, Editors. 1976, Cold Spring Harbor Laboratory. p. 891-914.

15.   Cibert, C. and J.-V. Heck, *Geometry drives the "deviated-bending" of the bitubular structures of the 9+2 axoneme in the flagellum.* Cell Motility and the Cytoskeleton, 2004. **59**: p. 153-68.

16.   Kasas, S., et al., *Mechanical properties of microtubules explored using the finite elements method.* Chemphyschem., 2004. **5**: p. 252-7.

17.   Kasas, S., et al., *Oscillation modes of microtubules.* Biol Cell, 2004. **96**(9): p. 697-700.

18.   Kis, A., et al., *Nanomechanics of microtubules.* Physical review letters, 2002. **89**(24): p. 1-4.

19.   Ricard, J., *Reduction, integration and emergence in biochemical networks.* Biol Cell, 2004. **96**(9): p. 719-25.

20.   Inaba, K., *Molecular architecture of the sperm flagella: molecules for motility and signaling.* Zoolog. Sci., 2003. **20**: p. 1043-56.

Regulation of Eukaryotic Flagellar Motility

David R. Mitchell

Department of Cell and Developmental Biology, SUNY Upstate Medical University, Syracuse, NY 13210, USA

Abstract. The central apparatus is essential for normal eukaryotic flagellar bend propagation as evidenced by the paralysis associated with mutations that prevent central pair (CP) assembly. Interactions between doublet-associated radial spokes and CP projections are thought to modulate spoke-regulated protein kinases and phosphatases on outer doublets, and these enzymes in turn modulate dynein activity. To better understand CP control mechanisms, we determined the three-dimensional structure of the *Chlamydomonas reinhardtii* CP complex and analyzed CP orientation during formation and propagation of flagellar bending waves. We show that a single CP microtubule, C1, is near the outermost doublet in curved regions of the flagellum, and this orientation is maintained by twists between successive principal and reverse bends. The *Chlamydomonas* CP is inherently twisted; twists are not induced by bend formation, and do not depend on forces or signals transmitted through spoke-central pair interactions. We hypothesize that CP orientation passively responds to bend formation, and that bend propagation drives rotation of the CP and maintains a constant CP orientation in bends, which in turn permits signal transduction between specific CP projections and specific doublet-associated dyneins through radial spokes. The central pair kinesin, Klp1, although essential for normal motility, is therefore not the motor that drives CP rotation. The CP also acts as a scaffold for enzymes that maintain normal intraflagellar ATP concentration.

Keywords: Central Pair, Cilia, Radial Spoke, Chlamydomonas, Flagella.
PACS: 87.16.Ka; 87.16.Nn; 87.17.Jj

INTRODUCTION

As motile organelles, cilia and flagella are highly conserved across most eukaryotic phyla, with only minor structural variations seen between ciliates and humans [1]. They are important organelles during the life cycle of such human parasites as *Plasmodium, Giardia, Trypanosoma*, and *Leishmania*. In vertebrates, ciliated epithelia provide essential transport functions in respiratory airways, brain ventricles, and fallopian tubes, and power spermatozoan motility. Defects in ciliary motility cause a high incidence of upper respiratory infections, and may be linked to hydrocephaly. Ciliary motility in the early embryo is also essential for normal left-right asymmetry during organogenesis [2]. The absence of sperm motility causes sterility in males, and recent evidence shows that regulation of sperm motility (capacitation and chemotactic attraction to the egg) is also important for successful transmission of the germ line in mammals [3]. Thus regulation of ciliary and flagellar motility has broad significance for eukaryotic organisms.

In mammalian systems, ciliary motility regulation is manifested as an increase in beat frequency to increase mucociliary clearance rates, and as changes in both

CP755, *ISIS: International Symposium on Interdisciplinary Science*
edited by A. Ludu, N.R. Hutchings and D.R. Fry
© 2005 American Institute of Physics 0-7354-0240-X/05/$22.50

frequency and waveform of sperm flagella during activation, capacitation, and chemotaxis. In the organism used most extensively for both genetic and biochemical dissection of flagellar motility, the single-celled alga *Chlamydomonas reinhardtii*, regulation of beat frequency, changes in waveform, and reversible quiescence are normal physiologic responses to environmental variables such as light intensity. Through the analysis of mutations that disrupt flagellar structures we and others have developed a detailed understanding of the *Chlamydomonas* flagellar structure and the contribution of many structural elements, including the central pair, to motility [4-8]. In this review we focus on recent advances in our understanding of the regulation of ciliary and flagellar motility by the central pair apparatus.

CENTRAL PAIR REGULATION

Although central pair microtubules are not part of the oscillatory bending mechanism and are even missing from some simplified "9+0" motile organelles, they contribute an essential regulatory function in typical cilia and flagella. Mutations that block central pair assembly in *Chlamydomonas* cause a complete lack of flagellar motility, and constitute a subset of paralyzed flagella (*pf*) mutations in the alga [9-11]. Human mutations that disrupt central pair structures, like those that disrupt dyneins, are associated with upper respiratory infections (due to lack of mucociliary clearance) and infertility (due to immotile sperm) [12-14], and mice with central pair defects have also been diagnosed with hydrocephaly from lack of brain ventricle ciliary activity [15].

Central Pair-Radial Spoke Interactions

Hypotheses about the mechanism of central pair regulation have focused on kinase-phosphatase cascades that require radial spokes and that regulate dynein-dependent microtubule sliding rates in simplified flagellar models (protease-treated axonemes) [16-18]. Based on these hypotheses dyneins are inhibited by the activity of a kinase or kinases associated with outer doublet microtubules, and activated when protein phosphatases dominate and kinase affects are reversed [5]. Each of the nine outer doublet microtubules has a row of radial spokes that are arranged as pairs within a 96 nm repeat. The two spokes in each pair are uniquely positioned relative to inner row dyneins and to a dynein regulatory complex [19]. In thin sections of rapidly fixed, actively beating cilia, radial spokes appear to attach to central pair projections and tilt as a result of dynein-induced sliding displacement of the doublet microtubules [20]. Because radial spokes are associated with all nine outer doublets, but dynein activity must be limited to a subset of doublets at any one location along a flagellum, specific rows of central pair projections are thought to interact with overlying radial spokes in a doublet-specific pattern to modulate spoke regulation of these kinases and phosphatases. Doublet microtubule sliding patterns in protease-treated *Chlamydomonas* axonemes support a model in which spokes that interact with the C1 central pair microtubule activate adjacent dyneins [21].

In *Chlamydomonas*, phototaxis requires the presence of inner row dynein I1, and has been linked to radial spoke-regulated phosphorylation of a 138 kD intermediate chain I1 subunit (IC138) [22]. Sale and colleagues have shown that IC138 phosphorylation is regulated by the activity of a cAMP-dependent protein kinase (PKA) and a PP1 isozyme, and that regulation of these enzymes in turn depends upon an intact radial spoke/central pair system [23]. At present, however, we still know little about the specific radial spoke proteins that interact with the central pair, or the specific central pair structures involved in those interactions. We recently completed a detailed structural analysis of the central pair [24] and a study of central pair orientation during the flagellar beat cycle [25] which formed the basis for experiments (summarized below) that clarified the relationship between dynein activity patterns and orientation of the central pair [26].

Central Pair Rotation and Twist

Although general features of central pair structure have been described for organelles from several different organisms, the most detailed studies have relied upon mutational analysis of *Chlamydomonas* flagella (Fig. 1). Mutations that partially disrupted CP structure, such as *pf6* [10], *pf16* [10], and *cpc1* [27], together with biochemical extractions that selectively solubilized portions of the CP complex [9], showed that the two CP microtubules are dissimilar and that each is associated with a unique set of associated proteins. To extend these structural studies, we analyzed longitudinal thin sections, transverse thin sections, and quick-freeze, deep-etch images to build a three-dimensional view of the *Chlamydomonas* CP [24,27]. We concluded that all projections directly associated with the C2 microtubule have a 16 nm periodicity, while several of the C1-associated projections have a 32 nm repeat period. We also found that many of the projections seen in transverse images serve as supports for elements that form an apparently smooth cylindrical surface apposed to radial spoke heads. Departures from these smooth surfaces occur primarily along the CP microtubule surfaces closest to radial spoke heads, such as the region to which the Klp1 kinesin has been localized. The recent cloning of several central pair proteins, including PF16p, essential for stability of the C1 microtubule [28], and the products of the *PF6* [29], *PF15* [30], *PF20* [31], and *CPC1* [32]genes, is gradually filling the blank spaces on our CP map.

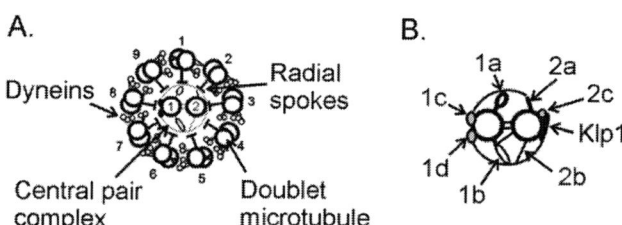

FIGURE 1. A, Diagram of a transverse section through a *Chlamydomonas* flagellar axoneme as viewed from inside the cell. Adapted from [25] by copyright permission of John Wiley and Sons, Inc. B, Diagram of the central pair apparatus with major densities labeled. The region labeled Klp1 is specifically depleted in axonemes from Klp1 knockdown strains [33].

Central Pair Kinesins

The central pair complex rotates during the bending cycle in the cilia of some organisms [34], including *Chlamydomonas* [35,36], but retains a fixed orientation in others [37]. Difficulty observing and recording central pair rotation has prevented a careful analysis of this phenomenon, which likely plays a significant role in the central pair regulatory process. The only motor proteins known to be in the central pair complex itself are kinesin-like proteins [38-40], so if rotation is driven by spoke-central pair interaction, kinesins are the likely motors. Although most kinesins act as translocating motor proteins, the functional capacity of central pair kinesins has not been determined and as yet no mutations have been characterized that selectively disrupt their function or prevent their assembly.

Thus far only two flagellar kinesin genes have been characterized in *Chlamydomonas*, *FLA10* which encodes one heavy chain of the kinesin for intraflagellar transport (IFT) [41-43], and *KLP1* which encodes a kinesin that has been localized to the C2 central pair microtubule [38]. As an initial step toward testing hypotheses about central pair kinesin activity, we characterized the phenotype that results from RNAi knockdown of Klp1 [33]. Previous immunolocalization showed that Klp1 is associated with the C2 central pair microtubule, but could not further determine which C2 structures might contain Klp1. Our thin section electron micrographs show that the only structure not seen in any of the klp1 knockdown flagella was a density associated with the C2 microtubule, part of projection 2c. A second structure, projection 2b, was also depleted in isolated, demembranated knockdown axonemes. For more detailed analysis of structural changes associated with knockdown of Klp1, cross section images of wild type and knockdown axonemes were analyzed by image averaging. These images confirmed that part of 2c is the only density completely missing from knockdown axonemes, and place Klp1 precisely on the surface of C2 in a position where it could interact with radial spoke heads (Fig. 1B).

Knockdown strains swim with reduced velocity as the result of an altered flagellar motility pattern. Preliminary analysis shows a reduction in beat frequency as the primary motility defect, with a small fraction of the population completely paralyzed. Therefore, Klp1 may be essential for flagellar motility. However (as summarized below) we have now shown that central pair rotation and twist in *Chlamydomonas* do not depend on interactions between radial spoke heads and central pair projections [26], so the role of central pair kinesins remains completely unknown. We hypothesize that CP kinesins act as mechanochemical switches, rather than as actual motors. Much like G proteins, they could undergo conformational changes based upon the presence of ATP vs ADP at the active site that resulted not in net movement, but in a change in conformation that alters the interaction between the kinesin and another flagellar component. ATP hydrolysis rates could respond to changes in flagellar bend angle or changes in the state of a chemical signaling pathway (e.g. phosphorylation) and each conformational state could in turn correspond to a different interaction between radial spokes and central pair projections.

Central Pair Orientation

In organisms with a fixed CP orientation the two central pair microtubules remain perpendicular to the bend plane, whereas one published report [44] suggested that, at least in bent regions, the central pair twists to become parallel to the bend plane. Resolving this puzzle, at least for *Chlamydomonas*, should help determine which central pair projections are in contact with radial spokes along active doublets, and which ones are in contact with spokes on inactive doublets, during formation and propagation of bends. To find out, we developed methods to fix actively swimming cells such that their waveforms were preserved, and then to embed and section those cells to reveal the orientation of the central pair in both straight and bent regions (Fig. 2). One tremendous advantage of the *Chlamydomonas* system for this study is the nearly planar shape of the waveform during both asymmetric, ciliary bends and symmetric, flagellar-style bends, and the added ability to induce cells to switch between these two bending modes by a simple shift in light intensity (photoshock).

Our conclusion, consistent with previous observations from other labs [34], is that the central pair in *Chlamydomonas* is twisted whereas surrounding doublet microtubules are not. In addition, the position of the twist propagates along with each bend, so that the *Chlamydomonas* central pair is always parallel to the bend plane within bent regions (Fig. 2B and 2D), but twists until it is perpendicular to the bend plane in straight regions (Fig. 2C) [25]. Propagation of a twist can account for the appearance of central pair rotation during each beat cycle. This rotation mechanism has interesting consequences for hypotheses about central pair-radial spoke interactions.

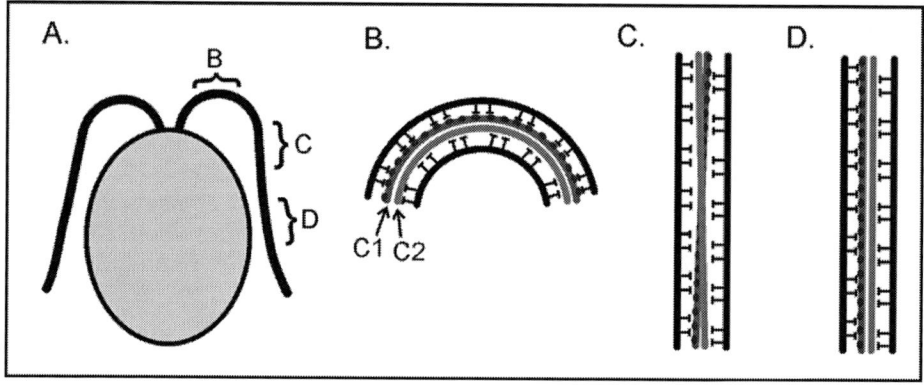

FIGURE 2. A, diagram of a *Chlamydomonas* cell during forward swimming. Brackets indicate the principal bend (B), the interbend region (C) and the reverse bend (D) and correspond to the three diagrams. B, CP orientation seen in thin sections through principal bends. The C1 CP microtubule is distinguished by a row of projections (dark ovals) repeating every 32 nm. C, central pair twist in an interbend region. D, CP orientation in a reverse bend region.

Forces Generating Central Pair Rotation and Twist

We recently determined that the central pair is inherently twisted and that neither twist nor bend-specific orientation depend on central pair-radial spoke interactions [26]. These conclusions are based on two types of experiments. First, quiescent (non-beating), straight flagella adhering to a flat surface (a coverglass) still have twists in their CPs, whether the cells assemble wild type radial spokes, or radial spokes that lack spoke heads. Second, when a mutation that blocks spoke head assembly is combined with a suppressor mutation that permits motility (bend propagation) in the absence of spoke heads, CP orientation in bent and straight regions follows the same pattern as seen in wild type flagella. This leads to the conclusion that CP orientation conforms to the bend, regardless of the bend plane. We hypothesize that central pair function evolved to regulate beat frequency and waveform independently of effective stroke orientation [1,26]. According to this hypothesis, in organelles that need to change effective stroke direction, the CP can passively re-orient to the beat direction. In organelles that always beat in the same plane, evolutionary selection for a system with fewer moving parts has resulted in a CP that maintains a fixed orientation, perhaps through semi-permanent attachments of modified spokes on opposite sides of the axoneme (doublets 3 and 8) to each CP microtubule.

CONCLUSIONS

In the single-celled alga *Chlamydomonas reinhardtii*, flagella beat with a nearly planar waveform and with a central apparatus that twists, although the surrounding doublet microtubules do not. We conclude that twist results from an underlying helical shape of the central apparatus, and that the orientation of the central pair within *Chlamydomonas* flagella depends upon the formation and propagation of bends. This orientation is independent of interactions between radial spoke heads and central pair projections, and therefore central pair rotation is not driven by these interactions but rather by the propagation of bends through forces generated by doublet-associated dynein motors. Central pair kinesins are important for normal motility, but cannot be essential motors for central pair rotation.

ACKNOWLEDGMENTS

Supported by grants from the National Science Foundation (MCB-9982062) and the National Institutes of Health (GM44228).

REFERENCES

1. D. R. Mitchell, *Biol. Cell* **96**, 691-696 (2004).
2. S. Nonaka, Y. Tanaka, Y. Okada, S. Takeda, A. Harada, Y. Kanai, M. Kido, and N. Hirokawa, *Cell* **95**, 829-837 (1998).
3. M. Eisenbach, *Dev Genet* **25**, 87-94 (1999).
4. D. R. Mitchell, *J. Phycol.* **36**, 261-273 (2000).

5. M. E. Porter and W. S. Sale, *J. Cell Biol.* **151**, F37-F42 (2000).
6. E. F. Smith and P. Yang, *Cell Motil. Cytoskeleton* **57**, 8-17 (2004).
7. M. E. Porter, *Curr. Opin. Cell Biol.* **8**, 10-17 (1996).
8. S. K. Dutcher, *Trends Genet.* **11**, 398-404 (1995).
9. G. M. W. Adams, B. Huang, G. Piperno, and D. J. L. Luck, *J. Cell Biol.* **91**, 69-76 (1981).
10. S. K. Dutcher, B. Huang, and D. J. L. Luck, *J. Cell Biol.* **98**, 229-236 (1984).
11. G. B. Witman, J. Plummer, and G. Sander, *J. Cell Biol.* **76**, 729-747 (1978).
12. C. Chapelin, A. Coste, P. Reinert, M. Boucherat, M. C. Millepied, F. Poron, and E. Escudier, *Ann. Otol. Rhinol. Laryngol.* **106**, 854-858 (1997).
13. H. Okada, H. Fujioka, N. Tatsumi, M. Fujisawa, K. Gohji, S. Arakawa, H. Kato, S. I. Kobayashi, S. Isojima, and S. Kamidono, *Hum. Reprod.* **14**, 110-113 (1999).
14. B. Baccetti, A. G. Burrini, A. Maver, V. Pallini, and T. Renieri, *Andrologia* **11**, 437-443 (1979).
15. R. Sapiro, I. Kostetskii, P. Olds-Clarke, G. L. Gerton, G. L. Radice, and I. I. I. Strauss, *Mol Cell Biol* **22**, 6298-6305 (2002).
16. A. R. Gaillard, D. R. Diener, J. L. Rosenbaum, and W. S. Sale, *J. Cell Biol.* **153**, 443-448 (2001).
17. E. F. Smith and W. S. Sale, *Science* **257**, 1557-1559 (1992).
18. P. Yang, D. R. Diener, J. L. Rosenbaum, and W. S. Sale, *J Cell Biol* **153**, 1315-1326 (2001).
19. L. C. Gardner, E. O'Toole, C. A. Perrone, T. Giddings, and M. E. Porter, *J. Cell Biol.* **127**, 1311-1325 (1994).
20. F. D. Warner and P. Satir, *J. Cell Biol.* **63**, 35-63 (1974).
21. M. J. Wargo and E. F. Smith, *Proc. Natl. Acad. Sci. USA* **100**, 137-142 (2003).
22. S. J. King and S. K. Dutcher, *J. Cell Biol.* **136**, 177-191 (1997).
23. G. Habermacher and W. S. Sale, *J. Cell Biol.* **136**, 167-176 (1997).
24. D. R. Mitchell, *Cell Motil. Cytoskeleton* **55**, 188-199 (2003).
25. D. R. Mitchell, *Cell Motil. Cytoskeleton* **56**, 120-129 (2003).
26. D. R. Mitchell and M. Nakatsugawa, *J Cell Biol* **166**, 709-715 (2004).
27. D. R. Mitchell and W. S. Sale, *J. Cell Biol.* **144**, 293-304 (1999).
28. E. F. Smith and P. A. Lefebvre, *J. Cell Biol.* **132**, 359-370 (1996).
29. G. Rupp, E. O'Toole, and M. E. Porter, *Mol. Biol. Cell* **12**, 739-751 (2001).
30. E. E. Dymek, P. A. Lefebvre, and E. F. Smith, *Eukaryot. Cell* **3**, 870-879 (2004).
31. E. F. Smith and P. A. Lefebvre, *Mol. Biol. Cell* **8**, 455-467 (1997).
32. H. Zhang and D. R. Mitchell, *J Cell Sci* **117**, 4179-4188 (2004).
33. R. Yokoyama, E. O'Toole, S. Ghosh, and D. R. Mitchell, *Proc Natl Acad Sci U S A* **101**, 17398-17403 (2004).
34. C. K. Omoto, I. R. Gibbons, R. Kamiya, C. Shingyoji, K. Takahashi, and G. B. Witman, *Mol. Biol. Cell* **10**, 1-4 (1999).
35. R. Kamiya, *Cell Motility [Suppl.]* **1**, 169-173 (1982).
36. H. J. Hoops and G. B. Witman, *J. Cell Biol.* **97**, 902-908 (1983).
37. S. L. Tamm and S. Tamm, *J. Cell Biol.* **89**, 495-509 (1981).
38. M. Bernstein, P. L. Beech, S. G. Katz, and J. L. Rosenbaum, *J. Cell Biol.* **125**, 1313-1326 (1994).
39. L. A. Fox, K. E. Sawin, and W. S. Sale, *J. Cell Sci.* **107**, 1545-1550 (1994).
40. K. A. Johnson, M. A. Haas, and J. L. Rosenbaum, *J. Cell Sci.* **107**, 1551-1556 (1994).
41. Z. Walther, M. Vashishtha, and J. L. Hall, *J. Cell Biol.* **126**, 175-188 (1994).
42. K. G. Kozminski, P. L. Beech, and J. L. Rosenbaum, *J. Cell Biol.* **131**, 1517-1527 (1995).
43. J. Rosenbaum, *Curr. Biol.* **12**, R125 (2002).
44. C. K. Omoto and C. Kung, *J Cell Biol* **87**, 33-46 (1980).

Flagellar Bend Dynamics in African Trypanosomes

Nathan R. Hutchings[*] and Andrei Ludu[†]

Interdisciplinary Experimentation and Scholarship (IDEAS) Program
*Department of Biological Sciences and †Department of Chemistry and Physics.
Northwestern State University of Louisiana. Natchitoches, Louisiana 71497

Abstract. The flagellated protozoa are highly dependent on flagellar dynamics for environmental sensing, reproduction, cell morphology, and disease progression. Functional and structural nuances amongst the different flagellated cells create the need to quantify the flagellar dynamics in each organism. In the African trypanosomes, specialized cellular and flagellar architecture sutures the flagellum to the cell body along its length, which results in a unique auger-like cell motility. In this paper, we provide a quantitative description of flagellar bends in procyclic *Trypanosoma brucei*. We used digital video microscopy to describe the geometry and dynamics of trypanosome flagellar bends. We present a formula that demonstrates trypanosome flagellar bends have a conserved and predictable shape that is related to conic functions. We also investigate the local dynamics of individual flagellar bends and show that trypanosome flagellar bends can dilate, constrict, and travel bidirectionally along the flagellum. The implications of this data in modeling trypanosome cell motility are discussed.

Keywords: Trypanosome, flagellum, axoneme, biophysics, cytoskeleton, waveform, conic
PACS: 87.15.He, 87.15.La, 87.15.Aa

INTRODUCTION

The eukaryotic flagellum is a structurally conserved yet functionally diverse organelle (for reviews see [1-4]). The movement of flagella and cilia is characterized by temporally and spatially regulated internally driven bends that produce dynamic undulations or oscillations along the filament [1]. More than a half century of research on the flagellum reveals that nearly every flagellated cell type exhibits nuances in the structure, use, and/or control of the flagellum. The molecular events that regulate flagellar dynamics are actively being studied, and although several experimentally supported models for flagellar bend regulation have been proposed [4-6], we can not yet speculate on any ubiquitous or fundamental regulatory mechanisms. Furthermore, although the force generating structure (axoneme) within the eukaryotic flagellum appears to be highly conserved, the dynamics and regulation of the flagellum that cooperatively determine flagellar function in each cell appear to be quite specialized. To date, flagellar movements have been described as a variety of shapes ranging from 2-dimensional sinusoidal-like waves and 2-dimensional planar oscillations to 3-dimensional helical-like waves [7,8]. Nonlinearities have been reported in both 2- and 3-dimensional flagellar waves [9,10], which further necessitates characterizing the

CP755, ISIS: International Symposium on Interdisciplinary Science
edited by A. Ludu, N.R. Hutchings and D.R. Fry
© 2005 American Institute of Physics 0-7354-0240-X/05/$22.50

parametric limits of each experimental system to derive the equations that most accurately describe the flagellar dynamics in that cell type.

In this study, we investigate the shape and dynamics of flagellar bends in the procyclic form of the African trypanosome, *Trypanosoma brucei brucei*. In addition to providing motility to the cell, the *T. brucei* flagellum is essential for viability and has also been implicated in cell morphology, organelle partitioning, cellular differentiation, and attachment to host tissues [11].

In contrast to the more commonly studied flagellated cells, such as *Chlamydomonas* or sperm, the trypanosome flagellum has several additional sub-structures that influence flagellar motion. The trypanosome flagellum contains a conventional 9+2 axoneme, but also contains: a second dense filament called the paraflagellar rod (PFR), a cytoskeletal suture to the cell body called the flagellar adhesion zone (FAZ) filament, and a specialized membrane trafficking organelle at the base of the flagellum called the flagellar pocket, (for review see [11,12]). Mutational analysis of proteins that function within these various structures has revealed that the collective integrity of the flagellar substructures is critical for productive trypanosome motility [13-15]. For example, *Leishmania sp.* lacking a functional PFR, exhibit decreased flagellar amplitude, wavelength, and beat frequency resulting in a 4-5-fold reduction in swimming velocity [16], and an African trypanosome lacking a complete PFR are nearly immotile [13,17]. The integrity of the FAZ in procyclic African trypanosomes is necessary for directional cell motility [18,19], and gamma tubulin, which is concentrated in the flagellar pocket/basal body region, was shown to be essential for central pair formation and motility [11,20].

A quantitative understanding of flagellar dynamics in all life stages of the African trypanosome is important for the following reasons. First, unlike most flagellated cells where the flagellum is a free appendage of the cell, the *T. brucei* flagellum is attached to the cell body along its entire length. This restricts flagellar motion and makes flagellar movement codependent on the physical properties of the cell body. Second, trypanosomes undergo significant changes in morphology, behavior, and metabolism during their life cycle. Each of these changes likely affects the structure, dynamics, and/or regulation of the flagellum [11], which ultimately affects the behavior of the parasite. Third, trypanosomes live in diverse and complex environments ranging from the midgut of an insect to the lymph fluid of a mammal. Each environment has significant differences in chemical composition and viscosity, which are known factors that influence flagellar dynamics in all flagellated cells [9,3,21,22]. Some global properties of trypanosome flagellar motion, such as beat frequency and wavelength, have been measured in *Crithidia* [3,23], *Leishmania* [16], and blood stream form *T. brucei* [22], but a quantitative description of local flagellar bend shape and dynamics has not been reported. Although the complex lifecycle and cyto-architecture of trypanosomes adds complexity to the experimental considerations, an understanding the natural genetic, behavioral, and environmental dynamics of the trypanosome flagellum will greatly increase our understanding of the basic biology of these, and related, protozoa.

To advance our understanding of how African trypanosome flagellar dynamics facilitate auger-like cell motility, we used digital video microscopy to quantify the shapes of trypanosome flagellar bends and describe the local dynamic properties within the trypanosome flagellum. In a related study, we have derived a novel solution to a generalized Korteweg-de Vreis (mKdV) equation that can accommodate the complex flagellar bend shapes described herein, which should enable us to create a mathematical model of trypanosome cell motility in the near future.

RESULTS AND DISCUSSION

Using digital video differential interference contrast (DIC) micrographs captured with either The Spot™ CCD (Diagnostic Instruments, Inc, Sterling Heights, MI) or a COHU 2600 CCD (Cohu, Inc., Sand Diego, CA), we measured the experimental amplitude, wavelength, and perimeter of trypanosome flagellar bends. Measurements were collected using The Spot Advanced (Version 4.0.2.0, Diagnostic Instruments, Inc, Sterling Heights, MI) or IP Labs (Scanalytics, Inc., Fairfax, VA) software calibrated to an Olympus 100X UPlanFl Objective (NA 1.3) on an Olympus BX60 microscope. To describe each flagellar bend, we measured the wavelength, amplitude, and perimeter, respectively. The wavelength (diameter of the bend, λ) was measured by drawing a line between the inflection-point on each side of the bend. Importantly, we can only measure the wavelength of a bend as the diameter of a single bend rather than the full period of the waveform (pair of sequential bends of opposite sign), because the wavelengths of two adjacent bends in the trypanosome flagellar are rarely symmetrical. The amplitude (a) was measured as a perpendicular from the midpoint of the wavelength line to the inner-diameter apex of the flagellar bend. The perimeter (p) was measured by tracing the inner-diameter arc-length between the inflection-points used to demark the wavelength.

From our analysis of over 250 flagellar bends, the average amplitude in procyclic *T .brucei* cells is 0.742 ± 0.3 microns, and the average wavelength is 3.25 ± 0.95 microns (Fig. 1). Generally, we can conclude that small wavelength bends have small amplitude, whereas, large wavelength bends can have a wide range of amplitudes. This suggests that the cytoarchitecture of the trypanosome flagellum imposes shape restrictions on the minimal bend shape of the flagellum.

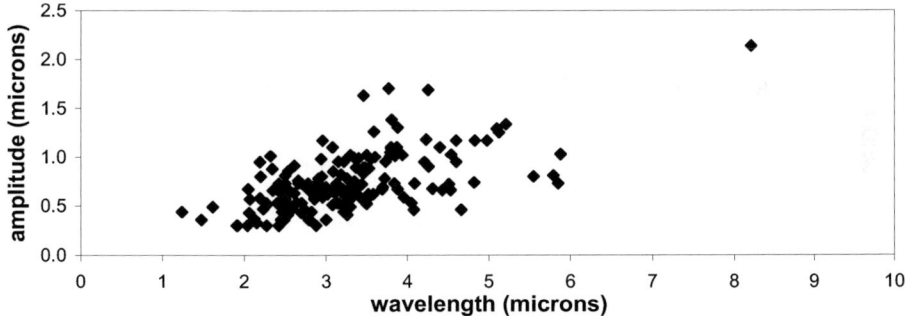

FIGURE 1. Amplitude and wavelength comparison for flagellar bends. The amplitude and wavelength of 164 flagellar waveforms are plotted. The average amplitude in procyclic *T .brucei* cells is 0.742 ± 0.3 microns with a range between $0.3 - 2.2$ microns, and the average wavelength is 3.25 ± 0.95 microns with a range between 1.24-8.21 microns

To assess the predominant shape of a trypanosome flagellar bend, we empirically modified the equation for elliptical perimeter to best fit the perimeter values in our experimental data set. We chose the equation for a semi-ellipse as our starting point since this is the closest basic geometric shape that resembles trypanosome flagellar bends. We determined the equation that most accurately describes the perimeter of trypanosome flagellar bends is:

$$p = \pi \frac{\left(\sqrt{(\frac{a}{2})^2 + (\frac{\lambda}{2})^2} \right)}{2} \tag{1}$$

To demonstrate the accuracy of Eq. 1, we compared the theoretical perimeters derived from Eq. 1 with the experimentally measured perimeters (Fig. 2). The calculated values for the theoretical perimeter have an average accuracy of 97 ± 6%. This result indicates that Eq. 1 well approximates the perimeter of trypanosome flagellar bends within the observed experimental ranges of amplitude and wavelength.

Interestingly, our analysis revealed that although the actual numerical values of the wavelength and amplitude are variable, many of the flagellar bends have a similarly proportioned shape. By plotting wavelength versus perimeter, our data revealed that flagellar wavelength and perimeter exhibit a strong linear relationship (Fig. 3). From the equation of the line in Figure 3 (p=1.12 λ), we can predict trypanosome flagellar bend perimeter nearly as well as Eq. 1 using only a single variable (λ). This result

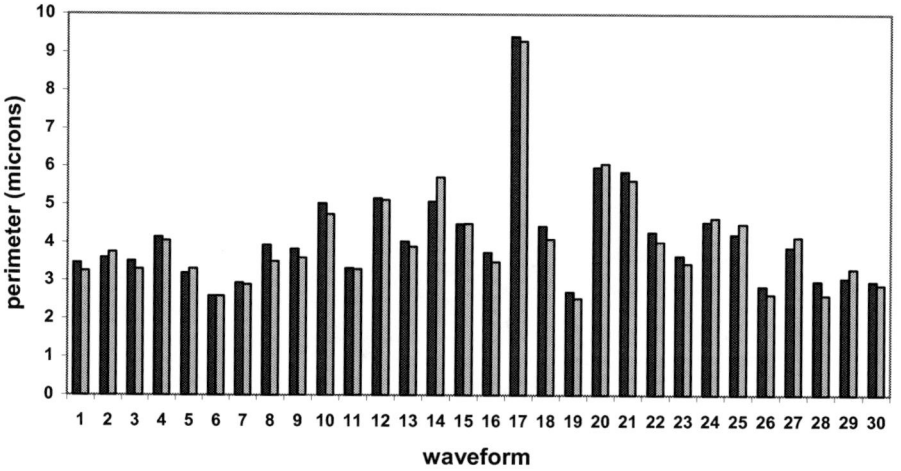

FIGURE 2. Equation 1 accurately predicts the perimeter of the flagellar bend shape based on the experimental amplitude and wavelength of the bend. The theoretical perimeter (black bars) and the experimentally measured perimeter (gray bars) of 30 example flagellar bends are shown. The values of the theoretical and experimental perimeters of each flagellar bend range from 87% to 100% accurate, with an average accuracy of 97%.

allows us to reasonably approximate the shape of a collective series of bends within the same flagellum based on the respective wavelength of each bend (not shown).

Formally, the amplitude of each bend is important to the exact geometric shape of the bend, but the relative scale of the amplitudes relative to the wavelengths apparently reduces the influence of the amplitude on the overall shape of the bend. For example, based on Eq.1 a 20% change in the average amplitude results in only a 3% change in the average perimeter. Although 80% of the bend perimeters in our data set can be predicted with >96% accuracy, the equation, p=1.12 λ, is exponentially less accurate as the bend shape approaches a circle or if the amplitude value is less than 0.3 microns. For example, the predicted perimeter of a bend with a wavelength to amplitude ratio greater than 3:1 is more than >98% accurate; whereas, the predicted perimeter of a bend with a wavelength to amplitude ratio near 2.1 is only ~92% accurate. However, by adding a second term to account for the wavelength to amplitude ratio in the linear approximation, Equation 2 accommodates 95% of the bends in our experimental data set with greater than 99% +/- 0.4% accuracy (Fig. 3). This term dramatically improves the predictability of flagellar bends with smaller eccentricities. From our analysis, the remaining inaccuracies are limited to bends with small amplitudes, and we are trying to further refine the equation to predict these bends with equal accuracy.

$$p = 1.12\lambda + \left[1.32 \frac{(a^{2.5})}{(\lambda^{2})} \right] \qquad (2)$$

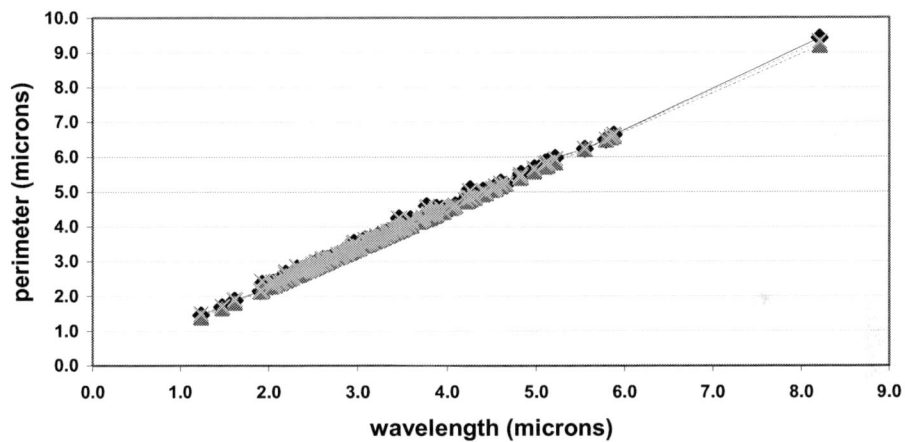

FIGURE 3. Experimental wavelength has a linear relationship to perimeter. The black diamonds represent experimental perimeter and experimental wavelength. Gray triangles represent experimental wavelength and perimeter predicted by p=1.12λ. Light gray X's represent experimental wavelength and perimeter predicted by Equation 2. Notice that the gray X's consistently fit more closely to the actual experimental point (black diamond). However all three series fit the equation p=1.12λ with an R^2 value above 0.99.

Historically, trypanosome flagellar dynamics have been measured using stroboscopic dark field microscopy, which allows for the global measurement of frequency and wavelengths. However, due to the nature of the trypanosome flagellum and its attachment to the cell body, we felt it was important to measure the local dynamics of individual flagellar bends.

The dynamic characteristics of individual bends were monitored by measuring the amplitude, wavelength, and velocity of a bend in sequential frames of a digital video micrograph of a trypanosome nonspecifically tethered to the glass slide. These conditions allow us to observe flagellar dynamics in different regions of the cell depending upon the orientation of the tethered cell. In many cases, the propagating bend velocities in the proximal regions of the flagellum were slow enough to accommodate the frame rate of our imaging system. Although the bend dynamics of tethered cells appear significantly slower than those of free-swimming cells, the three-dimensional swimming pattern in trypanosomes and limitations of our imaging system currently preclude us from monitoring individual flagellar bends in a 3-dimensionally rotating untethered cell. Nonetheless, our results clearly indicate that some flagellar bends in trypanosomes have a cohesive shape that translates along the filament, and some bends have degenerative shape that dilates or constricts as it translates (Fig. 4). These results beg the question whether the global shape and dynamics of the cell are significantly different from the local bend dynamics that occur within the trypanosome flagellum.

In our analysis, we also observed that the direction and velocity of bend

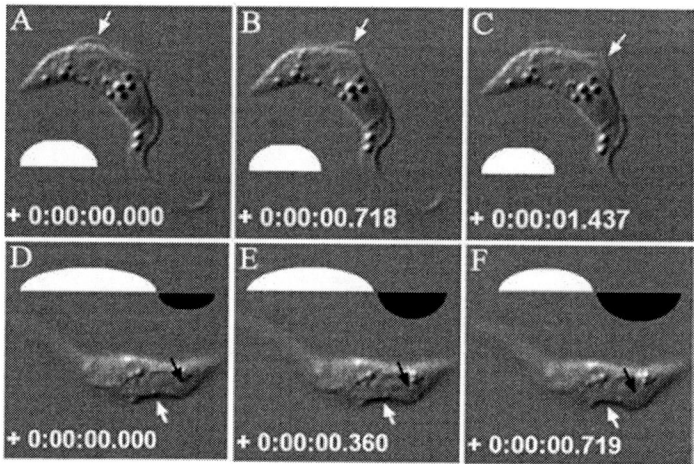

FIGURE 4. Trypanosome flagellar bends can be described as translating, dilating, and/or constricting conic functions. Panel A-C shows an example of a translating elliptical bend. The white arrow indicated the location of the bend in each frame, and the white semi-ellipse is a scaled representation of the shape of the bend. In this video, the waveform is translating from base-to-tip at approximately 1.5 microns per second. Panel D-F shows an example of a dilating/constricting pair of bends. The white arrow bend is constricting (decreasing wavelength over time) and the black arrow bend is dilating (increasing wavelength over time). The white and black semi-ellipses in each frame represent the shape of the respective bends. At the bottom of each frame, the elapsed time is indicated (hrs: min: sec. millisec).

propagation can change rapidly and sporadically, and a single trypanosome flagellum can simultaneously exhibit a complex variety of shapes within the same filament. Such bend dynamics can be stable for extended periods but can also suddenly destabilize and change into another pattern (not shown). We have generally observed that the dominant factor influencing the overall bend dynamics within a given cell is whether or not the anterior tip of the flagellum is free or restricted (not shown).

Although most flagellated cells flagellar bends propagate from base-to-tip [4], previous reports show that blood form *T. brucei* flagellar bends propagate from tip-to-base and *Crithidia* can propagate waves bidirectionally [15,22,23]. Our results clearly indicate that procyclic African trypanosome flagellar waveforms can travel bi-directionally along the flagellum (Fig 4). Although a free-swimming trypanosome does generate the overall impression of a tip-to-base cellular undulation and left-handed helical rotation (not shown), we have observed individual bends propagating along the flagellum in both directions in all regions of the flagellum (not shown).

Using the experimental waveform parameters described herein to solve a novel generalization of the modified Korteweg-de Vreis (mKdV) equation, strongly suggests trypanosome flagellar bends resemble nonlinear waves, solitary waves and even combinations of conic functions, like elliptical and hyperbolic arcs (not shown). Based on a novel mKdv equation, consideration of high-order nonlinear terms revealed the existence of piecewise connected conic functions as a real solution to the equation (not shown). Using the results herein in conjunction with material constants and force parameters from the literature [24], we can hypothesize that our novel mKdV equation can produce plausible scale dynamic flagellar bends that closely match our experimental observations.

Using a geometric approach to describe flagellar waveforms has proven to be a successful means to reduce the complexity of the whole flagellum into its definable parametric attributes since Brokaw fit complex flagellar shapes to a series of circular arcs connected by straight lines [25]. Subsequently, a geometric approach was used to describe the nature of the flagellar bend shapes in *Crithidia* [23], and a three-dimensional application of arcs and lines was used to describe the helical shape of a flagellum [26]. Most recently, the planar waveforms of sea urchin sperm were modeled as a series of circular tangential waveforms [6]. Thus, describing the geometry of flagellar bends is an effective means to deduce flagellar dynamics at the cellular level without having the molecular resolution of specific molecular structures and regulatory complexes. Having a quantitative standard and geometric description of trypanosome flagellar waveforms will allow us to integrate the conic function descriptions of trypanosome flagellar bends into a dynamic model of trypanosome flagellar motion, which will significantly enhance our ability to study the complex motility-dependent behavior of these cells.

ACKNOWLEDGMENTS

We gratefully acknowledge the support for this project from the Louisiana board of Regents grant number LEQSF (2004-2007)-RD-A25 to NRH and AL and the National Science Foundation grant number 0140274 to AL. We would like to thank Dr. Darrell

Fry and Ms. Anna Westergard for their contributions to this work. We also appreciate the support of The College of Science and Technology, the IDEAS Program (http://scietech.nsula.edu/ideas) and the IDEAS Program students.

REFERENCES

1   J. Cosson, Cell Biol Int **20** (2), 83 (1996).
2   S. K. Dutcher, Trends Genet **11** (10), 398 (1995).
3   M. E. Holwill, Sci Prog **61** (241), 63 (1974).
4   C. B. Lindemann and K. S. Kanous, Int Rev Cytol **173**, 1 (1997).
5   R. H. Dillon and L. J. Fauci, J Theor Biol **207** (3), 415 (2000); C. J. Brokaw, Cell Motil Cytoskeleton **42** (2), 134 (1999); M. E. Porter and W. S. Sale, J Cell Biol **151** (5), F37 (2000); D. R. Mitchell, Cell Motil Cytoskeleton **55** (3), 188 (2003).
6   C. Cibert, Cell Motil Cytoskeleton **51** (2), 89 (2002).
7   G. G. Vernon and D. M. Woolley, Cell Motil Cytoskeleton **42** (2), 149 (1999); D. M. Woolley and G. G. Vernon, Biophys J **83** (4), 2162 (2002).
8   F. Andrietti and G. Bernardini, Biophys J **67** (4), 1767 (1994); D. M. Woolley and I. W. Osborn, J Cell Sci **67**, 159 (1984).
9   M. Hines and J. J. Blum, Biophys J **25** (3), 421 (1979).
10   S. A. Koehler and T. R. Powers, Phys Rev Lett **85** (22), 4827 (2000).
11   K. Gull, Curr Opin Microbiol **6** (4), 365 (2003).
12   P. Bastin, T. J. Pullen, F. F. Moreira-Leite et al., Microbes Infect **2** (15), 1865 (2000); K. L. Hill, Eukaryot Cell **2** (2), 200 (2003); L. Kohl and K. Gull, Mol Biochem Parasitol **93** (1), 1 (1998); S. Vaughan and K. Gull, J Cell Sci **116** (Pt 5), 757 (2003).
13   P. Bastin, T. Sherwin, and K. Gull, Nature **391** (6667), 548 (1998).
14   J. A. Maga, T. Sherwin, S. Francis et al., J Cell Sci **112 (Pt 16)**, 2753 (1999).
15   J. A. Maga and J. H. LeBowitz, Trends Cell Biol **9** (10), 409 (1999).
16   C. Santrich, L. Moore, T. Sherwin et al., Mol Biochem Parasitol **90** (1), 95 (1997).
17   P. Bastin, T. H. MacRae, S. B. Francis et al., Mol Cell Biol **19** (12), 8191 (1999).
18   N. R. Hutchings, J. E. Donelson, and K. L. Hill, J Cell Biol **156** (5), 867 (2002).
19   D. J. LaCount, B. Barrett, and J. E. Donelson, J Biol Chem **277** (20), 17580 (2002).
20   P. G. McKean, Curr Opin Microbiol **6** (6), 600 (2003).
21   D. M. Woolley and G. G. Vernon, J Exp Biol **204** (Pt 7), 1333 (2001).
22   P. J. Walker, Naturwissenschaften **189**, 1017 (1961).
23   M. E. Holwill and J. L. McGregor, Nature **255** (5504), 157 (1975).
24   David H. Boal, *Mechanics of the cell*. (Cambridge University Press, Cambridge, UK ; New York, 2002); K. Kruse and F. Julicher, Phys Rev E Stat Nonlin Soft Matter Phys **67** (5 Pt 1), 051913 (2003); C. Cibert, Cell Motil Cytoskeleton **49** (3), 161 (2001).
25   C. J. Brokaw, J Exp Biol **43** (1), 155 (1965); D. Eshel and C. J. Brokaw, Cell Motil Cytoskeleton **9** (4), 312 (1988).
26   H. C. Crenshaw, Biophys J **56** (5), 1029 (1989).

MSP Dynamics and Retraction in Nematode Sperm

Charles W. Wolgemuth

Department of Cell Biology, University of Connecticut Health Center, Farmington, CT 06030-3505

Abstract. Most eukaryotic cells can crawl over surfaces. In general, this motility requires three distinct actions: polymerization at the leading edge, adhesion to the substrate, and retraction at the rear. Recent *in vitro* experiments with extracts from spermatozoa from the nematode *Ascaris suum* suggest that retraction forces are generated by depolymerization of the Major Sperm Protein (MSP) cytoskeleton. Combining polymer entropy with a simple kinetic model for disassembly I propose a model for disassembly-induced retraction that fit the *in vitro* experimental data. This model explains the mechanism by which deconstruction of the cytoskeleton produces the force necessary to pull the cell body forward and suggest further experiments that can test the validity of the model.

Keywords: MSP; cell motility; gel; major sperm protein; mathematical model.
PACS: 87.17.Jj, 82.33.Ln, 82.35.Pq, 82.35.Rs, 83.10.Ff, 83.80.Kn

INTRODUCTION

Imagine that you are a single cell . . . and you're hungry. You have two options. The first is that you can stay stationary, remain in place, and wait for food to come to you. By far, this method is the easiest: it requires no energy output. However, as you may guess, it is risky. If food does not come, you will die. On the other hand, you can figure out how to move and go in search of food. This method is harder but more productive. You, the cell, must somehow figure out how to exert force on your environment in such a way that you can produce and maintain directed motion. It is not surprising that many cells have chosen this route. The expense of energy is a small price to pay to stay alive. It is also not surprising that cells have figured out many different methods for achieving motility.

As with organisms at our scale, cells live in one of two environments. They are either immersed in a fluid or they live at a surface (here I loosely consider surrounding cells, such as in tissue, a surface as well). Swimming and flying are the types of motility possible for beings that are immersed. At the cellular scale, the predominant fluid is water. As the density of a cell is comparable to that of water, producing lift force is not an issue, and, indeed, cells are not known to fly. Swimming, however, is very common. As has been discussed in a number of the talks here at this symposium, many cells swim by waiving or rotating filamentary objects or composites of filamentary objects. The drag force that the fluid exerts back on these undulating cilia or flagella produces the thrust that pushes the cell forward.

CP755, *ISIS: International Symposium on Interdisciplinary Science*
edited by A. Ludu, N.R. Hutchings and D.R. Fry
© 2005 American Institute of Physics 0-7354-0240-X/05/$22.50

When in contact with a surface, a cell must somehow leverage friction to enable motility, just as we must to walk. Both eukaryotic cells and prokaryotic cells have figured out how to do this, though with quite different mechanisms. In this talk I will focus on the method employed by eukaryotic cells, which is generically called crawling. In general, this motility requires three distinct actions: polymerization at the leading edge, adhesion to the substrate, and retraction at the rear (**Figure 1**)[1-3]. A more detailed description of the model that I will discuss is presented elsewhere [4].

One of the main cytoskeletal components of eukaryotic cells is a cross-linked polymer network composed of actin filaments. Polymerization and addition of new actin filaments at the leading edge of the cell drives extension through either a polymerization ratchet mechanism [5-6] or swelling [7-9]. Transmembrane proteins, such as integrins, anchor cells to the substrate [10-12]. The mechanism by which force is generated to drive translocation of the cell body is still debated. Originally, this force was attributed to an actomyosin system similar to muscle [13]. However, Myosin II-null *Dictyostelium* cells are still capable of translocation [14-15]. Mogilner and Oster suggested that the depolymerization of an actin meshwork could generate a contractile force to pull up the cell rear [16]. Here we present a more detailed analysis of contractile force generation in a cell that lacks cytoskeletal protein motors. This problem has been addressed previously by finite element modeling [17] and continuum modeling [18-20]; the treatment here offers a microscopic explanation for the *in vitro* experiments on MSP force production [21] and its implications for nematode sperm locomotion.

MSP AND NEMATODE SPERM CRAWLING

Spermatozoa from nematodes, such as *Ascaris suum*, exhibit crawling motility strikingly similar to those of other crawling cells. Although they show all three characteristics of crawling, they do not possess an actin cytoskeleton. Rather, the nematode sperm utilizes a gel of an unrelated polymer, Major Sperm Protein (MSP). As in actin-based cells, polymerization of MSP at the leading edge of the lamellipod produces the force necessary to push out the front of the cell [22]. Unlike actin, MSP forms non-polar filaments [23], and molecular motors have not been identified. These results strongly suggest that the dynamics of the MSP network is responsible for both protrusive and retraction forces in crawling sperm cells. Recent *in vitro* experiments using cellular extracts from *A. suum* spermatozoa implicate disassembly of the MSP network as the force generating mechanism driving translocation of the cell body [21]. In these experiments, vesicles made from the membrane of *A. suum* sperm in the presence of sperm cytosol induce polymerization of a 'comet tail' cylinder of MSP that pushes the vesicle [22], similar to the motion of ActA coated beads in the presence of actin [24]. Retraction forces could be induced in the MSP gel by addition of *Yersinia enterocolytica* tyrosine phosphatase (YOP) to the cell-free extract of sperm (S100) in marked contrast to the behavior of the comet tails in buffer solution which only showed slight retraction [21].

In nematode sperm, MSP forms a network of interconnected charged filaments surrounded by cytosolic fluid, that constitutes a polyelectrolyte gel. This gel exhibits two coexisting forms: a distributed gel consisting of MSP filaments, and condensed regions of filaments (also called 'fiber complexes', 'bundles' or 'ribs') that span the lamellipod from the leading edge to the cell body. The arrangement of the filaments throughout the lamelipod appears to be mostly isotropic; however, images taken by electron microscopy show a bottle brush structure in the fiber complexes [25], suggesting that the filaments may be more ordered in these regions. The *in vitro* experiments on vesicles suggest that solation of this MSP gel induces contractile forces that pull the cell body forward.

THE DEPOLYMERIZATION WINCH

The MSP cytoskeleton is a gel: a crosslinked polymer network immersed in fluid. The volume of such gels is determined by the equilibrium between four forces [20,26]: (i) the entropic tendency for the gel filaments to diffuse outwards, (ii) the 'counterion pressure' that tends to inflate the gel, (iii) the entropic elasticity of the gel filaments that tends to resist expansion, and (iv) the attractive interactions between the filaments that also tend to hold the gel together. In gels, crosslinking of the filaments increases the rigidity of the overall structure and locks out entropic degrees of freedom. *Solation* involves breaking chains. When the structure solates, the rigidity of the structure decreases and the gain in filament entropic freedom drives retraction of the network

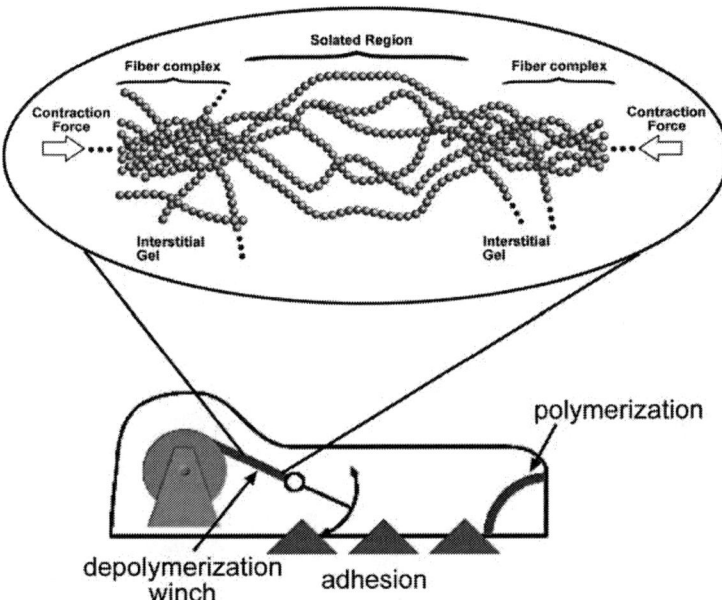

FIGURE 1. Schematic of the model for cell crawling. Polymerization of the leading edge pushes out the front of the cell. Adhesion to the substrate provides traction. We propose that depolymerization of the cytoskeleton produces the force necessary to haul the cell body forward.

(**Figure 1**). The energy sources for contractile work are the free energies of polymerization and crosslinking.

To quantitate this model, we denote by M_p the total mass of polymer in the gel. This mass can be related to the volume, V, of the gel through the volume fraction, ϕ, the ratio of polymer volume to the total mass of the gel (polymer plus solvent). $M_p = \rho_p \phi V$, where ρ_p is the density of a monomer of the polymer. Depolymerization of the polymer decreases the mass, but also can affect ϕ. Whether the gel contracts or not, depends on the ratio of these two effects.

Solation of the gel phase proceeds in two steps. First, chains are severed from the bulk gel creating chains with free ends. Second, the free ends depolymerize into monomers. Therefore, the polymer network contains two kinds of chains: those where both ends terminate in a crosslink, and those where one end is free. Let M_f be the mass of polymer in the free chains and M_c the mass in connected chains. The simplest model assumes that connected chains get broken and are transformed into free chains, which then depolymerize. The kinetics for this model are

$$\frac{dM_c}{dt} = -k_c M_c$$

$$\frac{dM_f}{dt} = k_c M_c - k_f M_f$$

(1)

with rate constants k_c and k_f. The total mass is $M_p = M_c + M_f$.

As mentioned above, solation increases the entropic freedom of the polymer in the gel and can lead to contraction. EM images show MSP filaments that are often bent at lengths of tens of nanometers, suggesting that this length is comparable to the persistence length of MSP [17]. Therefore, it is reasonable to treat the MSP filaments composing the cytoskeletal meshwork as flexible. Using a Flory-Huggins free energy [26] and assuming an isotropic and homogeneous gel, the stress, σ, is a function solely of the volume fraction:

$$\left(\frac{V_m}{k_B T}\right)\sigma_{ij} = \left(\underbrace{-\ln(1-\phi) - \phi}_{\text{mixing}} - \underbrace{\chi\phi^2}_{\substack{\text{polymer} \\ \text{interaction}}} + \underbrace{\frac{1}{N_e}\left(\frac{1}{2}\phi - \phi_0^{2/3}\phi^{1/3}\right)}_{\text{elasticity}} - \underbrace{2N_A V_m \left(C_{ion} - C_b\right)}_{\substack{\text{counterion} \\ \text{pressure}}} \right)\delta_{ij}$$

(2)

Here N_A is Avogadro's number, V_m is the volume of a monomer, k_B is Boltzmann's constant, and T is the temperature, and δ_{ij} is the identity matrix. χ is the Flory parameter which measures the interaction energy between polymer chains [26]. Solation of the gel breaks crosslinks, which changes effective number of monomers between crosslinks, N_e. As crosslinks are destroyed, N_e should increase. I assume a simple kinetics with a rate proportional to the rate that connected chains are broken,

$$\frac{dN_e}{dt} = \beta k_c M_c$$

(3)

where β is a constant

To compare the model derived above with the experiments in [21], length and optical density (OD) are converted to MSP mass and volume fraction using $M_p = \rho_p \phi V$ and Beer's law to relate the optical density to the volume fraction. The results are shown in **Figure 2**a, b. Both the mass and the volume fraction decrease as a function of time; however, the mass decreases faster than the volume fraction which requires an overall contraction of the MSP network. Using the mass vs. time plot, we fit the parameters k_f and k_c in KPM buffer and in S100 supplemented with YOP. We find good fits with a value of $k_f = 0.5$ min^{-1} and $k_c = 0.05$ min^{-1} in KPM buffer and $k_c = 0.14$ min^{-1} in S100 + YOP (**Figure 2**a,b). The depolymerization rates for MSP comet tails have not been measured; however, the value found for k_f is roughly comparable to the depolymerization rate for actin measured in crude extracts and *in vivo* [27,28].

Next we modeled the change in volume fraction, ϕ, with time using the determined values for k_c and the gel stress model for the solation of the MSP network. We find good agreement between the model and the data (**Figure 2**b). As shown in the figure, the volume fraction of the MSP gel decreases during the first 10 minutes and then tends to flatten out for both solution chemistries. In S100 supplemented with YOP, this decrease is more rapid than in KPM buffer. To compare the values calculated in this manner with the original data, we plot the change in length and optical density using the model. **Figure 2**c-d show that both solation models capture the disassembly and retraction of the MSP gel. The experiments that have been done so far show that disassembly can produce retraction in MSP fibers associated with vesicles. However, these experiments do not

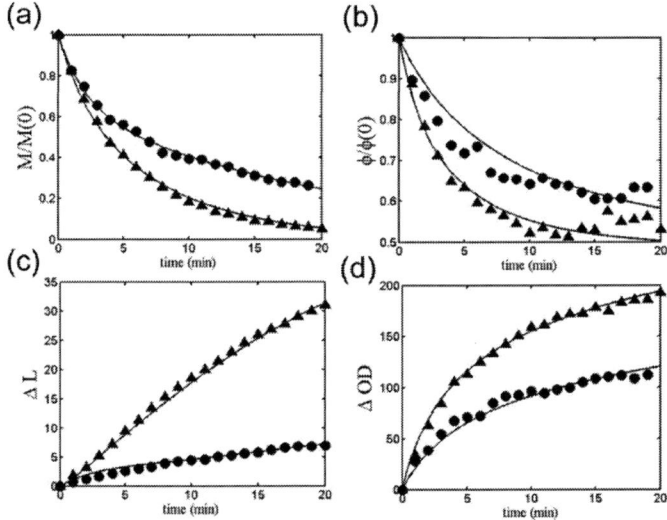

FIGURE 2. Model fits to data taken from [21]. (a) MSP polymer mass vs. time. (b) MSP volume fraction vs. time. (c) Cumulative loss in length of an MSP fiber vs. time. (d) Cumulative loss in optical density vs. time. (a-d) Circles (KPM buffer media) and triangles (s100 and YOP media) represent a replotting of the data from [21]. (a) Solid line is a fit to the mass kinetic model. (b-d) Solid lines are a fits using the gel retraction model.

show directly that this retraction can produce sufficient force to pull the cell body forward during crawling. The model suggests an experiment that can test the force production by disassembly of the MSP network. [21] observed that a bead could be attached to the MSP fiber and pulled along with the retracting fiber. If a bead is adhered to each end of the MSP fiber, the force required to prevent retraction can be measured using micromanipulation techniques such as flexible handles [29]. If we assume that under these conditions, the volume of the MSP fiber stays fixed, then $\phi = M/\rho_m V$. The force required to hold the ends is just the magnitude of the elastic stress times the cross-sectional area of the MSP tail.

Figure 3b shows that the maximum force on the comet tail produced by fiber depolymerization does not depend strongly on the presence of YOP. Both situations produce a maximum force ~ 30 nN. This force is comparable to the experimentally measured force required to halt crawling in keratocytes [30]. However, since a crawling cell traverses a cell length per minute, the physiological translocation force per bundle is more reasonably estimated by the force generated during the first minute. This force is found to be 5 nN in KPM buffer and 15 nN for S100 + YOP. Interestingly, the model predicts a slower rise for the force produced in the presence of YOP where the network is being disassembled faster. This result is somewhat counterintuitive since it seems that faster disassembly should lead to faster force production. However, the elastic strength of the network depends strongly on the crosslink density, whereas the stress depends strongly on the volume fraction, ϕ. In KPM buffer, the MSP mass that is contained in the free chains quickly depolymerizes, but crosslinks and connected chains stay intact. Therefore, the elasticity of the network remains strong, while entropic pressure from the free chains is removed driving network contraction. When YOP is added, crosslinks are broken more quickly. Therefore, the elasticity of the network decreases and free chain polymer is removed from the system at comparable rates; therefore, force production is slower. At longer times, the force decreases as the elasticity in the network is degraded.

This force dynamics may play a role in nematode sperm translocation. As the cell crawls, new polymer is added at the leading surface and old polymer gets progressively closer to the rear of the cell where disassembly induces the retraction necessary to pull the cell body forward. At the front of the cell, adhesion to the substrate is strong. Therefore, applying large forces at the leading edge are ineffective—or even counterproductive—if the force is large enough to break the adhesion to the substratum. Slower force production in the presence of YOP shifts the location of strong retraction towards the rear of the cell where it is most effective in pulling the cell body forwards.

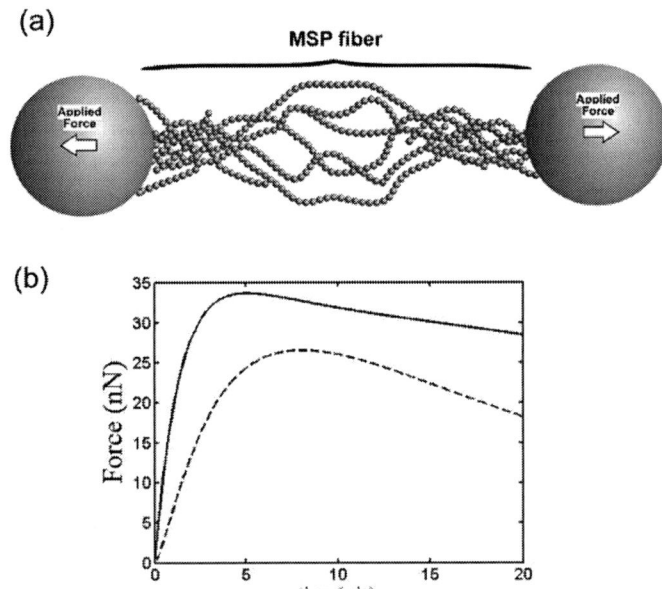

FIGURE 3. (a) Schematic of the proposed experiment. (b) Plot of force vs. time derived from the model when both ends of the MSP fiber are held fixed. The solid line shows the result for the MSP fiber in KPM buffer. The dashed line is the result for the MSP fiber in cell-free extract and YOP.

ACKNOWLEDGMENTS

The author thanks G. Oster, T. Roberts, O. Vanderlinde, and L. Miao for their contributions to this project.

REFERENCES

1. Abercrombie, M., *Proc. Roy. Soc. Lond. B* **207**, 129-147 (1980).
2. Lauffenburger, D.A., and Horwitz, A.F., *Cell* **84**, 359-369 (1996).
3. Mitchison, T.J., and Cramer, L.P., *Cell* **84**, 371-379 (1996).
4. Wolgemuth, C.W., Miao, L., Vanderlinde, O., Roberts, T., and Oster, G., (preprint).
5. Mogilner, A., and Oster, G., *Biophy. J.* **84**, 1591-1605 (2003).
6. Peskin, C., Odell, G., and Oster, G., *Biophys. J.* **65**, 316-324 (1993).
7. Herant, M., Marganski, W.A., and Dembo, M., *Biophys. J.* **84**, 3389-3413 (2003).
8. Oster, G., and Perelson, A., "The physics of cell motility," in *Cell Behavior: Shape, Adhesion and Motility*, edited by J. Heaysman CM, F. Watt, Cambridge, England: The Company Of Biologists Ltd., 1988, p. 35-54.
9. Oster, G., and Perelson, A., "Cell protrusions," in *Frontiers in Mathematical Biology*, edited by S. Levin, Berlin: Springer-Verlag, 1994, p. 53-78.
10. Gaudet, C., Marganski, W.A., Kim, S., Brown, C.T., Gunderia, V., Dembo, M., and Wong, J.Y., *Biophys. J.*, **85**, 3329-3335 (2003).
11. Koo, L.Y., Irvine, D.J., Mayes, A.M., Lauffenburger, D.A., and Griffith, L.G., *J. Cell Sci.*, **115**, 1423-1433 (2002).
12. Rahman, A., Tseng, Y., and Wirtz, D., *Biochem. Biophys. Res. Commun.*, **296**, 771-778 (2002).

13. Huxley, H.E., *Nature*, **243**, 445-449 (1973).

14. DeLozanne, A., and Spudich, J.A., *Science*, **236**, 1086-1091 (1987).

15. Knecht, D.A., and Loomis, W.F., *Science*, **236**, 1081-1086 (1987).

16. Mogilner, A., and Oster, G., *Euro. Biophys. J.*, **25**, 47-53 (1996).

17. Bottino, D., Mogilner, A., Stewart, M., Roberts, T., and Oster, G., *J. Cell Sci.*, **115**, 367-384 (2001).

18. Joanny, J.F., Julicher, F., and Prost, J., *Phys. Rev. Lett.*, **90**, Art. No. 168102 (2003).

19. Mogilner, A., and Verzei, D.W., *J. Stat. Phys.*, **110**, 1169-1189 (2003).

20. Wolgemuth, C.W., Mogilner, A., and Oster, G., *Eur. Biophys. J.*, **33**, 146-158 (2004).

21. Miao, L., Vanderlinde, O., Stewart, M., and Roberts, T.M., *Science*, **302**, 1405-1407 (2003).

22. Italiano, J.J., Roberts, T.M., Stewart, M., and Fontana, C.A., *Cell*, **84**, 105-114 (1996).

23. Bullock, T.L., Roberts, T.M., and Stewart, M., *J. Mol. Biol.*, **263**, 284-296 (1996).

24. Cameron, L.A., Footer, M.J., Oudenaarden, A.V., and Theriot, J.A., *Proc. Natl. Acad. Sci. USA*, **96**, 4908-4913 (1999).

25. Sepsenwol, S., Ris, H., and Roberts, T.M., *J. Cell Biol.*, **108**, 55-66 (1989).

26. Flory, P.J., *Principles of Polymer Chemistry*, Ithaca, NY: Cornell Univeristy Press (1953).

27. Theriot, J.A., Rosenblatt, J., Portnoy, D.A., Goldshmidt-Clermont, P.J., and Mitchison, T.J., *Cell*, **76**, 505-517 (1994).

28. Watanabe, N., and Mitchison, T.J., *Science*, **295**, 1083-1085 (2002).

29. Marcy, Y., Prost, J., Carlier, M.-F., and Sykes, C., *Proc. Natl. Acad. Sci. USA*, **101**, 5992-5997 (2004).

30. Oliver, T., Lee, J., and Jacobson, K., *Semin. Cell Biol.*, **5**, 139-147 (1994).

Divalent Cation Control of Flagellar Motility in African Trypanosomes

Anna M. Westergard and Nathan R. Hutchings

Interdisciplinary Experimentation and Scholarship (IDEAS) Program
Department of Biological Science
Northwestern State University of Louisiana. Natchitoches, Louisiana 71497

Abstract. Changes in calcium concentration have been shown to dynamically affect flagellar motility in several eukaryotic systems. The African trypanosome is a monoflagellated protozoan parasite and the etiological agent of sleeping sickness. Although cell motility has been implicated in disease progression, very little is currently known about biochemical control of the trypanosome flagellum. In this study, we assess the effects of extracellular changes in calcium and nickel concentration on trypanosome flagellar movement. Using a flow through chamber, we determine the relative changes in motility in individual trypanosomes in response to various concentrations of calcium and nickel, respectively. Extracellular concentrations of calcium and nickel (as low as 100 micromolar) significantly inhibit trypanosome cell motility. The effects are reversible, as indicated by the recovery of motion after removal of the calcium or nickel from the chamber. We are currently investigating the specific changes in flagellar oscillation and coordination that result from calcium and nickel, respectively. These results verify the presence of a calcium-responsive signaling mechanism(s) that regulates flagellar beat in trypanosomes.

Keywords: Divalent, trypanosome, flagellum, motility, axoneme.
PACS: 87.15.He, 87.15.La, 87.15.Aa

INTRODUCTION

The African trypanosome is a single-celled protozoan with a specialized cytoskeletal architecture that enables it to survive in the harsh conditions of the midgut of the tsetse fly and the mammalian bloodstream. The trypanosome is an extracellular parasite and the etiological agent of African sleeping sickness. This fatal disease affects millions of people in sub-Saharan Africa and has a dramatic impact on the agricultural and economic development in the endemic regions. There is no known cure for sleeping sickness. Cell motility is important for disease progression, yet very little is known about the mechanisms that regulate cellular motility in this parasite. Therefore, we have begun to characterize the influence of extracellular divalent cations on flagellar motility in procyclic African trypanosomes.

The trypanosome has a single flagellum that is attached along the entire length of the cell body, initiating at the flagellar pocket near the posterior end of the cell and extending past the anterior tip of the cell. The flagellum contains the typical nine-plus-two microtubule doublet arrangement that comprise the eukaryotic axoneme, but

CP755, *ISIS: International Symposium on Interdisciplinary Science*
edited by A. Ludu, N.R. Hutchings and D.R. Fry
© 2005 American Institute of Physics 0-7354-0240-X/05/$22.50

the trypanosome flagellum also contains a second filamentous structure known as the paraflagellar rod (PFR) and a connective region referred to as the flagellar adhesion zone (FAZ) (for review see [1]). The paraflagellar rod is a tripartite lattice-like filament that runs parallel to the axoneme. The exact role of the paraflagellar rod is unclear, but it is necessary for the productive cell motility. The flagellar adhesion zone contains the structures that suture the flagellum to the cytoskeleton of the cell body. All of the structures of the trypanosome flagellum are necessary for productive motility, and thus are necessary for the progression of the disease, as the parasite must migrate from the midgut of the tsetse fly to the salivary glands and through the blood stream of the mammalian host in order to cause disease. Furthermore the arrangement and composition of these structures creates a unique type of auger-like cell motility. Due to the lack of understanding of the precise nature of the motility of the African trypanosome, it is of specific interest to study the mechanics and regulation of flagellar dynamics in the trypanosome.

Recently, we determined that the swimming velocity of procyclic trypanosome brucei varies during different phases of growth in culture (not shown). Behavioral changes in culture, such as this difference in swimming velocity and those that occur during infection, suggest that the cells are able to sense and respond to changes in their environment. Although, these behavioral changes could result from many different signaling mechanisms, divalent cations are known chemical modulators of cellular motility in other flagellated systems. In *Chlamydomonas,* increases in the levels of intracellular calcium induce a change in waveform from an asymmetrical ciliary type to a symmetrical flagellar type [2]. In sea urchin sperm calcium concentrations above 100 micromolar increase the asymmetry of the waveform of the flagellum resulting in quiescence of flagellar beating [3]. In paramecium and tetrahymena, increases in the calcium concentration result in a directional reversal of swimming [4]. Furthermore, central pair regulation of microtubule sliding in extracted sea urchin sperm is linked to calcium concentration [5], and calcium reduces the sliding velocity of microtubules in extracted sea urchin sperm [6].

Nickel, another divalent cation, also affects the motility of flagellated cells. Nickel is a known antagonist of certain calcium channels [7]. For example nickel and cadmium reportedly block the flagellar calcium response in rat sperm [7], and nickel inhibits spontaneous wave generation by bull sperm [8]. Flagellar bend initiation is inhibited by nickel, which is independent from the sliding tubule mechanism responsible for wave propagation in bull sperm [8].

Although calcium and nickel have been shown to affect motility in many different flagellated cell types, the effects of these ions on trypanosomes has not been determined. Additionally, since the effects of calcium on different cell types is so varied, and because the architecture of the trypanosome is unique compared to other systems, we sought to directly measure the effect of calcium on trypanosomes. Nickel also has an assortment of effects on various flagellated systems, and the effects appear to be distinguishable form those caused by calcium. Therefore, we conducted the following experiments to determine what effect, if any, calcium and nickel would have on the flagellar motility of African trypanosomes.

Log phase cells (strain Ytat1.1) were collected by centrifugation and resuspended at 1×10^7 cells per milliliter in a solution of phosphate buffered saline with glucose (PBS-G [9]). Thirty micro liters of the cell suspension was placed in a flow through chamber on a glass microscope slide. The slide was then observed under oil emersion with a 100X UPlanFl objective on an Olympus BX60 microscope. A field of cells tethered to the glass but exhibiting obvious flagellar motility was selected, and a time-lapse digital video was recorded. Fifty microliters of PBS-G was then flowed through the chamber, and a second video of the same field of cells was recorded. This movie was used as a control to account for any affects that the chamber flow may have on the cells. Subsequently, fifty micro liters of PBS-G containing calcium or nickel (varying in concentration from 500 mM to 1 micromolar) was flowed through the chamber, and a "ion treated" video was recorded of the same field of cells. Finally, the chamber was flowed with excess PBS-G (>3 volumes) to remove the majority of the calcium or nickel from the chamber, and a 'recovery' video was recorded of the field. The movies were then analyzed on a cell-by-cell basis using the The Spot Advanced (Version 4.0.2.0, Diagnostic Instruments, Inc, Sterling Heights, MI). The maximum range of flagellar motion of each cell in the field was measured in each movie. The range of motility exhibited by each cell was determined in each of the four videos and compared to the control (PBS-G with flow) video. All experiments were conducted in at least triplicate.

RESULTS AND DISCUSSION

From these experiments, we determined that trypanosome flagellar motility is significantly inhibited by calcium concentrations as low as one hundred micromolar

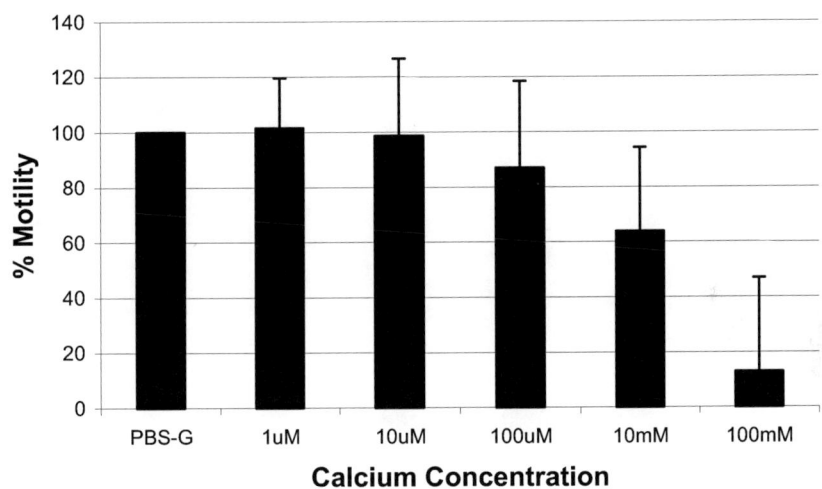

FIGURE 1. The average change in the range of motion is shown for each treatment group as the percent motility (compared to the PBS-G control). The inhibition is significantly different from the control state for all calcium concentrations of one hundred micromolar or higher (t=-3.376, P=0.002*).

(Fig. 1). We did not observe a significant inhibition in cells treated with less than 100 micromolar calcium (Fig 1). By comparing the range of motility in the calcium treated cells with the range of motility of the 'recovery' cells (after flooding the chamber with excess PBS-G buffer to remove the calcium from the chamber), we verified that the effects of the ions are non-lethal. All cells that experienced a decrease in motility after treatment with calcium recovered at least some motility when the system was flushed. Due to the lack of other obvious changes in cell morphology, these data suggest that a discrete and reversible signaling mechanism may be involved in regulating trypanosome flagellar dynamics.

The range of motion in the recovery videos does not return to the initial range of motion in the control video. This is due to the fact that when the cells are inhibited by calcium the amount of the cell body that is in contact with the glass surface increases. This results in a larger portion of the surface area of the cell becoming tethered to the glass, which prohibits the cell from fully recovering its original range of motion. We are exploring alternative means to quantify the flagellar movements that are not limited by this circumstance.

Our results indicate that nickel also inhibits the trypanosome flagellar motion down into the micromolar range (Fig. 2). All concentrations of nickel, except the 100 micromolar treatment group, experienced significant flagellar inhibition. Importantly, the magnitude of each respective concentration differs from that of calcium, which may indicate a difference in the mechanism through which each cation influences

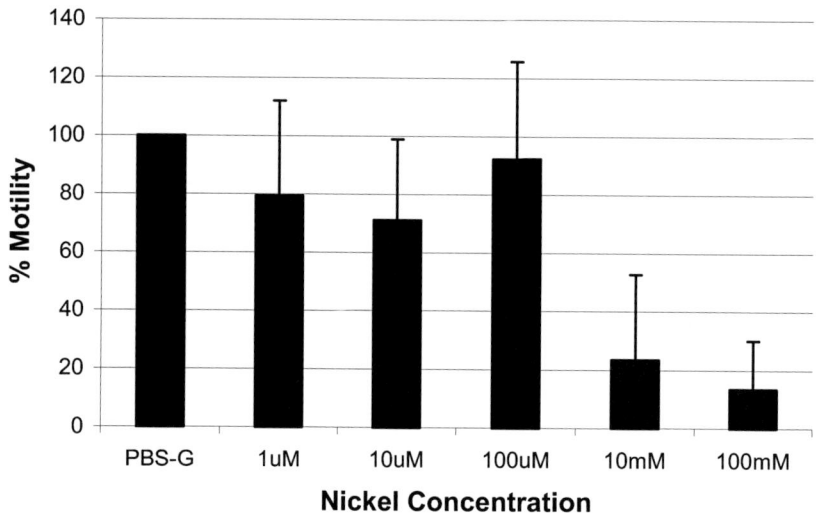

FIGURE 2. The average change in the range of motion is shown for each treatment group as the percent motility compared to the PBS-G control. The inhibition is significantly different from the control state for all nickel concentrations except one hundred micromolar ((t=-0.958, P=0.35). Additional experiments are being conducted to determine if the one hundred micromolar treatment group is an anomaly.

flagellar dynamics. Flushing the chamber with buffer to remove the nickel showed that the cells are able to recover at least some of the motion lost due to treatment with nickel; although the same problem interpreting the recovery data of calcium-treated cells also exists in this case.

The large standard deviations reported for the percent motility in our data suggest that either the individual cellular response to calcium or nickel is highly variable or there is a large variation in the initial range of motion values for each cell. Since our experimental system was designed to compare the range of motion of individual cells for each respective treatment, it did not take into account that each cell differed significantly in the initial range of motion. In fact, some cells exhibited a small initial range of motion while others exhibited a much greater range of motion in the control video. Thus, cells that demonstrated small initial ranges may have only numerically decreased slightly upon treatment relative to other cells that may have had a large initial range of motility (within the limits of our detection). Nonetheless, the mean results clearly indicate that both of the cations have an inhibitory effect on trypanosome flagellar motion.

The results of these experiments provide a foundational data set that confirms nickel and calcium are chemical modulators of trypanosome flagellar motility. Currently, additional experiments are being conducted to determine the exact range of effective concentrations for each chemical modulator on both live cells and extracted flagella. In the near future, we plan to describe the nature of the changes in motility experienced by the flagella, and to determine the effects of the cations on the swimming velocity of the cells in different stages of growth (i.e. stationary and log phase, respectively).

ACKNOWLEDGMENTS

We gratefully acknowledge the support for this project from the Louisiana board of Regents grant number LEQSF (2004-2007)-RD-A25 to NRH and AL. We also would like to thank Dr. Darrell Fry and Dr. Andrei Ludu for their contributions to this work. We appreciate the support of The College of Science and Technology, the IDEAS Program (http://scietech.nsula.edu/ideas), and the IDEAS Program students.

REFERENCES

1   K. L. Hill, Eukaryot Cell **2** (2), 200 (2003); L. Kohl and K. Gull, Mol Biochem Parasitol **93** (1), 1 (1998).
2   J. S. Hyams and G. G. Borisy, J Cell Sci **33**, 235 (1978); C. J. Brokaw and S. M. Nagayama, J Cell Biol **100** (6), 1875 (1985).
3   C. J. Brokaw, J Cell Biol **82** (2), 401 (1979).
4   Y. Naitoh, R. Eckert, and K. Friedman, J Exp Biol **56** (3), 667 (1972).
5   I. Nakano, T. Kobayashi, M. Yoshimura et al., J Cell Sci **116** (Pt 8), 1627 (2003).

6   H. Bannai, M. Yoshimura, K. Takahashi et al., J Cell Sci **113 (Pt 5)**, 831 (2000).
7   C. B. Lindemann and J. S. Goltz, Cell Motil Cytoskeleton **10** (3), 420 (1988); C. B. Lindemann, T. K. Gardner, E. Westbrook et al., Cell Motil Cytoskeleton **20** (4), 316 (1991).
8   C. B. Lindemann, I. Fentie, and R. Rikmenspoel, J Cell Biol **87** (2 Pt 1), 420 (1980).
9   Frederick M. Ausubel, *Current protocols in molecular biology*. (Greene Publishing Associates; J. Wiley, order fulfillment, Brooklyn, N. Y. Media, Pa., 1987).

Non-equilibrium Studies of Voltage-Gated Ion Channels

Armin Kargol

Loyola University New Orleans, Physics Department,6363 St. Charles Ave. New Orleans, LA 70118, USA

Anstract. Ion channels are proteins specializing in transport of certain ions across cellular membranes. They play a crucial role in physiological processes such as: excitability of brain or cardiac cells, cell volume control, messenger ion flow. Understanding and control of the ion channel functioning (their gating and selectivity) is one of the major goals of cellular biophysics. Ion channels are studied using molecular biology and electrophysiology techniques. Among the latter a standard method is patch clamping where currents across a whole cell membrane or a patch of it, sometimes containing only one channel, are recorded. These currents reveal changes in membrane conductance due to changes in gating variables. In our lab we study voltage-gated ion channels. We develop a new electrophysiological technique that involves applying specially designed, rapidly fluctuating voltage waveforms to the cell membrane, thus driving the channel molecules far from equilibrium. The method probes new details of the channel kinetics inaccessible to the standard techniques. It aids the development of new mathematical models of the ion channel gating and the testing and refinement of existing models.

INTRODUCTION

Voltage-gated ion channels are macromolecular assemblies in cell membranes that open ion-selective pores in response to electric fields. They determine cell electrical properties, excitability, and regulate a host of other processes [1]. One of the main goals in ion channel research is to develop molecular-kinetic models of their gating. Such models should reflect conformational rearrangements of the tertiary structure of the channels proteins and typically consist of a finite set of conformational states (closed C, open O, or inactivated I), with thermally activates transitions. More accurately, the channel gating ought to be viewed as a motion in a certain "energy landscape" subject to thermal fluctuations and governed by the Langevin or Fokker-Planck equations [2]. However, a picture of gating as a discrete Markov chains corresponding to the energy minima has been very useful, provided the internal structure of each energy well can be disregarded.

Electrophysiological studies of ion channels have been recently complemented by X-ray crystallography, fluorescence resonance and other imaging techniques [3,4]. However, the latter mostly offer a static picture of the molecular structure and a final test of any functional or structural model is still the agreement with electrophysiological data. Such data come mostly from the voltage clamp technique, in which the voltage across the cell membrane is controlled by a feedback circuit that measures the net current [5]. The technique originally based on measurements of the macroscopic ionic currents has been later refined to include gating currents and single channel recordings [6].

CP755, *ISIS: International Symposium on Interdisciplinary Science*
edited by A. Ludu, N.R. Hutchings and D.R. Fry
© 2005 American Institute of Physics 0-7354-0240-X/05/$22.50

Markov models of channel kinetics have been widely used and detailed models have been developed for several types of channels. Nevertheless, there are still substantial controversies as to the correct underlying physical picture of gating. Different models can often reproduce the data to within the experimental accuracy and only one of them can accurately depict the physical reality [7-9]. New methods are needed to resolve the ambiguities.

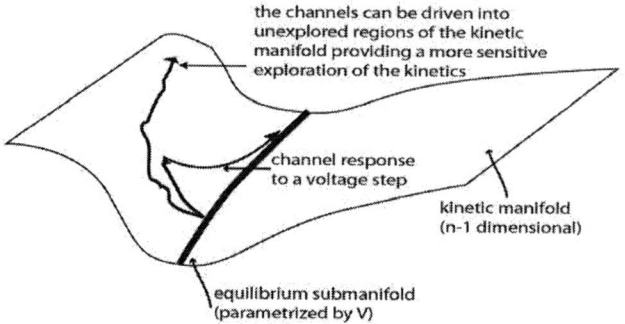

FIGURE 1. The kinetic manifold for a channel gating kinetics. For a channel Markov model with n kinetic states, the probability distributions are points in n-1 dimensional space (the kinetic manifold). All equilibrium distributions form a 1-dimensional submanifold (the equilibrium submanifold) paramertized by the holding voltage. The standard voltage-step experiments correspond to brief sojourns away from equilibrium followed by relaxation to a new point on the equilibrium submanifold determined by the test potential. Only the probability distributions near equilibrium are tested. In contrast, the NRS protocols drive channels into highly nonequilibrium distributions, thus exploring new areas of the kinetic manifold.

Since its inception in 1940's the ion channel electrophysiology has been based on voltage stepping. In this type of experimental protocols the membrane potential is varied in a stepwise fashion and the resulting currents are recorded. Recently a novel approach aimed at expanding the range of protocols used in voltage clamp and known as the nonequilibrium response spectroscopy (NRS) has been proposed [10-16]. It uses voltage waveforms oscillating on a time scale similar to the natural time scale of the gating kinetics. It has been argued that this uncovers new aspects of the channel kinetics, not observable under equilibrium conditions (see Fig. 1 caption).

The idea of subjecting voltage-sensitive channel molecules to such voltage fluctuations to study their non-equilibrium properties originates from physics [17-20]. Fluctuating stimuli can lead to new effects in nonlinear systems, not observed under stationary or deterministic perturbation. In particular, a non-equilibrium noise can lead to a coherent response known as the fluctuation-induced transport. Noise-induced phenomena appear promising in the context of biochemical reactions or biophysical transport processes, e.g. transport of large biomolecules in the absence of obvious potentials or thermal gradients. One of the examples of such molecule-electric field interaction is the selective activation of the K^+ and Na^+ transport modes by fluctuating voltages reported in the Na,K-ATPase [21]. In ion channels such nonequilibrium stimulation results in a new type of response, conflicting with current gating models.

MARKOV MODELS OF CHANNEL KINETICS

The physical picture of channel gating represented by a Markov model is that of thermally activated random jumps between a number of discrete states with certain transition probabilities. Fig. 2 shows examples of Markov models developed for the human heart sodium channels (hH1a). The voltage-dependent transition rates have the

(a)

form: $\alpha_i(V) = \alpha_i(0)exp(q_i^+V/kT)$ and $\beta_i(V) = \beta_i(0)exp(q_i^-V/kT)$, where $\alpha_i(0)$, $\beta_i(0)$ are the rates at 0 mV, and q_i^+, q_i^- are the (forward and backward) gating charges. The latter are often expressed as $q_i^+=q_i\delta_i$ and $q_i^-=-q_i(1-\delta_i)$, where the q_i is the total gating charge for a given transition and δ_i – the fractional electrical distance. The state occupancy probabilities form a vector $\vec{P}$ evolving according to the master equation:

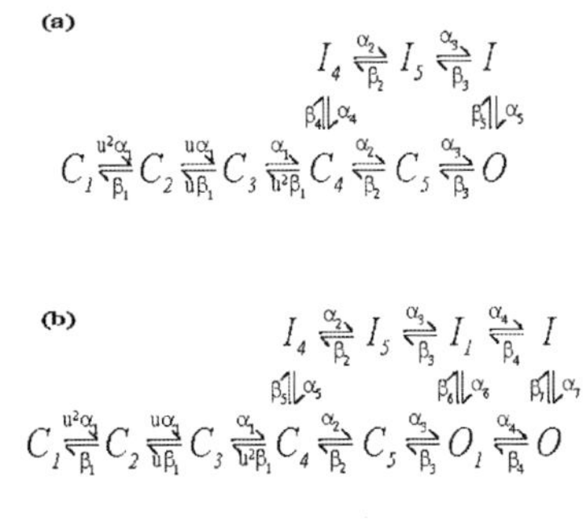

(1) $$\frac{d\vec{P}}{dt} = \tilde{W}[V]\vec{P}$$

FIGURE 2. Markov models for the human hear sodium channels. The Vandenberg-Bezanilla [22] model (a) has been proposed for the giant squid axon but has been used for the human heart sodium channels as well. The Millonas-Hanck [10] modification (b) introduces additional open and inactivated states.

where the matrix $\tilde{W}[V]$ consists of the transition rates $\alpha_i(V)$ and $\beta_i(V)$.

Based on the solution $\vec{P}(t)$ to the equation (1), a model macroscopic ionic current can be computed as:

(2) $$i(t) = g_0 g(V)(V - V_r)\vec{\Pi} \cdot \vec{P}(t)$$

where $\vec{\Pi}$ is the projection of the probability vector onto open states, V_r is the reversal potential, $g(V)$ represents nonlinear terms in instantaneous conductance due to effects other than the channel gating, and g_0 is the overall scaling factor related to the expression level of a particular cell. Similarly, the model gating current can be obtained as:

(3) $$i_g(t) = \frac{d}{dt}\vec{Q} \cdot \vec{P}(t) = \vec{Q} \cdot \tilde{W}[V] \cdot \vec{P}(t)$$

where: $\vec{Q} = \sum_{i-1}^{n} q_i$. These model currents are then compared to the experimental recordings and the model parameters (the transition rates and gating charges) are optimized for the best fit.

SINGLE-PULSE AND ENSEMBLE NRS PROTOCOLS

Kinetic models of channel gating are developed by fitting to various types of experimental data. If the data set is small, the number of compatible models is typically very large. Model refinement is achieved by introducing new types of data (e.g. gating currents or single-channel recordings) or by increasing the number of different voltage protocols. Besides the standard activation and tail transients, more complex protocols have been introduced [10-16]. They are not just a series of voltage steps, but rapid voltage fluctuations. Two types of such protocols have been proposed [12]: the single-pulse and ensemble protocols. Single-pulse protocols are based on few complex voltage

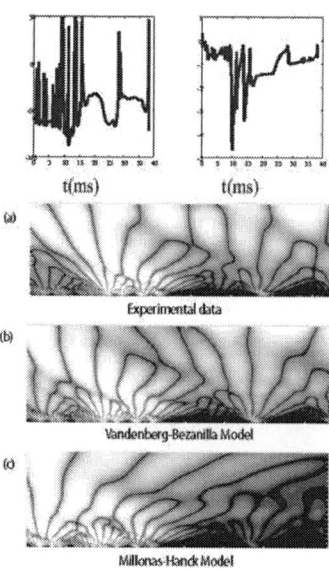

t(ms)   t(ms)

(a)

Experimental data

(b)

Vandenberg-Bezanilla Model

(c)

Millonas-Handk Model

FIGURE 3. An example of the single-pulse protocol applied to the human heart Na$^+$ channel. The voltage pulse (top left) and ionic current response (top right) are shown. Voltage is measured in mV and the current in nA. Voltage pulse was designed for maximal model sensitivity for the VB and MH models (cf. Fig. 2). We compared the continuous wavelet transforms of the experimental data (a), the VB (b) and MH (c) model currents. The plots show spectro-temporal patterns in the data. The horizontal axis represents the time during the signal (0-40 ms in this case). The vertical axis shows the wavelet levels. They are related to the frequency content of the signal with the lowest frequencies on the top and the highest at the bottom. The absolute values of the wavelet transform coefficients are plotted on a gray scale. Dark areas correspond to near-zero values and the bright areas to large values of the coefficients. The significance of the bright areas is that at these times (read on the horizontal scale) specific frequencies dominated the signal. (reproduced from [13] with permission). Copyright © Springer Verlag

waveforms designed for specific purposes. Design algorithm is based on the dyadic wavelet bases and allows development of pulses with specific spectral and temporal properties. If the goal is the comparison of two models proposed for the same channel, the pulses are designed for maximal model sensitivity:

(4) $$S_m = \sum_t \left(i_1(t) - i_2(t)\right)^2$$

where $i_1(t)$ and $i_2(t)$ are the model currents from the two competing models. Once the optimal model topology is determined, the best estimates for model transition rates can be obtained. To that end the pulses maximizing the parametric sensitivity can be used. Parametric sensitivity has been defined as the change in model response (the occupation probabilities or the currents, ionic or gating) caused by a small variation in model parameters. For instance, for the macroscopic ionic currents defined by equation (2) the parametric sensitivity in a generic parameter α (which can be one of the $\alpha_i(0)$, $\beta_i(0)$ or q_i^+, q_i^-'s) is:

(5) $$S_\alpha = \frac{1}{\delta\alpha}\left[\int_0^t \left(i[\alpha + \delta\alpha](t) - i[\alpha](t)\right)^2\right]^{1/2}$$

where $\delta\alpha$ is a small parameter variation and $i[\alpha](t)$ is the current computed from equation (2) for a specific value α of the model parameter. The sensitivity can be computed for each kinetic parameter independently or jointly. In the latter case the sensitivity is directionally nonuniform in parameter space. Typically there are well defined manifolds along which the sensitivity gradient is the least steep. Joint variation of parameters along this manifold is not easily detectable by currently used experimental protocols.

Fig. 3 shows an example of a single-pulse applied to the human heart Na$^+$ channel. The pulse maximizing the model sensitivity for the VB and MH models has been applied using standard whole-cell patch clamp technique, described elsewhere [13]. The model currents and the experimental data were analyzed using the continuous wavelet transform. It can be noticed that the while the VB model reproduces the experimental data well, the MH model overestimates the rate of inactivation.

As seen in the graphs the MH signal decays in time faster than the VB or the experimental data and exhibits a different spectro-temporal pattern. This is an argument against the inclusion of the second open and inactivated states in the model.

The ensemble protocols are based on a repeated application of a random voltage waveform from a statistical ensemble with certain properties. An example can be the dichotomous noise (DN) applied to the human heart Na^+ [10, 11] and to Shaker K^+ channels [14]. The DM consists of random switching between two voltage values: V^+ and V with a certain frequency f and the temporal asymmetry ε. The latter expresses the fraction of time spent in either of the voltages. The experimental ensemble average (the average over a number of realizations of this stochastic process) can be compared to the average model currents. The latter can be easily computed by embedding the channel Markov model into a higher-dimensional model combining the channel gating with the DN voltage switching [10]. In [10,11] the DN ensemble protocols have been applied to the human heart sodium channel (hH1a). The results show that in fact neither of the two models, the Vandenberg-Bezanilla model or the Millonas-Hanck model, can fully reproduce the experimental currents. For low frequency dichotomous noise the models perform well, but for DN frequencies higher than 1 kHz there is an increasing difference between the experimental and model currents, a difference that cannot be corrected by simple reoptimization of model parameters. That shouldn't be surprising, since the low-frequency DN is a set of few voltage steps and the models were developed based on such protocols.

The VB model was further studied using the parametric sensitivity [12,15]. It is a variation of the single-pulse NRS technique where the pulses were designed for maximal sensitivity in selected model parameters. With this method we reduced the uncertainty in model parameters but at the same time we provided an independent verification that the model is not fully compatible with the experimental data.

CONCLUSIONS

Models of ion channel gating kinetics are hypotheses, subject to experimental verification. Model refinement is achieved by introducing more types of that or by increasing the number of voltage protocols. In standard electrophysiology the types of data are the macroscopic or single-channel ionic currents and gating currents. The voltage protocols are typically based on few voltage steps applied at different times and aimed at uncovering various aspects of channel kinetics. However, once a number of such protocols has been used to generate experimental data and a model developed based on these data, a further increase in the number of stepped-voltage protocols will not result in model refinement. A model that reproduces a certain basic sets of stepped-voltage data will reproduce all other data of this type.

In this and other reports we argue in favor of a nonequilibrium approach to studies of channel gating kinetics. We propose a new experimental method, the nonequilibrium response spectroscopy, based on driving ion channels into nonequilibrium distributions with fluctuating voltage waveforms. The novelty of the NRS method is not in the new type of data (a new physical quantity being measured) but in the type of voltage input and the conditions it forces upon the channels. The method has been tested on whole-cell macroscopic ionic channels, however the same paradigm can be applied to single-channel

163

currents as well as gating currents. We have shown that the dichotomous noise protocols as well as the single-pulse protocols designed for maximal sensitivity can be very effective in selecting the best of models that were undistinguishable with traditional techniques. We developed the single-pulse optimal sensitivity method to target specific transition rates and to obtain better estimates of the model parameters.

The method has been so far applied to two well-studied ion channels, the human heart sodium channel and the Shaker potassium channel. Applications to other channels as well as the extension to gating currents is currently in progress.

REFERENCES

1. B. Hille, *Ionic channels of excitable membranes*, 3rd ed. Sinauer Assoc., Sunderland, MA, 2001
2. N. van Kempen, *Stochastic processes in physics and chemistry*, Elsevier, New York, 1981
3. Y. Jiang, A. Lee, J. Chen, V. Ruta, M. Cadene, B.T. Chait, and R. MacKinnon, *Nature*, **423**, 33-41 (2003)
4. A. Cha, G.E. Snyder, P.R. Selvin, and F. Bezanilla, *Nature*, **402**, 809-816 (1999)
5. A. Hodgkin, A.F. Huxley, and B. Katz, *J. Physiol (London)*, **108**, 37-77 (1952)
6. B. Sakmann and E. Neher (eds.), *Single channel recording*, Plenum, New York 1983
7. F. Bezanilla, E. Peroso, and E. Stefani, *Biophys. J.*, **66**, 1011-1021 (1994)
8. N.E. Schoppa and F.J. Sigworth, *J. Gen. Physiol.*, **111**, 295-311 (1998)
9. W.N. Zagotta, T. Hoshi, and R. Aldrich, *J. Gen. Physiol.*, **103**, 312-362 (1994)
10. M.M. Millonas and D.A. Hanck, *Biophys. J.*, **74**, 210-229 (1998)
11. M.M. Millonas and D.A. Hanck, *Phys. Rev. Lett.*, **80**, 401-404 (1998)
12. A. Kargol, B. Smith, and M.M. Millonas, *J. Theoret. Biol.*, **218**, 239-258 (2002)
13. A. Hosein-Sooklal and A. Kargol, *J. Membrane Biol.*, **199**, 199-212 (2002)
14. A. Kargol, *Cell. Mol. Biol. Lett.*, **9**, 375-388 (2004)
15. A. Kargol and A. Hosein-Sooklal, *J. Membrane Biol.*, **199**, 113-118 (2004)
16. A. Kargol, A. Hosein-Sooklal, L. Constantin, and M. Przestalski, *Gen. Physiol. Biophys.*, **23**, 53-75 (2004)
17. C.R. Doering, W. Horsthemke and J. Riordan, *Phys. Rev. Lett.*, **72**, 2984-2987 (1994)
18. A. Fulinski, *Phys. Lett A.*, **193**, 267-273 (1994)
19. W. Horsthemke, R. Lefever, *Biophys. J.*, **35**, 415-432 (1980)
20. M.C. Menconi, M. Pellegrini, M. Pellegrino, D. Petracci, *Eur. Biophys. J.*, **27**, 299-304 (1998)
21. B. Liu, R.D. Astumian, T.Y. Tsong, *J. Biol. Chem.*, **265**, 7260-7267 (1990)
22. C.A. Vandenberg, F. Bezanilla, *Biophys. J.*, **60**, 1499-1510 (1999)

Phenomenological Energetics for Molecular Motors

Takahiro Harada

Department of Physics, Graduate School of Science, Kyoto University, Kyoto 606-8502, Japan.

Abstract. The phenomenological consideration on the energetics of molecular motor that are working at non-equilibrium conditions, is presented in this paper. First, a phenomenological equation of motion for a motor molecule is suggested, in which the physical properties of the motor are incorporated. Based on this equation of motion, the energetics of the system were considered. It is found that the energy fluxes can be expressed in terms of several mechanical observables, so that we can directly calculate the energetic efficiency of the motor experimentally. The present framework can adequately explain the recent experimental data on the energetic efficiency of kinesin. In order to gain further insight, we examined the theory by employing a well-known ratchet model, which demonstrated that the present framework is consistent with an already known framework of energetics, referred to as stochastic energetics, provided that the nonlinearity of the system is not so high. The present framework provides an easy procedure to estimate the energy fluxes on a molecular motor, with mechanical experiments at a single-molecule level.

Keywords: molecular motors, energetics, fluctuations, response, phenomenology
PACS: 05.40.Jc, 05.70.Ln, 87.16.Nn

1. INTRODUCTION

Molecular motors, which are present in several types of cells, play important roles in various aspects of cellular functions such as intracellular transport or cellular motility. Diverse types of motor proteins are already known, and their most significant function is to transduce chemical energy of hydrolyzed substrate such as Adenosine Tri-Phospate (ATP) to mechanical work. Intensive studies, concerning their transduction mechanism, have been performed by many groups. In particular, techniques to observe their motion at a single-molecule level with resolutions in the order of ms and nm have been developed, these are often called single-molecule-measurement techniques [1, 2]. Although detailed data on their motion has been acquired and analyzed using such techniques, the mechanism of their chemo-mechanical energy transduction, unfortunately, still remains obscure. One of the reasons may be the difficulty in directly obtaining information on the amount of chemical energy that is input at a single-molecule level. Although T. Yanagida and co-workers succeeded in monitoring attachment and detachment of an ATP molecule to a motor by use of near-field microscopy [2], the amount of energy accompanying the ATP molecule could not be directly determined. Therefore, in order to understand the mechanism of the chemo-mechanical energy transduction, a novel strategy to obtain the information on the chemical energy input at a single-molecule level has to be developed.

As mentioned above, a molecular motor receives an amount of chemical energy and transduces it to several forms of mechanical energy. Since the motor itself is an enzyme,

CP755, *ISIS: International Symposium on Interdisciplinary Science*
edited by A. Ludu, N.R. Hutchings and D.R. Fry
© 2005 American Institute of Physics 0-7354-0240-X/05/$22.50

the internal energy remains unchanged after each cycle of transduction. Therefore, being aware of the law of energy conservation, we should be able to estimate the input energy by measuring all the mechanical work done by the motor. In this context, the term "mechanical work" implies not only the work accomplished in executng the directed movement but also the fluctuation around it. This provides an easier method to estimate the energy input as compared to using a lot of tricks of biochemistry, because we already have been able to obtain sufficient detailed information on the motion or the mechanical response of a motor molecule.

A phenomenological procedure to estimate the energy input on a molecular motor from purely mechanical measurements, is discussed in this paper. It is a virtue of phenomenology that we can apply the theory without regard to the specific mechanism of the motor, although we must be careful about the applicability of the framework. For example, we demonstrated that the prediction by the present framework is consistent with the recent experiment of the energetic efficiency of kinesin. We also examined whether the present theory can correctly estimate the energy input of a well-known Brownian ratchet model, in order to clarify the applicability of the theory. The result of our examination proves that we can provide an accurate prediction with the present framework unless the internal dynamics of the system is highly nonlinear.

2. PHENOMENOLOGICAL ENERGETICS

In this paper, we consider a system in which a small motor particle is confined to a one dimensional track in an aqueous solution. We first introduce several observable quantities regarding this system. When a perturbative force, $f(t)$, is applied to the particle, the linear response of the velocity, $v(t)$, of the particle can be characterized as

$$\langle v(t) \rangle = v_0 + \int_{-\infty}^{t} \chi(t - t') f(t') \mathrm{d}t', \tag{1}$$

where $\langle \cdots \rangle$ denotes the ensemble average, v_0 is the mean velocity of the steady system without perturbation, and $\chi(t)$ is the response function. The Fourier-Laplace transformation of $\chi(t)$, given as

$$\chi[\omega] \equiv \int_{0}^{\infty} \chi(t) e^{i\omega t} \mathrm{d}t, \tag{2}$$

represents susceptibility. Another important quantity is the correlation function of the velocity fluctuation, given as $\Phi(t) \equiv \frac{1}{2} \langle \{v(t) - v_0\}\{v(0) - v_0\} \rangle$, or its Fourier transformation, given as

$$\Phi[\omega] \equiv \int_{-\infty}^{\infty} \Phi(t) e^{i\omega t} \mathrm{d}t. \tag{3}$$

We then develop a phenomenological model for the motion of the molecular motor under a constant small load, $f(t) = -f_L$. In order to describe the stochastic motion of the motor, we adopt a generalized Langevin equation. Although we do not have any information on the detailed mechanism of the motor, such an equation can incorporate various properties of the motor and it can be considered as a coarse-grained model of more detailed models. The phenomenological model is presented by the following equation:

$$m\dot{v}(t) + \int_{-\infty}^{t} \Gamma(t-t')v(t')\mathrm{d}t' = -f_{\mathrm{L}} + \Xi(t) + F(t), \tag{4}$$

where m, $\Gamma(t)$, $\Xi(t)$, and $F(t)$ are the mass of the motor, a memory kernel, a thermal noise, and an unknown fluctuating force that is regarded as being produced by the motor accompanied with energy input, respectively. In this model, susceptibility is calculated as follows:

$$\chi[\omega] = (-im\omega + \Gamma[\omega])^{-1}, \tag{5}$$

where $\Gamma[\omega]$ is the Fourier-Laplace transformation of $\Gamma(t)$. $\Xi(t)$ should satisfy both, $\langle \Xi(t) \rangle = 0$ and the second type of FDT as $\langle \Xi(t)\Xi(t') \rangle = T_0 \Gamma(t-t')$, where T_0 is the temperature of the solvent.

Subsequently, we take the average of each term in eq. (4), after multiplication by $v(t)$, and obtain the following expression:

$$f_{\mathrm{L}}\langle v \rangle = \langle -\int_{-\infty}^{t} \Gamma(t-t')v(t)v(t')\mathrm{d}t' + \Xi(t)v(t) \rangle + \langle F(t)v(t) \rangle. \tag{6}$$

The left-hand side, $\dot{E} \equiv f_{\mathrm{L}}\langle v \rangle$, represents the power against the external load. The first term on the right-hand side, $Q \equiv \langle -\int_{-\infty}^{t} \Gamma(t-t')v(t)v(t')\mathrm{d}t' + \Xi(t)v(t) \rangle$, is interpreted as the rate of heat transfer between the system and the heat bath [3], while the second term, $P \equiv \langle F(t)v(t) \rangle$, represents the rate of energy input from the external energy source. Therefore, eq. (6) represents conservation of energy $\dot{E} = Q + P$.

The expression of Q can be further transformed using eqs. (3), (4), and (5). After calculation, we obtain [6]

$$Q = \frac{\langle v \rangle^2}{\chi[0]} - \int_{-\infty}^{\infty} \frac{2}{\chi[\omega]} \left(\Phi[\omega] - T_0\chi'[\omega] \right) \frac{\mathrm{d}\omega}{2\pi}, \tag{7}$$

in the derivation of which we have assumed that there is no correlation between $\Xi(t)$ and $F(t)$. $\chi'[\omega]$ denotes the real part of $\chi[\omega]$. The rate of energy input, P, can be determined by $P = \dot{E} - Q$, as follows:

$$P = F_0 \langle v \rangle + \int_{-\infty}^{\infty} \frac{2}{\chi[\omega]} \left(\Phi[\omega] - T_0\chi'[\omega] \right) \frac{\mathrm{d}\omega}{2\pi}, \tag{8}$$

where $F_0 \equiv f_{\mathrm{L}} + \chi[0]^{-1}\langle v \rangle$ is found to be independent of f_{L}, and is interpreted as an effective force of the motor. Finally, the energetic efficiency, which is defined as the ratio of the power against the load to the rate of energy input $\eta \equiv \dot{E}/P$, can be obtained using eq. (8).

3. EXPERIMENTAL VERIFICATION

In this section, we verify the result obtained above by comparison with the experimental data of conventional kinesin. Although, in principle, the efficiency, η, can be calculated only from measurable quantities, the present data available for molecular motors

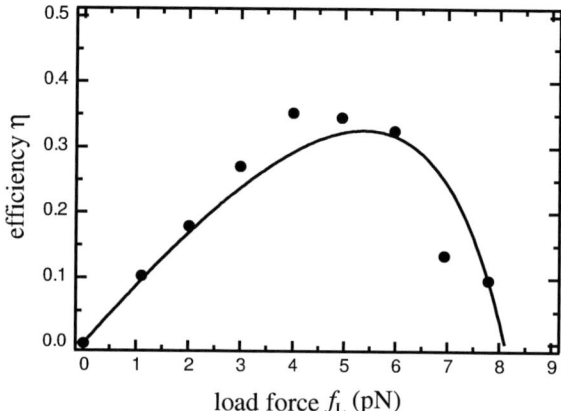

FIGURE 1. Energetic efficiency against a load force for kinesin. The closed circles denote the experimental data from ref. [8]. The solid line indicates the theoretical prediction in eq. (9), which is fitted to the experimental data by the least-square method. The resulting parameters are shown in TABLE I.

is, unfortunately, insufficient. Therefore, we adopt several assumptions in order to compensate for the lack of data. First, we assume that the fluctuation-dissipation relation (FDR) between $\Phi[\omega]$ and $\chi[\omega]$ [4] holds only for frequencies larger than $2\pi/\tau$, where τ is a positive constant. Thus, within this range, the relation between them is given as $\Phi[\omega] = T_0\chi'[\omega]$. On the other hand, for frequencies smaller than $2\pi/\tau$, FDR can be violated in general, since the system is not in an equilibrium state. Such a tendency has already been observed in the case of Brownian ratchets [5, 6, 7]. In this case, we assume their relation with Lorentz type as $\Phi[\omega] \approx \chi'[\omega](D/\mu + T_0\omega^2\tau^2)/(1 + \omega^2\tau^2)$, where $D \equiv \Phi[0]$ is the diffusion coefficient and $\mu \equiv \chi'[0]$ is the mobility. Subsequently, we assume that the inertia effect is sufficiently small because of high friction. This assumption results in $\chi'[\omega]\chi[\omega]^{-1} \approx 1$, since the imaginary part of $\chi[\omega]$ is related to inertia. Under these assumptions, the integral in the expression of P can be simplified as $\int_{-\infty}^{\infty} 2\chi'[\omega]^{-1}(\Phi[\omega] - T_0\chi'[\omega])\mathrm{d}\omega/2\pi \approx (D/\mu - T_0)\tau^{-1}$. Therefore, the efficiency is expressed as follows:

$$\eta \approx \frac{f_\mathrm{L}\langle v\rangle}{F_0\langle v\rangle + (D/\mu - T_0)\tau^{-1}},\tag{9}$$

with $\langle v\rangle = \mu(F_0 - f_\mathrm{L})$.

We fitted this expression to the experimental data that Nishiyama *et al.* obtained by using a nanometry technique with optical trapping [8]. Figure 1 shows that the above expression (the solid line) is successfully fitted to the experimental data (the closed circles). The parameters obtained from this fitting are shown in table 1 along with the experimental data from refs. [8, 9]. The values of the fitting parameters were found to be consistent with the independently obtained experimental data. It should be noted that the timescale of the energy input, τ, was found to be close to the mean first passage time (dwell time) of kinesin. This fact implies that the energy input is tightly coupled to the motion in the case of conventional kinesin.

TABLE 1. Parameter values in eq. (9). μ, D, τ, and F_0 represent the mobility, the diffusion coefficient, the time constant for the effective temperature, and the effective force, respectively. The upper row represents values obtained from the least-square fitting of eq. (9), with the Boltzmann constant recovered, to the experimental data (Fig. 1). The lower row shows the experimental value obtained from refs. [9, 8] for kinesin under a saturated ATP concentration (≥ 1 mM) at room temperature. For the value of τ, the mean first passage time (dwell time) is shown.

	μ (nm/s·pN)	D (nm^2/s)	τ (ms)	F_0 (pN)
Fitting	0.89×10^2	1.6×10^3	6.7	8.1
Experimental	1.0×10^2	1.3×10^3	10	8.0

4. EXAMINATION WITH RATCHET MODELS

We need further discuss the applicability of the theory because, in the above example of kinesin, the amount of analyzed data was limited and we have introduced several assumptions. Thus, in this section, we examine whether we can correctly estimate the input energy of a well-known Brownian ratchet model; here we focus on the flashing ratchet model suggested by J. Prost et al. [10]. In this model, we suppose that a particle that has several discrete internal states is undergoing Brownian motion in one dimensional space. According to its internal state, the particle experiences different periodic potentials. The dynamics of the internal states can be either deterministic or stochastic. It is well known that the particle can have a non-zero mean velocity if all the periodic potentials are not symmetric with respect to reflection. At present, since we are interested in energy input, we simply use the model in which the particle is assumed to have two states and the transition from one state to another is assumed to be stochastic at a uniform rate σ. We can, then, describe the motion of the particle by the following two-state Fokker-Planck equation:

$$\partial_t \begin{pmatrix} W_0(x,v,t) \\ W_1(x,v,t) \end{pmatrix} = \begin{pmatrix} L_0 - \sigma & \sigma \\ \sigma & L_1 - \sigma \end{pmatrix} \begin{pmatrix} W_0(x,v,t) \\ W_1(x,v,t) \end{pmatrix}, \qquad (10)$$

where $L_i \equiv -\partial_x v - \partial_v(-\gamma v - \partial_x U_i(x) - \gamma \partial_v)$ $(i = 0, 1)$ are Kramers operators for each state. With regard to the periodic potential, we adopt that

$$\begin{cases} U_0(x) &= \Delta \cos 2\pi x \\ U_1(x) &= const. \end{cases} \qquad (11)$$

We use a system of units in which the mass of the particle, the temperature, and the period of the potential energy are set to unity.

As already considered in literature mentioned earlier [3, 11], the energy corresponding to the potential difference, $U_j(x) - U_i(x)$, is supposed to be supplied on the occasions of transition from state i to state j. The probability per unit time for the transition from state 0 to state 1 is given as $\sigma \overline{W}_0(x, v)$, where $\overline{W}_i(x, v)$ $(i = 0, 1)$ are the stationary distributions, while that for the transition from state 1 to state 0 is given as $\sigma \overline{W}_1(x, v)$. Therefore, the

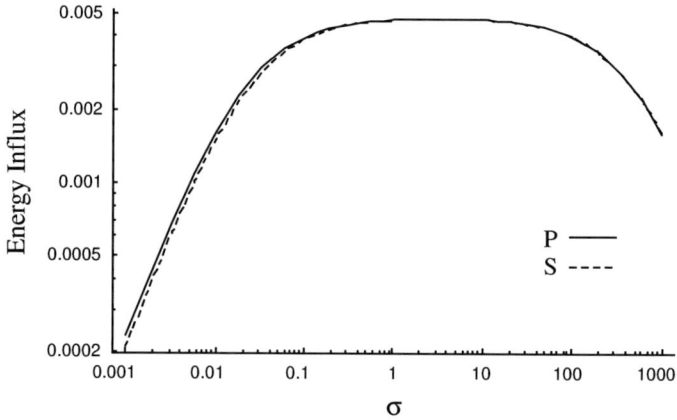

FIGURE 2. The rate of energy input for the flashing ratchet model. The green line shows the energy input determined from the model that has been calculated according to eq. (12) as a function of the transition rate σ. The red line represents the energy input calculated using the phenomenological procedure (13). Here, parameter values are chosen as $\Delta = 1$ and $\gamma = 10^{-3}$.

rate of energy supply is given as

$$S \equiv \int dx \int dv \left\{ \sigma(U_1(x) - U_0(x))\overline{W_0}(x,v) + \sigma(U_0(x) - U_1(x))\overline{W_1}(x,v) \right\} \quad (12)$$

On the other hand, according to the procedure described in Sec. 2, the rate of energy input is estimated as

$$P = \int_{-\infty}^{\infty} 2\chi[\omega]^{-1} \left(\Phi[\omega] - T_0\chi'[\omega] \right) \frac{d\omega}{2\pi}. \quad (13)$$

Note that, in the present system, $v_0 = 0$. Since all the quantities on the right hand side of (13) and (12) can be calculated on the basis of eq. (10) [12], we can directly compare the estimation, P, with the input energy, S, which is determined by the model. As seen in Fig. 2, the estimate P, well reproduced S. Unfortunately, the coincidence between them is not perfect when σ is small. Although this difference becomes larger if we increase the height of the potential, Δ, we can obtain a good agreement for the entire range of σ in the case of a small value of Δ.

5. DISCUSSION AND CONCLUSION

In this paper, we suggested a phenomenological procedure to estimate chemical energy input to a molecular motor on the basis of mechanical measurements. It has been shown that energetic efficiency of kinesin can be quantitatively explained with this framework. We have also shown the applicability of the present phenomenological framework by investigating a simple ratchet model. From an experimental point of view, direct determination of chemical energy input is a much harder problem than obtaining measurements of mechanical quantities, as mentioned in Sec. 1. The present procedure provides a simple way to substitute the measurement of chemical energy with mechanical energy. This procedure is expected to be applied to various species of motor proteins. Although the applicability of the framework is rather limited at present, it is practical enough for application to molecular motors. The possible extension of the theory should be studied in future.

ACKNOWLEDGMENTS

The author acknowledges the useful discussions with S. Sasa, K. Hayashi, H. Linke and K. Yoshikawa. This work is supported in part by Research Fellowships for Young Scientists, of the Japan Society for the Promotion of Science (No. 05494).

REFERENCES

1. K. Svoboda and S. M. Block, *Cell*, **77**, 773 (1994).
2. T. Funatsu, Y. Harada, M. Tokunaga, K. Saito, and T. Yanagida, *Nature*, **374**, 555 (1995).
3. K. Sekimoto, *J. Phys. Soc. Jpn.*, **66**, 1234 (1997).
4. R. Kubo, M. Toda, and N. Hashitsume, *Statistical Physics II: Nonequilibrium Statistical Mechanics* (Springer, Berlin, 1991).
5. T. Harada and K. Yoshikawa, *Phys. Rev. E*, **69**, 031113 (2004).
6. T. Harada, cond-mat/0310547.
7. K. Sasaki, *J. Phys. Soc. Jpn.*, **72**, 2497 (2003).
8. M. Nishiyama, H. Higuchi, and T. Yanagida, *Nat. Cell. Biol.*, **4**, 790 (2002).
9. K. Svoboda, P. P. Mitra, and S. M. Block, *Proc. Natl. Acad. Sci. USA*, **91**, 11782 (1994).
10. A. Ajdari and J. Prost, *C. R. Acad. Sci. Paris II*, **315**, 1635 (1992).
11. F. Jülicher and J. Prost, *Phys. Rev. Lett*, **75**, 2618 (1995).
12. H. Risken, *The Fokker-Planck Equation: Methods of Solution and Applications* (Springer, Berlin, 1988).

Chemical Master Equation Reduction Methods

Rui Zhu and Marc R. Roussel

Department of Chemistry and Biochemistry, University of Lethbridge, Lethbridge, Alberta T1K 3M4, Canada

Abstract. We study invariant manifold methods for reducing chemical master equations using the Michaelis-Menten mechanism as an example. We try Fraser's functional iteration method first, but find that it is difficult to use for master equations of high dimension. Using the insights gained from Fraser's method, we develop a technique to produce reduced chemical master equations directly from the eigenvectors of the state-to-state transition rate matrix. The dimension of the original chemical master equation grows quadratically with number of molecules, while the dimension of the reduced one we obtain is linear in the number of molecules. Additionally, a simple, effective way is developed to generate initial conditions for the reduced models.

Nowadays, the kinetics of biochemical reactions occurring in cells has attracted considerable attention. In this context, stochastic models are required to give a proper description due to the small numbers of certain biomolecules in cells. Stochastic models are often studied by simulation of chemical master equations, which is firmly based on the collision theory [1,2], and treats homogenous reactions as discrete Markov jumps in the space of the populations for every species [3,4]. It is hard to directly solve chemical master equations if the number of the molecules involved in the reaction system is large. This is because the dimension of chemical master equation grow as the i'th power of the number of molecules, where i is the number of independent chemical species in the system. Is it possible to considerably reduce the dimension of the original chemical master equation to get a reduced version which can still produce the results we need within an acceptable precision range? We will explore this possibility in this work. A full report appears in Ref. [5].

We study the classical enzyme model system, the irreversible Michaelis-Menten (MM) mechanism as an example:

$$E + S \xrightarrow{k_1} C,$$

$$C \xrightarrow{k_{-1}} E + S, \tag{1}$$

$$C \xrightarrow{k_2} E + Q.$$

Here E is an enzyme molecule, S a substrate molecule, C an enzyme-substrate (ES) complex, and Q the product; k_1, k_{-1} and k_2 are rate constants. The corresponding chemical master equation is [6-8]

CP755, *ISIS: International Symposium on Interdisciplinary Science*
edited by A. Ludu, N.R. Hutchings and D.R. Fry

$$\partial p(s,c)/\partial t = -[k_1 s(e_0 - c) + (k_{-1} + k_2)c]p(s,c)$$
$$+ k_1(s+1)(e_0 - c + 1)p(s+1, c-1)$$
$$+ k_{-1}(c+1)p(s-1, c+1) \tag{2}$$
$$+ k_2(c+1)p(s, c+1),$$

where e, s, and c are the numbers of E, S, and C molecules, e_0 the initial number of E, and $p(s,c)$ the probability that the realizable state (s,c) exists at time t; k_1, k_{-1} and k_2 are now the probability rate constants. Note that there is one such equation for each possible pair of values (s,c) consistent with the mass conservation constraints $0 \le c \le e_0$ and $0 \le s + c \le s_0$, where s_0 is the initial number of molecules of S. This chemical master equation therefore consists of a set of linear ODEs, which can be written as a product of a matrix $\mathbf{R}$ and a vector of state probability variables $\mathbf{p}$,

$$d\mathbf{p}/dt = \mathbf{R}\,\mathbf{p}. \tag{3}$$

Fraser proposed an invariant manifold method to reduce the ODE system [9-12]. If q variables of the ODEs are used to parameterize the other variables, a q-dimensional manifold will be obtained. The evolution of the system on the manifold is only determined by the evolution of the q parameterizing variables. Thus, the ODEs for the evolution of the parameterizing variables provide a q-variable reduced model which exactly describes the evolution after the q-dimensional manifold has been reached. We apply the Fraser method to the chemical master equation (2). In this case, we have found that we can use $q = s_0$, provided we pick the s_0 parameterizing variables properly [9]. For the parameters of Fig. 1, panel d for instance, this reduces the dimension of the system from 18 to 5. The comparison of the mean values from the original master equation and the reduced one in four cases is illustrated in Fig. 1. Obviously, the original system can be well described by the s_0-dimensional reduced model. Additionally, comparing the four cases in Fig.1, we can tell that the more steadily the statistical mean of enzyme-substrate complex decreases, the earlier the reduced model can well describe the original kinetics.

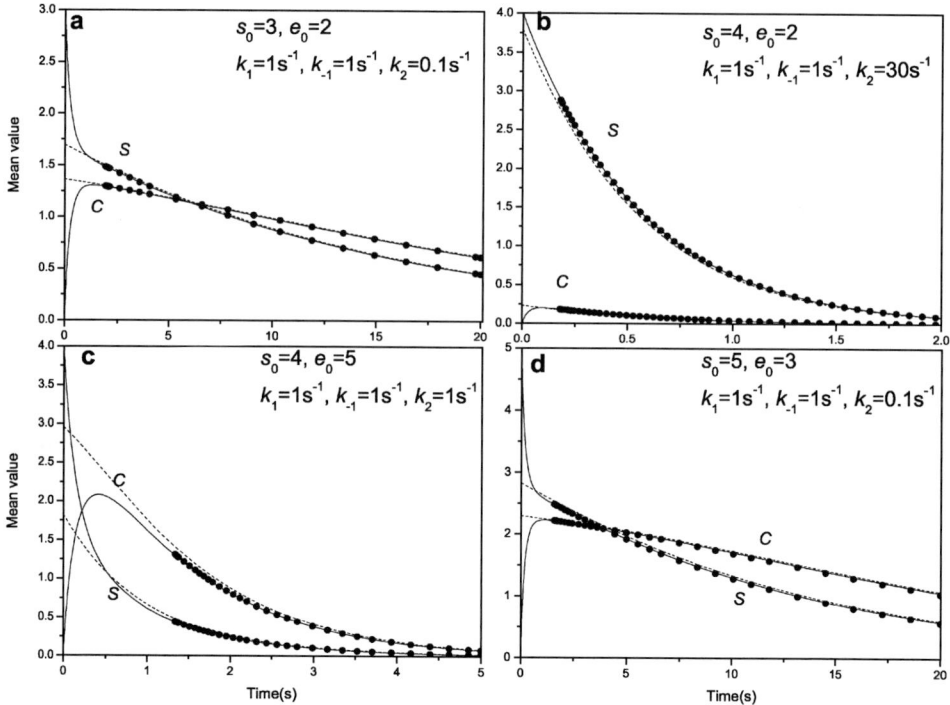

FIGURE 1. Comparison of the statistical means from the original master equation and the reduced one in four cases. Solid lines are the time series of the mean values of S and C based on the probabilities from the original chemical master equation. Symbols are the corresponding results based on the probabilities from the reduced s_0-dimensional models. The reduced-model trajectories (symbols) were initialized from the solutions of the full model at the time when the latter solutions approached the computed manifold. Dashed lines are the results of simulations using the initialization technique for the reduced model described in the text. The rate constants and numbers of molecules are given in each panel.

In order to illustrate how well the reduced models can approximate the original kinetics, we initialized in Fig. 1 the reduced-model trajectories from the solutions of the full model at the time when the latter solutions approached the computed manifold. Integrating the full model of course defeats the purpose of model reduction. So, we need a way to generate reasonable initial conditions for the reduced models. We developed a simple, but surprisingly effective way to generate initial conditions for them based on the physical insight that the first state populated from the typical initial experimental condition $(s,c) = (s_0,0)$ is $(s_0-1,1)$. The probability of the latter state should be high at early times, but cannot be so high as to make $\sum_{(s,c)} p(s,c) > 1$ on the computed manifold. Choosing the greatest $p(s_0-1,1)$ consistent with this constraint and setting all the other parameterizing variables to zero generally yields a very satisfactory initialization of the reduced model, as seen in Fig. 1.

By Fraser's method, we have been able to obtain s_0-dimensional reduced models for master equations of dimension up to 17. When trying the cases with more molecules, we meet convergence problems.

The procedure for generating manifolds can be recast into a more robust form using matrix algebra. This method is briefly stated as follows:

1. Calculate the c leading eigenvalues of the matrix $\mathbf{R}$ appearing in Eq.(3), along with the corresponding eigenvectors. These can be computed efficiently using iterative methods.

2. Define a matrix $\mathbf{N}$ whose columns consist of the eigenvectors arranged in an order of descending real parts of the c eigenvalues. Let $\mathbf{T}_1$ be a $c \times c$ sub-matrix of $\mathbf{N}$ whose rows correspond to c independent variables, and $\mathbf{u}$ be the vector of the c independent variables.

3. Now if $\mathbf{T}_2$ is another matrix obtained by eliminating the rows corresponding to the c independent variables from $\mathbf{N}$, then the remaining variables, collected in a vector $\mathbf{m}$, can be calculated from the independent variable vector $\mathbf{u}$ by

$$\mathbf{m} = \mathbf{T}_2\mathbf{T}_1^{-1}\mathbf{u}. \qquad (4)$$

By this method, we reduce a 395-dimensional master equation ($s_0 = 40$ and $e_0 = 10$) to a 36-dimensional reduced one, while 37-, 38-,…, 40-dimensional reduced models are not successfully obtained as $\mathbf{T}_1$ becomes extremely ill-conditioned.

In this work we develop a few techniques to reduce the master equation of the MM mechanism. By these methods, we can get very good reduced models which can well approximate the original kinetics. The most important thing is that the dimension of the original chemical master equation is substantially reduced. Specifically, the dimension of the original chemical master equation grows quadratically with number of molecules, while the dimension of the reduced one is only linear in the number of molecules. However, there remain several issues. The matrix $\mathbf{T}_1$ which arises in the matrix-oriented method can be ill-conditioned, leading to a difficult inversion process. It may be possible to avoid some of the problems associated with ill-conditioning of this matrix by using a Schur basis, as is done in the Maas-Pope method [13]. Some of the speed of our technique arises from the computation of only the leading eigenvectors of the matrix $\mathbf{R}$ using an appropriate iterative method. Ideally, we would have a corresponding method for the iterative computation of the leading Schur vectors. We plan to investigate this possibility.

Secondly, how large a system can be handled effectively by manifold methods? The reason that there has been relatively little work done directly with the master equation is that, all other things being equal, the number of ODEs in this set grows as the i'th power of the number of molecules, where i is the number of independent chemical species in the system. Thus, the number of states for any realistic system will tend to be very, very large. Since the state-to-state transition rate matrix $\mathbf{R}$ contains terms for each pair of states connected by the reactions, the number of nonzero entries in this matrix scales as a multiple of the number of states. Thus, brute-force approaches to solving the master equation are unlikely to get very far. For this particular problem, we found that we could reduce the number of states to s_0, a quantity which scales linearly with system size, unlike the total number of states. Moreover, by using an iterative method, we can compute only

s_0 of the eigenvectors. Clearly, as the size of **R** grows, the amount of computational power required to compute these eigenvectors will also grow. While we have not carried out a detailed study of the computational complexity of this problem, our calculations thus far are easily feasible on a desktop computer so that we may hope that they will scale well to more complex problems.

ACKNOWLEDGEMENTS

The authors are grateful to Catharine Roussel for valuable assistance. This work was supported by an Alberta Ingenuity Fund Fellowship to R.Z., while M.R.R.'s research is supported by the Natural Sciences and Engineering Research Council of Canada.

REFERENCES

1. D. T. Gillespie, J. Comput. Phys. **22**, 403-434 (1976).
2. D. T. Gillespie, J. Phys. Chem. **81**, 2340-2361 (1977).
3. D. T. Gillespie, Physica A **188**, 404-425 (1992).
4. I. Oppenheim, K. E. Shuler, and G. H. Weiss, *Stochastic Processes in Chemical Physics: The Master Equation.* (MIT, Cambridge, 1977).
5. M. R. Roussel and R. Zhu, J. Chem. Phys. **121**, 8716-8730 (2004).
6. A. F. Bartholomay, Biochemistry **1**, 223-230 (1962).
7. C. J. Jachimowski, D. A. McQuarrie and M. E. Russell, Biochemistry **3**, 1732-1736 (1964).
8. C. C. Heyde and E. Heyde, J. Theor. Biol. **25**, 159-172 (1969).
9. S. J. Fraser, J. Chem. Phys. **88**, 4732-4738 (1988)
10. A. H. Nguyen and S. J. Fraser, J. Chem. Phys. **91**, 186-193 (1989).
11. M. R. Roussel and S. J. Fraser, J. Chem. Phys. **94**, 7106-7113 (1991).
12. M. R. Roussel and S. J. Fraser, Chaos **11**, 196-206 (2001).
13. U. Maas and S. B. Pope, Combust. Flame **88**, 239-264 (1992).

Automated cell tracking tools for quantitative motility studies

Christophe Zimmer*, Bo Zhang*, Samantha Blazquez†, Elisabeth Labruyère†, Freddy Frischknecht**, Robert Ménard**, Nancy Guillén† and Jean-Christophe Olivo-Marin*

*Unité d'Analyse d'Images Quantitative, Institut Pasteur,
25 rue du Docteur Roux, 75015 Paris, France
†Unité de Biologie Cellulaire du Parasitisme, Institut Pasteur,
28 rue du Docteur Roux, 75015 Paris, France
**Unité de Biologie et Génétique du Paludisme, Institut Pasteur,
25 rue du Docteur Roux, 75015 Paris, France

Abstract. Optical microscopy in 2 or 3 dimensions allows extensive observations of the motility and morphology of living cells, in culture or in tissue. This leads to an exploding accumulation of imaging data and shifts the bottleneck from data acquisition to data analysis. Manual image analysis is often either impossible or exceedingly time-consuming and subject to uncontrollable user bias and errors. Computerized methods promise to ensure fast, accurate and reproducible processing, but the basic image analysis functions available in standard commercial software are generally not adapted to the complexity of biological images. For this reason, we develop methods based on active contours, a powerful and flexible technique to segment and track objects, that has become very popular in computer vision research. Here, we describe the main benefits and limitations of active contours for our application, and our efforts to adapt and improve these methods for the analysis of cellular dynamics.

Keywords: cell motility, cell morphology, image processing, segmentation, tracking, active contours
PACS: 07.05.Kf, 07.05.Pj, 42.30.Tz, 42.30.Va, 87.17.Jj, 87.64.Rr, 87.64.Tt, 87.64.Vv, 87.80.Tq

1. CELL MOTILITY, IMAGING, AND QUANTIFICATION

Cell motility and morphology play central roles in a variety of processes of major importance to biological and biomedical research. Examples are the polarization and homing of immune cells in response to antigens, deformation and migration of metastatic tumour cells, the coordinated displacements and organization of cells in developing tissue, or phagocytosis and invasion by parasites. Thanks to standard or new optical microscopy techniques, living cells can be observed routinely and in great detail over prolonged periods of time and in 3 spatial dimensions. To go beyond mere visual inspection, it is increasingly important to perform quantitative analyses of large amounts of imaging data. This is particularly true for the study of cell dynamics: Since cell motion is largely stochastic, determining underlying parameters, such as the persistence time of correlated random walks [1] requires averaging over large ensembles of cells and/or times. Another reason is that the cell cytoskeleton, which governs many aspects of cell morphology and motion, is regulated by a very large number of proteins, thus most alterations have only limited effects and are likely to remain unnoticed in the absence of quantification. Fi-

CP755, *ISIS: International Symposium on Interdisciplinary Science*
edited by A. Ludu, N.R. Hutchings and D.R. Fry
© 2005 American Institute of Physics 0-7354-0240-X/05/$22.50

nally, quantification of experimental data is crucial to test theoretical models of cell locomotion.

Quantification may often be performed manually, but the task is tedious, time-consuming and consequently expensive. In addition, it is prone to an unknown and variable human bias and lacks reproducibility. In contrast, computerized methods are appealing because processors are increasingly fast and cheap, and thus accessible to most biology labs; fully automated methods yield reproducible results, and although they may still produce errors, these are systematic and can potentially be estimated and controlled. However, no existing computer program comes close to reaching the performances of the human vision system for images in general, even for seemingly simple tasks such as identifying and tracking distinct objects. Nevertheless, specialized computerized methods can replace and outperform human vision in specific applications where images have relatively constant properties.

In the past few years, we have developed methods for outlining and tracking cells using the technique of active contours popular in the general computer vision literature [2, 3], and adapted it to the needs of cellular video microscopy data. Here, we present a non-technical overview of some of these methods, focusing on the practical difficulties posed by cellular imaging and on how the standard methods can be improved to overcome them. Technical aspects can be found in the cited literature.

2. CELL SEGMENTATION AND TRACKING METHODS

Quantifying cell morphology and motility generally requires segmenting and tracking individual cells from large image sequences. Segmentation means identifying the boundaries of meaningful objects (here: cells) in an image; tracking means linking these objects across different time frames. In practice, a segmentation and tracking algorithm typically produces a data file that contains a description of the boundary of each cell at each time point, consisting for example of the (x, y) coordinates of a finite list of points along that boundary (Fig. 1, center).

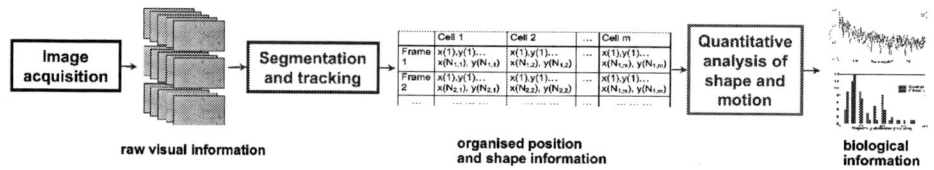

FIGURE 1. Image analysis flowchart: segmentation and tracking transforms the raw images into a file that encodes the boundaries of every cell at every time point, directly allowing computation of shape and motion parameters.

These 'extracted' data can readily be used to quantify parameters characterizing cell shape and motion, generally without the need to go back to the original images. Examples of such quantifications, which depend on the biological question at hand, will be shown in section 3. Most of this paper is concerned with the segmentation and tracking stage, which is very often the bottleneck of the workflow that begins with image acquisition and ends with the analysis of quantitative results.

2.1. Limitations of basic methods

Most commercial microscopy software distributions now contain various image processing tools, including functions for segmentation and tracking. These tend to rely on basic operations, such as thresholding, edge detection or correlation tracking. Unfortunately, these techniques are generally successful only for rather restricted types of images, and fail frequently on typical data from biological research, thus requiring substantial manual intervention to correct the errors. Limitations of common processing methods are illustrated in Fig. 2: simple thresholding fails even on highly contrasted images when objects touch, or if object portions have intensities similar to the background (Fig. 2a); an edge detection filter fails because the image gradients are weak along parts of the cell boundaries, and because gradients at the intercellular interface are not larger than intracellular gradients (Fig. 2b); template matching fails because the cells change their shape while moving (Fig. 2c).

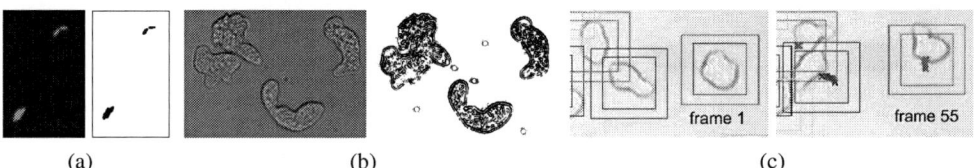

(a)   (b)   (c)

FIGURE 2. Shortcomings of basic image processing functions. (a) Thresholding of an image of fluorescent *Listeria* bacteria: the bacterium on top is incorrectly split into two objects, while the two touching bacteria at the bottom are identified as a single object. (b) An edge detector filter is applied to a phase contrast microscope image of four amoebae. Note that the detected points do not form closed contours, and do not allow separation of touching cells. (c) A template matching algorithm initialized on the first frame of an image sequence is loosing track of the cells on frame 55.

The limitations of such methods, which are based on local pixel properties only, can be overcome with techniques based on global models of the objects to be detected and tracked. One such set of techniques is provided by the framework of active contours.

2.2. Active contours and adaptations to cellular imaging

Since their introduction in [4], active contours have become a hugely popular technique in computer vision research, with a very large number of applications and adaptations [2], including for cellular imaging [5, 6, 7]. Active contours are closed mathematical curves that are used to represent the boundaries of an object in an image. The term 'active' refers to the fact that these curves evolve iteratively under the action of 'forces' computed from the image and from their internal geometrical properties. A segmentation is obtained when the evolving contours reach an equilibrium (see Fig. 3a). Tracking can be achieved by propagating the contours to the following frames and letting them evolve according to the new image content (Fig. 3b). In the following, we discuss the main difficulties and technical choices involved in adapting active contours to the needs of cellular imaging.

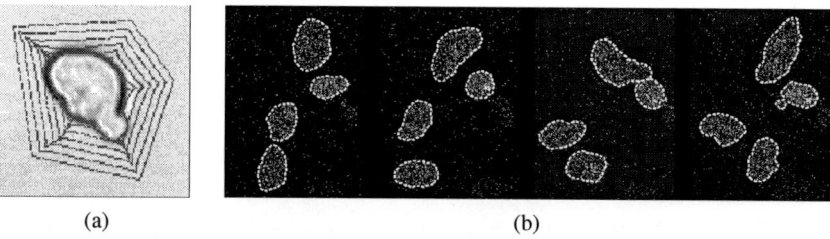

(a)　　　　　　　　　　　　　　　　　(b)

FIGURE 3.　Segmentation by parametric active contours. (a): Superposed on the original phase contrast image, successive iterations of an evolving contour are shown in blue; the final contour is shown in red; the outermost polygon was traced by the user. (b): Active contours tracking moving cells in a time series, observed with fluorescence microscopy. User intervention is required for the first image only.

2.3. Segmenting cells without edges

The forces that govern active contour evolution are often derived from a cost functional, commonly called 'energy'. In contrast to biophysical models, this energy is generally based on image processing techniques or assumptions about the mathematical properties of images rather than physical principles. As a result, a large variety of energies have been proposed.

Following [4], most active contour methods use an energy where the image data appears as an integral of the image gradient along the curve. Since the image gradient is often high at the object boundary, the energy is minimized when the curve lies along it, as desired. However, the 'attraction basin' of object boundaries is small, which means that the initial curve must already be located close to the object for the contour to move in the right direction. In addition, the image gradient is not always largest along the cellular membrane. In phase contrast microscopy, membrane protrusions such as pseudopods often have weaker gradients than structures inside the cell (Fig. 2a, [7]). In fluorescence microscopy, where limited exposure time often leads to low signal-to-noise ratios, cell boundaries may have no significant intensity gradients at all (Fig. 3b, [8]).

In [7], we have described a method to segment pseudopods using an elaborate variation of the gradient-based energy. This modification yields better results on phase contrast microscopy, but involves several parameters that make the generalization to other image types, such as fluorescence microscopy, difficult. Therefore, we have now replaced the gradient-based energy by the region-based energy of the 'active contours without edges' proposed by [9]. This energy takes into account the full information contained in the whole image, not just along the contour. As a result, the segmentation is much less dependent on good placement of the initial contour, and performs well on fluorescence images with low signal-to-noise ratios (Fig. 3b). This property was put to use in analyzing amoebae imaged inside hamster livers using two photon microscopy [10].

2.4. Automatic detection of multiple incoming or dividing cells

Another essential aspect of the technique is the mathematical representation of contours. There are essentially two representations: (i) the earlier parametric contours, where curves are represented explicitly as a finite list of points [4, 2], and (ii) the level set technique -originally developed to simulate complicated moving interfaces such as propagating flames, and later used for imaging- where curves are represented implicitly as zero level sets of a scalar function ϕ defined over the whole image [3].

The choice of curve representation has important consequences for the performance of the segmentation method. The parametric representation allows fast computations, but has severe limitations: the initial placement of each contour must be manually specified (e.g. using the mouse to draw a polygon around each cell as in Fig. 3a). The number of contours is then essentially fixed, because parametric curves cannot easily change their topology (they cannot split or merge) without the help of complicated 'surgical' schemes [11]. This in turn makes it impossible or difficult to automatically handle dividing cells or detect cells that enter the field of view in the course of the sequence. Most importantly, the parametric approach is ill suited to handle 3D data [3].

FIGURE 4. Segmentation by implicit (level set based) active contours, illustrating the ability of contours to split and merge automatically. The image shows 3 *Plasmodium* sporozoites. Leftmost image: initial contours. Rightmost image: Final contours. Other images show intermediate stages of contour evolution.

For these reasons, we also develop methods based on the level set approach, which places no constraints on contour topology. Thanks to the implicit representation, the contours can freely split and merge without any additional schemes. This allows fully automatic initialization (Fig. 4), and detection of dividing, or entering cells (not shown). Also, this approach is much more easily generalized to 3D data (see section 2.6).

An important drawback of the level set approach, however, is the considerably larger computation time. Another limitation, related to the topological flexibility itself, will be discussed in section 2.5.

2.5. Handling touching cells

A recurrent problem in cellular imaging is posed by touching cells. Cells are often easily discriminated from uniform backgrounds based on their fluorescence intensity or granular appearance (texture). However, it is much less obvious to define pixel properties allowing to automatically discriminate touching cells, even when this is done easily by the human eye. In practice, active contours that succeed in segmenting and tracking isolated cells will usually fail as soon as the cells come close or touch. Most parametric active contour models will fail, because the two independently evolving contours will tend to encompass both cells [7, 12]. Most implicit active contour models will also fail, because the topological flexibility will cause previously distinct contours to merge into a single one [8].

To address these problems for parametric active contours, we described a modification where contours are coupled to each other in such a way that overlaps of contours are discouraged [7]. Our evaluations showed that such coupling substantially increases tracking performance in image sequences where cells are in partial contact. Nonetheless, additional work is needed to reduce the number of remaining tracking failures, which becomes unacceptably large when cells are so dense that the visible background is significantly reduced. As a new step in this direction, we have recently devised a coupling scheme for parametric active contours, based on introducing a penalty against contour overlaps into a global energy functional ([12], submitted). This improvement will allow better control of the boundaries of touching cells and provide a better basis for other extensions of the model. For implicit active contours, we have proposed a scheme that uses several multiple level set functions ϕ_i to represent different cells, and, again, couples these functions through an overlap penalty [8]. These multiple coupled level sets can track cells that transiently touch each other in cases where standard level set methods would fail.

2.6. Going to 3D

Cellular motility and morphology are strongly influenced by the embedding medium and are now increasingly studied inside real or synthetic 3D tissue rather than on 2D substrates. This creates a need for segmentation in 3D and tracking in 3D+time. The technique of choice to segment 3D data is level sets [3] (see above). However, the large computation time of implicit methods rapidly becomes prohibitive for 3D or 3D+time data. We have therefore adopted a fast optimization scheme recently described in [13]. Fig. 5 shows a segmentation of an amoeba in 3D using this technique.

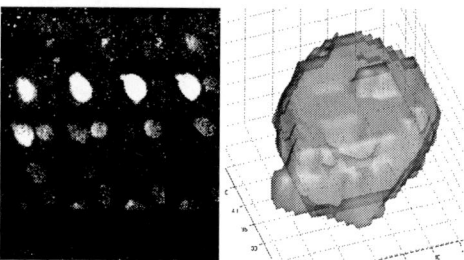

FIGURE 5. Segmentation of a cell in 3D. Left: montage of the 20 slices composing a z-stack taken from two photon microscopy observations of amoebae inside explanted hamster liver (fragment). Right: volume rendering of a segmented cell. The computed cell volume is 15000 μm^3.

We recently improved and extended this technique for segmenting and tracking multiple cells in 3D+time microscopy method using coupled active surfaces ([14], submitted).

3. APPLICATIONS

Successful cell segmentation and tracking tools allow to perform a variety of quantitative analyses of motility and morphology, examples of which are illustrated in Figs. 6 and 7. See also [15, 10, 16, 17] for published work that includes quantitative analyses obtained with our automated methods.

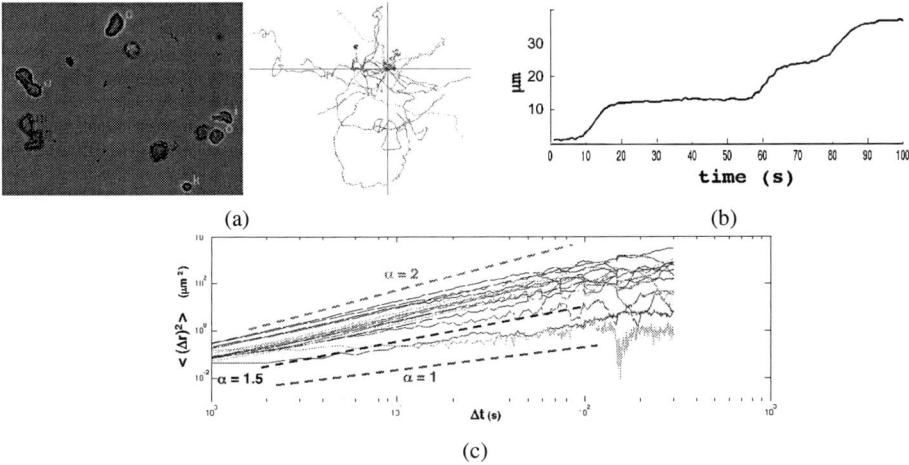

(a)   (b)

(c)

FIGURE 6. Quantitative analyses of cell motion. (a) Tracks (right) of individual amoebae (left, fragment), aligned so that they share a common starting point for better visualization of migration direction. Polar histograms (not shown) reveal no preferential direction. (b) Curvilinear abscissa of a *Plasmodium* sporozoite moving along the salivary duct of an *Anopheles* mosquito, indicating intermittent motion; see [16]. (c) Mean-square displacements from individual amoeba tracks plotted as function of time interval (log-log scale). The slope α of these curves is comprised between 1 and 2, consistent with super-diffusive motility (dashed lines are power laws $<(\Delta r)^2>\propto (\Delta t)^{\alpha}$); see also [10].

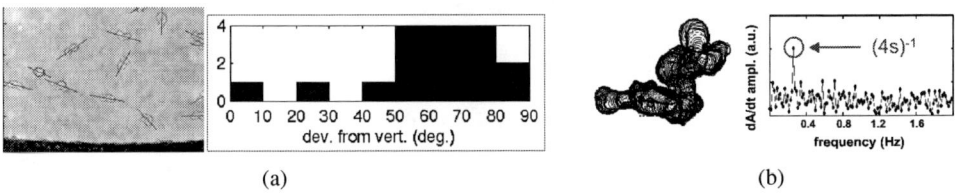

(a)   (b)

FIGURE 7. Quantitative analyses of cell shape. (a) Elongation directions (left, red bars) of amoebae in a chemotactic gradient. Histogram of angles (right) suggests preferential elongations perpendicular to the gradient. (b) Frequency spectrum (right) of the apparent area of a crawling amoeba (superposed outlines shown on left), suggesting the existence of a 4s periodicity.

4. CONCLUSION

We have described some of our work on developing cell segmentation and tracking techniques based on the framework of active contours. We have emphasized chief difficulties of cellular imaging, such as cells 'without edges', touching cells, dividing or incoming cells, and described how active contours can be adapted to address them. Our methods can be used for various biological projects, as illustrated. Nevertheless, substantial work remains necessary to improve and extend these methods to more complex images and ensure faster and more reliable processing. An important goal is to develop methods without arbitrary parameters, or methods that learn these parameters from examples. Another one is to automatically estimate the errors, in order to reduce the need for visual quality check. These efforts will likely require probabilistic frameworks that incorporate more precise knowledge about the object properties and the optics of image formation.

We will report on such improvements in future papers.

ACKNOWLEDGMENTS

We thank Stéphanie Seveau of Institut Pasteur for the *Listeria* image of Fig. 2 and François Amblard of Institut Curie for initiating the analysis shown in Fig. 6c. The work reported here was funded by Institut Pasteur, partly through a 'Programme Transversal de Recherches' and the Anopheles 'Grand Programme Horizontal'. F.F. was supported by a Human Frontier Science Program postdoctoral fellowship.

REFERENCES

1.   R. B. Dickinson, and R. T. Tranquillo, *AIChE J.*, **39**, 1995–2010 (1993).
2.   A. Blake, and M. Isard, *Active contours*, Springer, 1998.
3.   J. Sethian, *Level set methods and fast marching methods*, Cambridge University Press, New York, 1999.
4.   M. Kass, A. Witkin, and D. Terzopoulos, *International Journal of Computer Vision*, **1**, 321–331 (1988).
5.   F. Leymarie, and M. D. Levine, *IEEE Trans. Pattern Anal. Machine Intell.*, **15**, 617–634 (1993).
6.   N. Ray, S. Acton, and K. Ley, *IEEE Trans. Med. Imaging*, **21**, 1222–1235 (2002).
7.   C. Zimmer, E. Labruyère, V. Meas-Yedid, N. Guillén, and J.-C. Olivo-Marin, *IEEE Trans. Med. Imaging*, **21**, 1212–1221 (2002).
8.   B. Zhang, C. Zimmer, and J.-C. Olivo-Marin, *Proceedings of IEEE International Symposium on Biomedical Imaging (ISBI 2004)*, pp. 476–479 (2004).
9.   T. Chan, and L. Vese, *IEEE Trans. Image Processing*, **10**, 266–277 (2001).
10. E. Coudrier, F. Amblard, C. Zimmer, P. Roux, J.-C. Olivo-Marin, M.-C. Rigothier, and N. Guillén, *Cellular Microbiology*, **7**, 19–27 (2005).
11. C. Zimmer, V. Meas-Yedid, E. Glory, E. Labruyère, N. Guillén, and J.-C. Olivo-Marin, "Active contours applied to the shape and motion analysis of amoeba," in *Proc. of SPIE's 46th annual meeting, Vision Geometry X*, 2001, vol. 4476, pp. 124–134.
12. C. Zimmer, and J.-C. Olivo-Marin, *IEEE Trans. Pattern Anal.* (2005), submitted.
13. B. Song, *Topics in Variational PDE Image Segmentation, Inpainting and Denoising*, Ph.D. thesis, University of California, Los Angeles (2003).
14. A. Dufour, V. Shinin, S. Tajbakhsh, N. Guillén, J.-C. Olivo-Marin, and C. Zimmer, *IEEE Trans. Image Processing* (2005), submitted, Dec. 2004.
15. E. Labruyère, C. Zimmer, V. Galy, J.-C. Olivo-Marin, and N. Guillén, *Journal of Cell Science*, **116**, 61–71 (2003).
16. F. Frischknecht, P. Baldacci, B. Martin, C. Zimmer, S. Thiberge, J.-C. Olivo-Marin, S. L. Shorte, and R. Menard, *Cell Microbiol.*, **6**, 687–94 (2004).
17. V. Das, B. Nal, A. Roumier, V. Meas-Yedid, C. Zimmer, J.-C. Olivo-Marin, P. Roux, P. Ferrier, A. Dautry-Varsat, and A. Alcover, *Immunological Reviews*, **189**, 123–135 (2002).

Differential Tethering of Log Phase *Trypanosoma brucei* onto Chemically Distinct Surfaces

Lydia Archuleta, Ashley Dunham, Justin Rains and Darrell Fry

Northwestern State University, Natchitoches LA

Our long-term objective is to understand and model the motility of *T. brucei*. Obtaining high quality images of T. brucei that allow one to differentiate between cell body movement and flagallar movement is difficult with *T. brucei* because the flagellum is attached along the cell body. Currently, our approach is to tether T. brucei onto a microscope friendly surface. The contributions to the ISIS proceedings summarize our progress to date. Specifically, we look at the adhesion density of *T. brucei* to numerous microscope friendly surfaces and at the optimum adhesion conditions for *T. brucei*.

MODELING MOTILE CELLS REQUIRES IMAGING MOTILE CELLS

Generally speaking, two approaches are being used to model the motility of flagellated organisms. The first modeling approach begins with the fundamental components of a flagellum. The much cited, geometric clutch model proposed by Lindemann began with the fundamental components of "9+2" axoneme and was used to describe the beating of cilia and flagella [1-3]. Moreover, the model has been modified to describe multiple aspects of boar sperm motility [4]. Other researchers are taking using fundamental components to build models that describe the motility of organisms. In contrast, the second modeling approach uses both static and dynamic images of the flagellum as the initial parameters. For instance, Gueron and Liron used images of the beat pattern of a cilium of a *Paramecium* to calculate the parameters of an internal engine that depends only upon the geometry of the cilium [5]. Obviously the two approaches are complimentary and not mutually exclusive. Ultimately, any model that describes the motility of an organism must be compared with the actual motility of the organism. Most often this comparison is accomplished by imaging the organism as it moves. For these reasons, imaging motile cells is important.

AN ORGANISM OF INTEREST: *TRYPANOSOMA BRUCEI*

In our case, we are interested in the unicellular flagellated parasite, *Trypanosoma brucei*. *T. brucei* is the causative agent for African Sleeping Sickness, which threatens

CP755, *ISIS: International Symposium on Interdisciplinary Science*
edited by A. Ludu, N.R. Hutchings and D.R. Fry

approximately 60 million individuals in sub Saharan Africa.[6] Although the flagellum is a highly conserved structure, the attachment of the flagellum to the cell body of *T. brucei* is significantly different than other commonly modeled flagellated organisms. Unlike other flagellated cells, the flagellum of T. brucei is attached along the length of the cell body.[7] Furthermore, the flagellum of *T. brucei* is attached in a left-handed helix as it extends from the posterior to the anterior of the cell body.[7] Parallel to the axoneme and the cell body is a highly organized lattice-like structure called the paraflagallar rod. The axoneme is attached to the paraflagallar rod through the outer doublets four through seven. In turn, the paraflagallar rod and axomeme structure is attached to the cell body through the flagallar adhesion zone.

The unique architecture of *T. brucei* has several consequences. Since the flagellum is attached in a helical pattern to the cell body, the flagellum naturally moves in a three dimensional pattern. [7, 8] Furthermore, *T. brucei* uses it's flagellum to pull itself through a medium; whereas other commonly modeled flagellated organisms use their flagellum to push themselves through a medium.[8] Finally, distinguishing between movement and cell body movement is more difficult with *T. brucei* since the flagellum is attached along the length of the cell body.

The unique architecture of *T. brucei* requires novel approaches to understanding and modeling its motility, which in turn requires new imaging strategies. Currently, our strategy is to tether *T. brucei* onto microscope friendly surfaces. In these conference proceedings, we communicate the optimum conditions for immobilizing log phase *T. brucei* onto several different chemically distinct chemical surfaces. Furthermore, we compare the adhesion density of *T. brucei* on several chemically distinct chemical surfaces.

METHODOLOGY

All chemicals were purchased from Sigma-Aldrich unless otherwise stated. Two of the four surfaces are commercially available. (Sigma-Aldrich sells the Silane Prep microscope slides while Electron Microscopy Sciences sells the poly-L-lysine microscope slides.) The other two surfaces, Sigma-cote and acid washed, are simple to prepare. For the Sigma-cote, a 15-minute incubation in fresh Sigma-Cote, followed by a thorough rinsing, produces slides with consistent hydrophobicity (as measured by contact angle meter). The acid washed slides are prepared by incubating slides in chromic acid for 24 hours followed by ample rinsing.

Briefly, tethering is accomplished by first determining the concentration (and hence growth phase) of *T. brucei* using a hemocytometer. The appropriate volume of log phase cells is centrifuged to a pellet, rinsed with PBS, and re-suspended in PBS so that the concentration of cells is $1*10^7$ cells/mL. Fifteen microliters of the $1*10^7$ cells/mL was applied to the various microscope slides.

An Olympus BX50 fluorescent, DIC-equipped microscope equipped with SPOT software was used to image the tethered cells. Cell counts were obtained at a 400X magnification; the approximate imaging size was about 300 microns by 300 microns. Live and dead cells were both counted.

RESULTS

Characterization of Microscope Surfaces

The four microscope surfaces were characterized using a contact angle meter. Table 1 details the results. Briefly, chromic acid washed slides were used as a control because of the ease of their preparation. As one would expect, water placed onto a chromic acid washed slide did not bead up. Poly-L-Lysine slides are often used to mimic cell / protein interactions. Hypothetically the surface contains charged $-NH_3^+$ moieties along with methylene (CH_2) units. Silane prep slides are more hydrobyobic, but still contain the ionizable $-NH_2$ group. Finally, Sigma-cote treated slides are very non-polar, as indicated by the contact angle measurment. Sigma-cote is a mixture of short chain chlorosilanealkanes and is often used to render glass surfaces hydrophobic.

TABLE 1) Contact Angle, which serves as a measure of hydrophobicity, for 4 different chemical surfaces.

Chemical Treatment	Contact Angle
Chromic Acid Washed	$<5^0$
Poly-L-Lysine	48^0
Silane Prep	68^0
Sigma-Cote ™	84^0

Optimum Incubation and Temperature Conditions for Tethering

Figure 1 shows the results of incubation time and temperature on tethered cell density for Poly-L-lysine slides. Incubation time refers to how long the cells were allowed to sit on the microscope slide prior to viewing. Temperature does not play a significant role in the tethering density. However, it was observed that at elevated temperatures ($26.2^{\circ}C$) be tethered cells were most likely to be dead. However, the living cells were more motile than at lower temperatures. Incubation time does play a role in how many cells are tethered to a surface. For simplicity, we choose ten minutes as the immobilization time. Similar results were obtained for chromic acid slides, Sigma-cote slides, and silane prep slides (data not shown).

Tethering Density on Different Surfaces

The different surfaces tethered the cells with a different density and observed characteristics. The chromic acid washed slides adhered the most cells, with 26 cells per field. In general, the cells tethered on the chromic acid had a variety of tethering orientations (flagellum tethered, cell body tethered, tip of flagellum and posterior end of cell body tethered, etc…). The silane slides also demonstrated no preference for tethering orientation, and 23 cells per field were immobilized. The Sigma coated slides and the poly-L-lysine adhered approximately 15 cells per field. Interestingly, the cells tethered to the Sigma-coated slides were able to crawl along the surface.

In future work, we plan to quantify the observed orientations of cells and compare log and stationary phase cells tethering density.

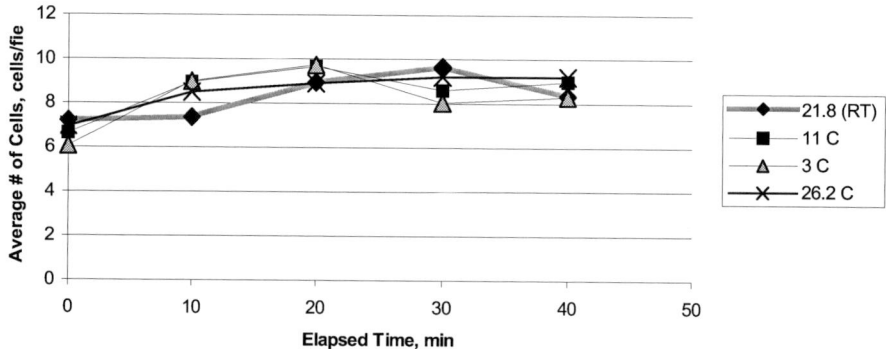

Figure 1: Number of cells tethered onto Poly-L-Lysine slides at four temperatures and with varying incubation times. The number of cells immobilized does not vary greatly with temperature; however, above room temperature, the cells were more likely to be destroyed although the living cells did exhibit more motility. Increasing incubation time does increase the number of cells immobilized.

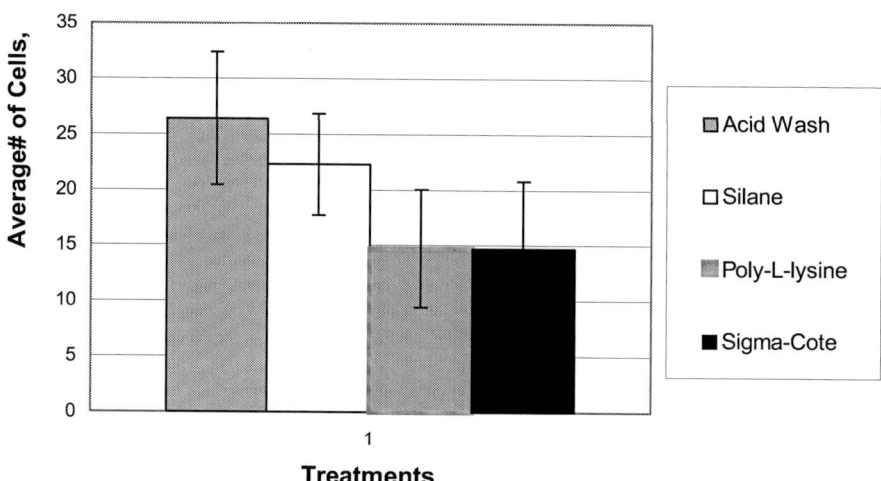

Figure 2: Number of cells tethered onto four different chemical surfaces. Various slide treatments yield a different number of tethered cells. Chromic acid washed slides (acid washed) adhere the most cells per field then silane treated slides. Finally, poly-l-lysine and sigma-cote slides adhere about the same number of cells.

ACKNOWLEDGMENTS

The authors acknowledge Northwestern State University JOVE, IDEAS and CURIA programs for supply and scholarship monies. The authors would also like to thank Ms. Anna Westergard, Dr. Nathan Hutchings, and Dr. Andy Ludu for their encouragement and assistance.

REFERENCES

1. Lindemann CB. A geometric Clutch hypothesis to explain oscillations of the axoneme of cilia and flagella. J Theor Biol 168:275-189, 1994.
2. Lindemann CB. A model of ciliary functioning which uses the forces transverse to the axoneme as the regulator of dyenine activation. Cell Motil Cytoskeleton 29:141-154, 1994.
3. Lindemann CB, Kanous KS. 1995. "Geometric Clutch" hypothesis of axonemal function: Hey issues and testable Predictions. Cell Motil Cytoskeletron 31:1-8, 1995.
4. Schmitz KS, Holcomb-Wygle DL, Oberki DJ, Lindemann CB. Measurement of the force producted by and intact bull sperm flagellum in isometric arrest and estimation of the dynein stall force. Biophysics Journal 79:468-478, 2000.
5. Gueron, S. and Liron, N. Simulations of three-dimensional ciliary beats and cilia interactions. Biophys. J. 65, 499-507. 1993.
6. World Health Organization, "Tropical disease, including Pan African tsetse and trypanosomiasis eradication campaign" Report by the Seretriat. April 14, 2003. Available at http://www.who.int/gb/ebwha/pdf_files/WHA56/ea569.pdf.
7. Hill, Kent, Biology and Mechanism of Trypansome Cell Motility. Eukaryotic Cell, Vol. 2, No. 2 p[ages 200-208, April 2004.
8. Fry, D.; Hutchings, N.; Ludu, A. Dynamics of Immobilized Flagella. Los Alamos National Archives, Identifier oai:arXiv.org:physics/0309026 (20040731). September 2003.

Microbial Biofilms: Persisters, Tolerance and Dosing

N. G. Cogan

Mathematics Department Tulane University New Orleans, LA. 70118 e-mail: cogan@math.tulane.edu FAX: 504 865 5063

Abstract.
   Almost all moist surfaces are colonized by microbial biofilms. Biofilms are implicated in cross-contamination of food products, biofouling, medical implants and various human infections such as dental cavities, ulcerative colitis and chronic respiratory infections. Much of current research is focused on the recalcitrance of biofilms to typical antibiotic and antimicrobial treatments. Although the polymer component of biofilms impedes the penetration of antimicrobials through reaction-diffusion limitation, this does not explain the observed tolerance, it merely delays the action of the agent. Heterogeneities in growth-rate also slow the eradication of the bacteria since most antimicrobials are far less effective for non-growing, or slowly growing bacteria. This also does not fully describe biofilm tolerance, since heterogeneities arr primairly a result of nutrient consumption.
   In this investigation, we describe the formation of 'persister' cells which neither grow nor die in the presence of antibiotics. We propose that the cells are of a different phenotype than typical bacterial cells and the expression of the phenotype is regulated by the growth rate and the antibiotic concentration. We describe several experiments which describe the dynamics of persister cells and which motivate a dosing protocol that calls for periodic dosing of the population. We then introduce a mathematical model, which describes the effect of such a dosing regiment and indicates that the relative dose/withdrawal times are important in determining the effectiveness of such a treatment. A reduced model is introduced and the similar behavior is demonstrated analytically.

Keywords: persister, tolerance, biofilm
PACS: 87.17.Ee 87.17.Aa

INTRODUCTION

It has been estimated that 99% of all bacteria live in structured communities termed biofilms [1]. Recently the US National Institutes of Health announced that, "Biofilms are medically important, accounting for over 80% of microbial infections in the body". These two observations, in conjunction with the observation from numerous sources [2, 3, 4, 5, 6, 7] that typical antimicrobial treatments fail to eradicate the bacteria, leads one to the conclusion that understanding resistance mechanisms for bacterial biofilms is of paramount importance in treating bacterial infections. There are several hypotheses concerning resistance mechanisms which can be placed into three broad categories: transport limitation, physiological tolerance and phenotypic resistance.

   There are many mechanisms by which a biofilm life-style can confer resistance to bacteria. Some of these mechanisms include delayed penetration, variation of growth rates due to heterogeneous nutrient distribution and the expression of extremely tolerant phenotypes termed persisters. The focus of this paper is on phenotypic resistance. The physiology of persister cells is not currently well understood, although their existence

CP755, *ISIS: International Symposium on Interdisciplinary Science*
edited by A. Ludu, N.R. Hutchings and D.R. Fry
© 2005 American Institute of Physics 0-7354-0240-X/05/$22.50

has been demonstrated [8, 9, 4]. As yet there is no consensus as to what phenotypic variations a primarily responsible for biofilm tolerance [4].

Here we incorporate current biological observations into a mathematical model of bacterial tolerance with the aim of analyzing the dynamics of a population of bacteria which contains both persister cells and susceptible cells when exposed to periodic doses of an antimicrobial.

Bacterial tolerance to antibiotics has been well established although the specific mechanisms are still being investigated. In two papers [8, 9], the susceptibility of different species of bacteria at different stages of growth is investigated. It is shown, experimentally, that the expression of the persister phenotype is linked both to the application of the antimicrobial and the growth rate of the bacteria. Thus, the dynamics of the population of persister cells is intrinsically linked to the dynamics of the nutrient.

In the following sections we describe our mathematical model which incorporates the experimental results. The model consists of three ordinary differential equations (ODE) which govern the dynamics of the susceptible and persistent bacteria and one growth limiting substrate. In the absence of antibiotic, the susceptible bacteria consume substrate and reproduce. When antibiotic is added, a fraction of the the susceptible cells are killed while another fraction convert to persister cells. Persister cells are not effected by the antibiotic, nor do they grow. Instead, if there is no antibiotic, persister cells revert to susceptible cells at a fixed rate.

MODEL ASSUMPTIONS AND DESCRIPTION

We denote the two phenotypes as B_s and B_p for susceptible and persister density, respectively. We assume that there is one growth-limiting substrate, S and one antibiotic, denoted A. The population of susceptible bacteria changes due to growth, death due to antibiotic action, loss due to transition to persister cells and gain as the persistent cells revert back to susceptible cells. We assume that there is one growth limiting substrate, oxygen, and that the growth is described by Monod kinetics with maximum specific growth rate, Monod coefficient and yield denoted μ_{max} and K_s and Y, respectively. Thus the growth term is

$$g(B_s, S) = \frac{\mu_{max}}{Y} \frac{S}{K_s + S} B_s. \tag{1}$$

Disinfection of the susceptible population depends on the type of antibiotic used. If the antibiotic is a beta-lactam then the disinfection rate is assumed to be proportional to the growth rate. If the antibiotic is a fluoroquinolone we allow for disinfection in the absence of growth, although at a reduced rate. We assume that the disinfection term is

$$d(B_s, S, A) = k_d(A, t) \mu_{max} \frac{S + \alpha}{K_s + S} B_s, \tag{2}$$

where α is zero for beta-lactam and non-zero for fluoroquinolone. The function $k_d(A, t)$ depends on the antibiotic concentration. In particular, $k_d = 0$ if $A = 0$ and is nonzero otherwise. Because we are interested in dosing strategies which are time dependent the disinfection rate is necessarily time dependent.

The loss of susceptible cells to the persister population is assumed to occur at a rate proportional to the growth rate and depends on the antibiotic concentration. Thus

$$l(B_s, S, A) = k_l(A, t)\mu_{max}\frac{S}{K_s + S}B_s.$$ (3)

Again, the transition from susceptible to persister is assumed to be caused by exposure to the antibiotic so this rate is also time dependent.

Persister cells can revert to susceptible cells as long as the applied antibiotic concentration is zero. Otherwise, there is no reversion back to susceptible cell type. Mathematically we have

$$r(B_p, S, A) = k_g(A, t)B_s,$$ (4)

where k_g is zero if there is antibiotic present and non-zero otherwise.

Putting these together gives the equation governing the dynamics of the susceptible population as

$$\frac{dB_s}{dt} = \frac{\mu_{max}}{Y}\frac{S}{K_s + S}B_s - k_d(A, t)\mu_{max}\frac{S + \alpha}{K_s + S}B_s$$
$$-k_l(A, t)\mu_{max}\frac{S}{K_s + S}B_s + k_g(A, t)B_p.$$ (5)

The persister cells are not killed by the antibiotic. The population changes by conversion to and from susceptible cells. The governing equation is,

$$\frac{dB_p}{dt} = k_l(A, t)\mu_{max}\frac{S}{K_s + S}B_s - k_g(A, t)B_p.$$ (6)

Here we have assumed that the persister population does not grow; either growth is being inhibited by the antibiotic or persister cells revert to susceptible cells.

The substrate is being consumed by the susceptible population,

$$\frac{dS}{dt} = -\mu_{max}\frac{S}{K_s + S}B_s.$$ (7)

Equations (5) - (7) describe the dynamics for suspended populations of susceptible and persisting bacteria and substrate. In the next section, we describe simulations which yield results comparable to experimental results shown in [9] and [8]. Then we show results from a simulated dosing protocol entailing application of a constant concentration of antibiotic for a specified length of time, withdrawing the antibiotic and allowing the population to regrow. We see that results depend dramatically on the length of dose/withdrawal times. In particular, we will show that for short withdrawal times, we generate a persister population which is eliminated extremely slowly. If the treatment is terminated before the persister population is cleared, there is rapid regrowth of the bacterial population. If the withdrawal time is very long, the susceptible population is only transiently eliminated. The persister cells, generated from the dosing step are a source for the susceptible population, which will regrow given enough time. There is a dosing regiment for which neither of these cases occurs; instead, both bacterial populations are eliminated.

TABLE 1. Parameters used in the simulations.

Parameter	Symbol	Units	Value	Source
Maximum Specific Growth Rate	μ_s	h^{-1}	0.417	[10]
Yield Coefficient	Y		0.2	[10]
Monod Coefficient	K_s	$mg\ l^{-1}$	0.2	[10]
Maximum Disinfection Rate	k_d	h^{-1}	40	Estimated
Non-growing Disinfection	α	$mg\ l^{-1}$	0.07	Estimated
Rate of Loss	k_l	h^{-1}	0.001	Estimated
Rate of Gain	k_g	h^{-1}	0.05	Estimated

Simulated Dosing Experiments

In this section we describe results from a simulated dosing experiment. In [4], Lewis describes a possible treatment to eradicate all persisters. The treatment requires applying an antibiotic for a period of time, killing all susceptible bacteria while generating small population of persister cells. Withdrawing the antibiotic allows the persister cells to grow. As the persisters grow they lose the persister phenotype, reverting back to susceptible cells. At this point a second application will remove virtually all of the bacteria.

In this section, we describe simulations to test this hypothesis. We begin our simulations with a population of susceptible cells. The cells are exposed to nutrient, initially twice the half-saturation value, and an antibiotic for a fixed length of time, termed dose period and denoted T_d. This kills the susceptible cells quickly and generates a small population of persister cells. We then remove the antibiotic and allow the cells to grow for a fixed length of time, termed withdrawal period and denoted T_w. Persister cells revert to susceptible cells, which then consume nutrient and grow. This completes one dose/withdrawal period. After one dose/withdrawal period, fresh nutrient is added and the cycle is repeated. Results from the experiment, with a dose period of 10 hours and varying withdrawal period are shown in Figures (1)- (3). If the bacterial population is not eliminated within approximately 30 days, the treatment is not effective and the simulation is stopped.

Here we see a very interesting result. For very short withdrawal periods, the treatment fails to eliminate the persister population within the alloted time period (see Figure (1)). If the treatment is discontinued, the population will quickly regrow. For extremely large withdrawal periods, the susceptible population is not eliminated. Instead, it responds to the treatment quickly, but the persister population is a source of susceptible cells, which regrow to the carrying capacity before the next dose period. Again this results in an ineffective treatment (see Figures (3)). However, there is an intermediate dose/withdrawal pair, for which the treatment is effective in clearing both the susceptible and persister population. This indicates that periodic dosing may be effective, but the dosing regiment must be bacteria specific. Specifically, the withdrawal period must be long enough to allow the persister cells to revert to susceptible cells, but not long enough for the susceptible cells to reach the exponential growth stage.

The effectiveness of the dosing strategy depends on the number of susceptible and

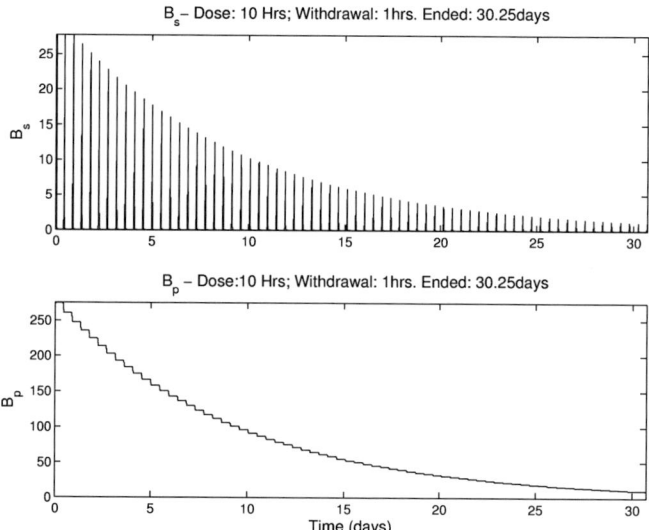

FIGURE 1. Periodic dosing experiment with withdrawal period of one hour. We see that, while the susceptible cells are killed, the persister population is being killed very slowly. This treatment was not successful.

persister cells that are not killed. Clearly this depends on the dosing strategy. To find the optimal withdrawal length, we measure the change in the population as a function of the this length. We first define the envelope of the survival curves as the maximum, during each cycle, of the respective populations. An example of the survival curve and the envelope is given in Figure (4). The slope of the envelope on a logarithmic scale gives the exponential rate of decrease or increase of the population. For successful treatments (i.e. those for which the each of the populations tends to zero), the maximum rate is negative. In Figure (5), we show the maximum slope of the logarithm of the envelope as a function of the withdrawal time. Minimizing this curve gives the optimal strategy which for the parameters given in Table I and $Td = 10$ hours, is to withdraw the antibiotic for approximately 7.5 hours.

DISCUSSION

We have presented a mathematical model of bacterial tolerance based experimentally observed 'persister cells', which are assumed to be a dormant phenotype that is expressed at a rate that depends on the population growth rate and the antibiotic concentration. Parameters of the model have been chosen to yield results comparable to experiments. The model is used to show that alternating dose/withdrawal of the antibiotic can eliminate the bacteria. Moreover, the optimal dose/withdrawal times can be computed.

Because we assume that the rate of persister formation is proportional to the growth rate, we find that persister cells are formed throughout the growth-cycle. This is not reflected in experimental results (Dr. Kim Lewis personal communication). Instead, it seems that there is no persister formation until the mid to late exponential phase. This may indicate that the expression of the persister phenotype is regulated by an auto-inductive signal; however, in the absence of any direct physiological evidence, we feel that our results are indicative of persister dynamics.

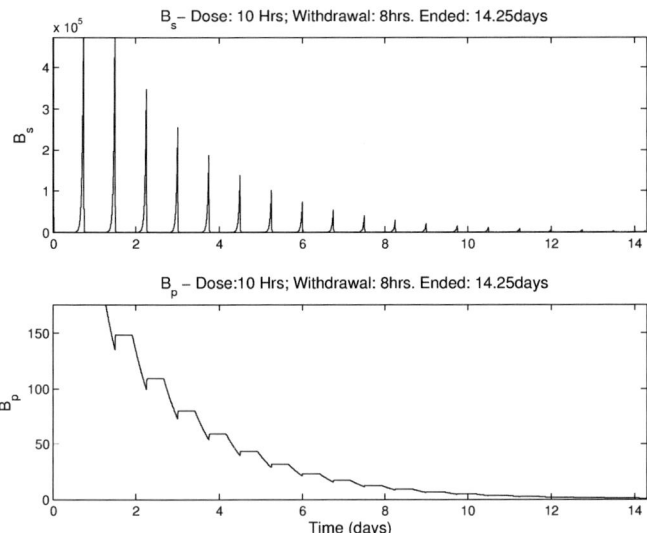

FIGURE 2. Periodic dosing experiment with withdrawal period of eight hours. We see that both the susceptible and persister cells are killed indicating a successful treatment.

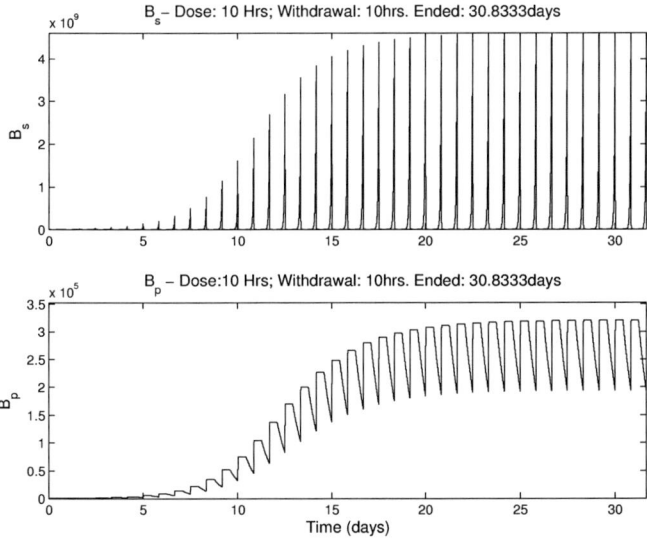

FIGURE 3. Periodic dosing experiment with withdrawal period of ten hours. We see that neither the susceptible nor the persister cells are eliminated. This treatment was not successful.

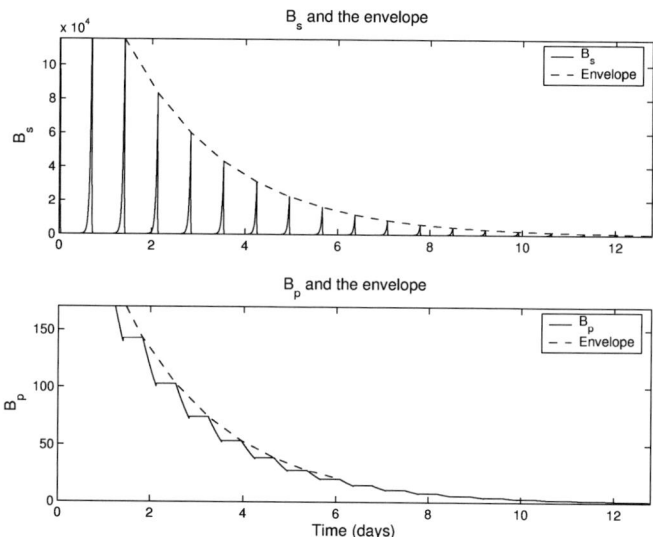

FIGURE 4. Survival curves for susceptible and persister cells along with the envelopes of the curves. The envelopes are given by the maximum of the populations during the dosing cycle. This gives an overall view of the effect of the dosing strategy.

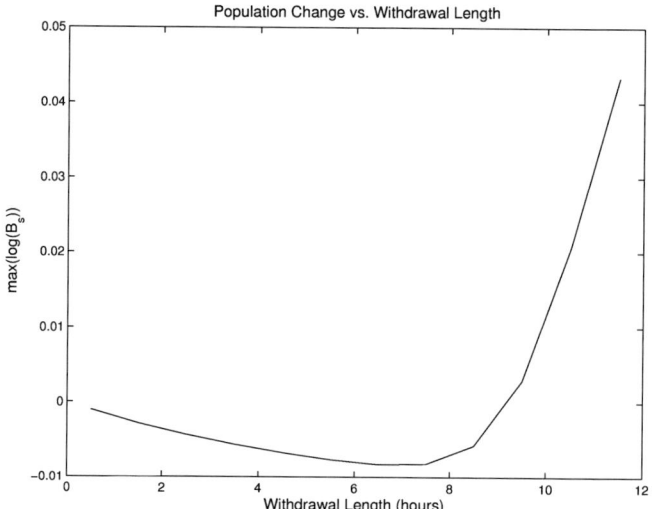

FIGURE 5. Maximum rate of change for the population of cells as a function of the withdrawal length. The maximum rate of change of envelope of the population is calculated for various withdrawal lengths. We see that for withdrawal time shorter than approximately 9 hours, the treatment is effective. For withdrawal periods longer than this, the overall population actually increases. The minimum if this curve gives the optimal withdrawal time of approximately 7.5 hours for the parameters used.

REFERENCES

1. M. Ashby, J. Neale, and I. Critchley, *Journal of Antimicrobial Chemotherapy*, **33**, 443–452 (1994).
2. J. Costerton, P. S. Stewart, and E. P. Greenberg, *Science*, **284**, 1318–1322 (1999).
3. D. Davies, *Nature Reviews Drug Discovery*, **2**, 114–122 (2003).
4. K. Lewis, *Antimicrobial Agents and Chemotherapy*, **45**, 999–1007 (2001).
5. D. Allison, and P. Gilbert, *Journal of Industrial Microbiology*, **15**, 311–317 (1995).
6. J. G. Elkins, D. J. Hassett, P. S. Stewart, H. P. Schweizer, and T. R. McDermott, *Applied and Environmental Microbiology*, **65**, 4594–4600 (1999).
7. H. M. Lappin-Scott, and J. W. Costerton, editors, *Microbial Biofilms*, Cambridge University Press, Cambridge, 1995, chap. Mechanisms of the Protection of Bacterial Biofilms from Antimicrobial Agents, pp. 118–130.
8. M. Desai, T. Buhler, P. Weller, and M. Brown, *Journal of Antimicrobial Chemotherapy*, **42**, 153–160 (1998).
9. I. Keren, N. Kaldalu, A. Spoering, Y. Wang, and K. Lewis, *FEMS Microbiology Letters*, **230**, 13–18 (2004).
10. M. E. Roberts, and P. S. Stewart, *Antimicrobial Agents and Chemotherapy*, **48**, 48–52 (2004).

Computational Model of Population Dynamics Based on the Cell Cycle and Local Interactions

Sorinel Adrian Oprisan[1], Ana Oprisan[2]

[1]Department of Psychology, University of New Orleans
[2]Department of Physics, University of New Orleans
2000 Lakeshore Dr., New Orleans, LA 70148

Abstract. Our study bridges cellular (mesoscopic) level interactions and global population (macroscopic) dynamics of carcinoma. The morphological differences and transitions between well and smooth defined benign tumors and tentacular malignat tumors suggest a theoretical analysis of tumor invasion based on the development of mathematical models exhibiting bifurcations of spatial patterns in the density of tumor cells. Our computational model views the most representative and clinically relevant features of oncogenesis as a fight between two distinct sub-systems: the immune system of the host and the neoplastic system. We implemented the neoplastic sub-system using a three-stage cell cycle: active, dormant, and necrosis. The second considered sub-system consists of cytotoxic active (effector) cells – EC, with a very broad phenotype ranging from NK cells to CTL cells, macrophages, etc. Based on extensive numerical simulations, we correlated the fractal dimensions for carcinoma, which could be obtained from tumor imaging, with the malignat stage. Our computational model was able to also simulate the effects of surgical, chemotherapeutical, and radiotherapeutical treatments.

Keywords: cellular automata, cancer model, fractal dimensions
PACS: 87.17.Aa, 87.18.Bb, 89.75.Fb, 82.40.Qt

INTRODUCTION

In humans, cancer denotes a collection of more than a 100 forms of diseases involving excessive, autonomous and non-homeostatic cell growth. In human beings, cancer and osteosarcoma were found in a mandibular fragment of a Pleistocene hominid with approximately 100 thousand years [1]. Cancer cells accumulate some genetic and phenotypic changes [2,3] and are endowed with the same resistance and survival abilities conferred by the clonal adaptation. Each type of cancer is unique due to the complexity in factors of development such as the variety of tissues, the numerous combinations of genetic alterations, and the succession of events such as vasculogenesis and angiogenesis [4]. However, several critical events are common to the evolution of each cancer, which are potential targets for therapeutic actions. Among them ere the suppression of apoptosis (programmed cellular death) and the deregulation of cell proliferation, we refer to [5] for a review of this aspect.

In 1932, Mayneord [6] proposed the so-called "living layer model" based on experimental observations that solid tumors often have a dead kernel surrounded by a shell of viable tumor cells. Mayneord's model introduced for the first time the concept of a tumor's structural heterogeneity. Solid tumors have two growth phases. During the initial phase, the tumor relayed on surrounding healthy tissue to obtain nutrients and oxygen by diffusion. Most models describing the

CP755, *ISIS: International Symposium on Interdisciplinary Science*
edited by A. Ludu, N.R. Hutchings and D.R. Fry
© 2005 American Institute of Physics 0-7354-0240-X/05/$22.50

avascular growth are variations of the diffusion-limited growth models. These models assume that nutrients must be taken up across the surface of the tumor. Uptake of nutrients at the surface, together with their use by the tumor's tissue creates a nutrient gradient within the tumor that determines the proliferation and death of cells [7]. For large tumors, these models predict structurally heterogeneous tumors with an outer living layer and a necrotic core [8]. Basic diffusion-models have been extended to account for more realistic biological details such as the presence of growth inhibitors [9], non-uniform nutrient consumption [10], and tumor-immune system interactions [11]. The avascular phase of the tumor's growth entered the vascular phase growth when the tumor was no longer able to obtain sufficient nutrients by diffusion alone. It was experimentally found that the tumor cells were able to produce factors to stimulate angiogenesis [12,13] and form tumor blood vessels. After vascularization, the tumor may have become larger.

Quantitative description of a tumor in terms of its spatial size was only possible after the tumor reached a detectable size (a few million cells). The experimental data on human tumor growth was sparse and mainly referred to growth of tumors in vitro (tumor spheroids), tumors inoculated in animal models, or natural tumors in vivo for the short duration of their growth ontogeny. It was also experimentally found that all tumors were characterized by deceleration of the growth process in a late phase [14,15]. The growth deceleration has been attributed to several factors such as increase in cell loss, increase in cell-cycle time, and decline in the growth fraction [16,17]. The growth fraction, defined as the ratio of proliferating cells to total cells, is a concept frequently used to compare tumors in terms of their growth capacity. Another concept used for this purpose is the tumor doubling time, which is the time a tumor needs to double its size.

Theoretical studies on tumor dynamics are by far outnumbered by experimental studies. The existing mathematical models belong to one of the following four classes: kinetic (describes the relationship between exposure to a (pro)carcinogen and an internal dose of carcinogen), tumor induction (describes the toxico-dynamic mechanisms through which the carcinogen induces the transformation of normal cells into tumor cells), tumor growth (relates to the clonal expansion of a tumor), and effects (involves the consequences of tumor development for the organism) [18]. Our model belongs to the tumor growth phase category and was recently extended to include different cell cycle stages and local interactions between the solid tumor and the immune system.

COMPUTATIONAL MODEL

Mathematical models of tumor dynamics have several benefits. First, a model helps improve the testability of the hypotheses by means of quantitative formulations about the interactions. Second, a mathematical model complements the individual experimental studies because the model asks for an overall view of a number of processes and their interrelationships. Finally, such a mathematical model is convenient for switching on or off particular hypothetical mechanisms in order to evaluate their impact on and relevance for the expected outcome.

The avascular growth phase of in vivo tumors is modeled by the multicellular tumor spheroids of the *in vitro* model system. The initial exponential growth phase of avascular tumor models include Monte-Carlo computational models [19], cellular automata models [20,21], and differential equation models [22,23]. These models, however, are limited in their purpose. For instance, differential equation models cannot easily represent either the individual and spatial heterogeneity of tumor cells or their adaptive behaviors. Cellular automaton models cannot account for all the detailed representations of individual tumor cell behaviors and states, and lack the flexibility to represent the adaptive behavior of individual tumor cells. In addition, none of the aforementioned approaches allow the detailed study of nutrient and waste transport through a spheroid.

The biophysics of the computational model developed by us is as follows. In vitro tumor cells consume oxygen and nutrients and release metabolic byproducts. Under optimal environmental conditions tumor cells undergo mitosis, producing new tumor cells, which in turn leads to an initial exponential growth phase and a multicellular spheroid. As the spheroid's size increases, internal oxygen and nutrient availability are reduced, and metabolic byproducts accumulate within the spheroid leading to necrosis of cells near the center of the spheroid. The byproducts of necrosis can be cytotoxic, and their presence can further inhibit tumor cell proliferation. Tumor cells near the surface of the spheroid can continue to proliferate because of favorable environmental conditions. Cells located between this outer proliferating layer and the inner necrotic regions become quiescent since the environmental conditions are adequate to their survival, but not for proliferation [15,21]. This dynamics interplays with the immune system's effector cells (EC) which attack, modulate, and control the spatio-temporal dynamics of tumor growth.

The computational model stands for the most representative and clinically relevant features of oncogenesis, viewed as a fight between two distinct sub-systems: the immune system of the host and the neoplastic system [24,25]. The cells of the neoplastic sub-system (A) could be in three different states depending on their spatial position. If the distance between the tumor cell and the nutrient medium is less than a given number, δ, of cell layers (we used three cell layers), then the tumor cell is in the "active" (mitosis) state. The cells in the "active" state could undergo cell division if they are on the tumor's outer boundary. Otherwise, the "active" inner cells force outer layer diffusion. Some authors [26] also consider the possibility of tumor cell division even if there is no room left around the mother cell leading to an irregular spatial distribution of tumor cells. If the distance between the tumor cell and the nutrient medium is more than δ cell layers, then the tumor cell will enter the resting, or dormant, phase. Finally, tumor cells resting in the dormant state for more than a critical number of time steps, τ, enter the necrosis phase and form the necrotic region of the tumor.

A major simplification in our model refers to cell phase duration which is supposed to take place during a single time step. We are aware that, for example, cell division has a series of sub-phases (G1 in which proteins and ribonucleic acid are synthesized; S in which DNA is replicated; G2 in which cell synthesizes RNA and proteins molecules required for transit through this phase; M leading to cell division [26]). Our current implementation of the computational model lumps all of the above stages into one phase but has the flexibility of independent phase duration setup for more realistic simulations.

In our computational model of tumor – immune system interaction, the second considered sub-system consists of cytotoxic active cells, or effector cells (EC), with a very broad phenotype, ranging from NK cells to CTL cells, macrophages, etc. We abstracted the dynamics of these cells to the following properties. Every EC cell can randomly walk through the extracellular matrix and suffers an aging effect depending on its interaction history. The age of a free EC increases by one unit every time we update the clock. To mimic the effect of tumor-EC interaction, we assumed that any EC interacting with a tumor cell increases its age with different amounts, depending on the tumor's cell phase. Finally, the EC will die after a given age and we implemented two different schemes for EC removal. In the deterministic removal scenario, all EC older than a given limit age disappear. In a stochastic removal approach, an EC has a survival probability depending on its age $P_{surv} = \dfrac{1}{1 + e^{-\gamma(t-t_0)}}$, where t_0 is the limit age of the immune cells and γ is the characteristic-aging factor of the immune cells.

We implemented the aforementioned mechanism into an ANSI C language program. The computer experiment simulating in vitro growth of a tumor starts with a single tumor cell, in the mitotic phase, placed in the center of a nutrient medium of a two-dimensional cell space. The model is based on the cell cycle regulation mechanisms described above.

RESULTS

The simulations which were run started with a single tumor cell that resided in the state of mitosis. We assumed an infinite reservoir of oxygen and nutrients. Under such favorable conditions, the tumor cell underwent cell division and the tumor grew due to cell division of the outer layer and cell diffusion due to continuous pressure from inner active cells. The increase in tumor size was the cumulative effect of two processes: cell division and cell migration. If the cell division process dominated over the cell diffusion, then the shape of tumor spheroid would have been more regular, but the necrotic core occurred earlier due to intense nutrient depletion. On the other hand, if the cell diffusion was the dominant process then the shape of the tumor would have been much more irregular and most of the tumor cells would have been in active or dormant state with a very small necrotic core. We performed numerical simulations using two different setups for the immune system. In one set of simulations, we turned off the immune system and studied the dynamics of the tumor spheroid under different cell division/diffusion rates. In another set of numeric simulations, we turned the immune system on and studied influence of the EC life time duration on the geometry of the tumor spheroid.

Consistent with clinical observations, we deemed all patterns roughly imperfect circles bounded by an irregular perimeter. Therefore, the geometry of the simulated tumors can be quantitatively described by the gyration radius, the fractal dimension, and the number of cells on the tumor's periphery [27]. We evaluated the scaling exponents of the cross section of tumor and the total perimeter of the cross section (Fig. 1). It was assumed that the cross section of the tumor spheroid scales with tumor cells number as follows: $\Sigma \propto n^{\alpha}$, and the perimeter $L \propto n^{\beta}$. The scaling exponents α and β include global information on cell cycle phase and they can be measured from computer tomography or other medical imaging techniques. We found that the capacity dimension of the tumor section with the immune system turned off (Fig. 1A) was 1.95±0.05, which suggests a compact structure filling almost uniformly the available space. The capacity dimension of the tumor spheroid decreases to 1.15±0.05 when the immune system is turned on, which reflects the effectiveness of the EC cells of the immune system in killing the tumor cells on the outer boundary.

A

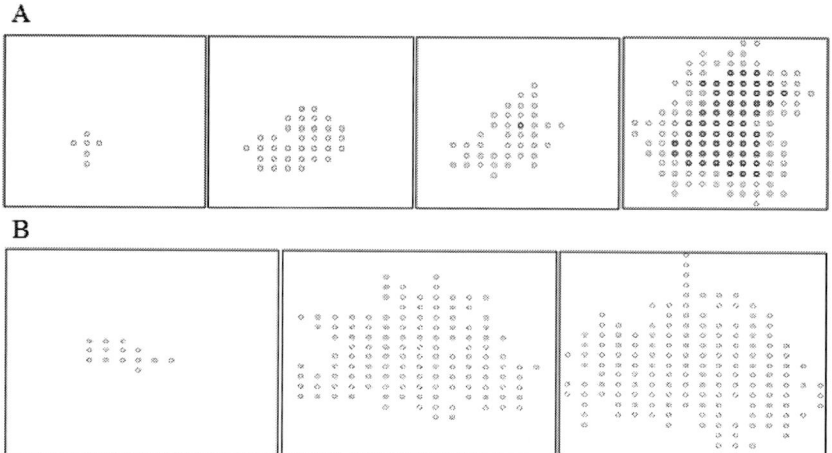

FIGURE 1. Time course of tumor spheroids during avascular growth. Successive snapshots of the 2D lattice space shows cancer cells in the mitotic phase by empty circles and the necrotic cells by thick circles. A. The immune system is turned off and the tumor develops freely, expanding into the extracellular matrix. B. Turning the immune system on leads to fractured borders of the tumor.

DISCUSSION AND CONCLUSIONS

Despite continuous efforts to improve prevention, diagnosis, and treatment, cancer strikes one in three people, and one in four will eventually die of the disease [28]. The tumor response to radiation therapy and chemotherapeutic drugs is significantly influenced by the cell-cycle phase. Therefore, we developed a phase-oriented model of tumor growth to gain a more realistic analysis of different treatment protocols.

The specialists in oncology provide a diagnosis of the malignancy of a tumor based on the physical appearance of the tumor spheroid. This empiric knowledge refers to the degree of irregularity showed by the tumor's contour. We proposed a computational model capable of mimicking the observed fractal structure of the malign tumor. The calibration of the model required high quality images of similar tumors. The relevant parameters used to calibrate the mode were the fractal dimensions.

REFERENCES

1. J. O'D. McGee, P.G. Isaacson, N.A. Wright (Eds.), Oxford Textbook of Pathology, Oxford University Press, Oxford, 1992.
2. R. S. Cotran, V. Kumar, S.L. Robbins (Eds.), Robbins Pathologic Basis of Disease, 5th ed., W.B. Saunders Co., London, 1994.
3. J. A. Anderson (Ed.), Muir's Textbook of Pathology, 12th ed., Edward Arnold, London, 1987.

4. A. R. A. Anderson and M. A. J. Chaplain, *Appl. Math. Lett.*, **11**(3), 109-114 (1998).
5. G. I. Evan and K. H. Vousden, *Nature*, Vol. **411**, 342-348 (2001).
6. W. V. Mayneord, *American Journal of Cancer,* **16**, 841-846 (1932).
7. K. Groebe and Mueller-Klieser W, *European Biophysics Journal,* **19**, 169-181 (1991).
8. R. H. Thomlison and L. H. Gray, *British Journal of Cancer*, **9**, 539-549 (1955).
9. H.P. Greenspan, *Studies in Applied Mathematics,* **51**, 317-340 (1972).
10. D. L. S. McElwain and P. J. Ponzo, *Mathematical Biosciences,* **35**, 267-279 (1977).
11. J. A. Adam N. and Bellomo (editors) A survey of models for tumor-immune system dynamics. Modeling and simulation in science, engineering and technology. Birkhauser, Boston, 1997.
12. J. Folkman and M. Klagsbrun, *Science*, **235**, 442-447 (1987).
13. R. S. Kerbel, *Carcinogenesis,* **21**(3), 505-515 (2000).
14. G. G. Steel, Growth kinetics of tumors. Clarendon Press, Oxford, 1977.
15. R. M. Sutherland, *Science*, **240**, 177-182 (1988).
16. P. K. Lala and H. M. Patt, *Proceedings of the National Academy of Sciences U.S.A.,* **56**, 1735-1742 (1966).
17. I. F. Tannock and R. P. Hill (editors), The basic science of oncology, McGraw-Hill, New York, third edition, 1998.
18. I.M.M. van Leeuwen, *Mathematical Models in Cancer Risk Assessment*, PhD-thesis, Vrije Universiteit, Amsterdam, 2003.
19. D. Drasdo, "A Monte-Carlo approach to growing solid nonvascular tumors", in Dynamical networks in physics and biology, edited by G. Beysens and G Forgacs, Springer, New York, 1998, pp 171-185.
20. S. Dormann, A. Deutsch, *Silico Biol.*, **2**(3), 393-406 (2002).
21. A. R. Kansal, S. Torquato, G. R. Harsh, E.A. Chiocca, T. S. Deisboeck, *J. Theor. Biol.,* **203**, 367-382 (2000).
22. J. J. Casciari, S. V. Sotirchos, R. M. Sutherland, *J Cell Physiol.*, **151**(2), 386-94, (1992).
23. J. P. Ward, J. R. King, *J. Math. Appl. Med. Biol.*, **14**(1), 39-69 (1997).
24. S. A. Oprisan, A. Ardelean, and P.T. Frangopol, *Bioinformatics*, **16** (2), 96 – 100 (2000).
25. S. A. Oprisan, *Journal of Physics A: Mathematical & General*, **34**, 10013 – 10028 (2001).
26. W. Duchting, *J. Cancer*, **32A**, no.8, 1293-1292 (1996).
27. S.C. Ferreira Junior, M.L. Martins , M.J. Vilela, *Physica A*, **261**, 569-580 (1998).
28. L. A. G. Ries, C. L. Kosery, B. F. Hankey, B. A. Miller, L. Clgg, and B. K. Edwards, *SEER Cancer Statistics Review 1973-1996*, National Cancer Institute, Bethesda, 2000.

Extensions of Self-Organizing Maps

Marjan Trutschl*†, Urška Cvek*†

*Department of Computer Science, LSU Shreveport, One University Place, Shreveport, LA 71115
† Center for Bioinformatics and Computational Biology, LSU Health Sciences Center, Shreveport, LA

Abstract. We present two novel methods we developed that take advantage of the large power of the Self-Organizing Map (SOM) technique developed by Teuvo Kohonen. SOM is an unsupervised neural network mapping a set of n-dimensional vectors to a *two*-dimensional topographic map. We first present a method that combines the analytic SOM with the scatter plot into an interpolated model. Second, we introduce an interactive technique that intelligibly organizes occluded or overlapped points, a self-organizing map-based Smart Jittering algorithm. Large and high-dimensional data sets mapped to low-dimensional visualizations often result in perceptual ambiguities. We address this ambiguity and present a method to systematically organize the occlusion.

Keywords: Visual data exploration, visualization formalism, Self-Organizing Map.

PACS: 89.20.Ff; 02.70.–c; 89.75.Fb.

INTRODUCTION

Large quantities of multivariate data generated in various scientific, engineering, business and other fields have triggered exciting developments in visualization, data mining and knowledge discovery. One key objective of the discovery process is to identify and describe previously unknown and interesting properties as hypotheses to be explored. Due to their size and high dimensionality, these data sets represent a challenge for analysis with standard data analysis tools and methods.

To handle the analysis of such complex and large-dimensional data sets from diverse sources, Kohonen [1] in the 1980's developed Self-Organizing Maps (SOMs). Unlike supervised learning (with known class), the interpretation of results generated by unsupervised learning requires familiarity with the data sets. Moreover, even when a class is known, the analyst and the domain expert may not discover possible sub-clusters simply because of the selected presentations of the predefined classes. What remains as one of the major challenges is finding clusters, patterns and relationships in the data under any clustering or classification.

How can we tell what the clustered data represents, especially when the clustering is non-Euclidean in nature? What is the relationship of the Euclidean points and the new clustered representation? What if there are overlapping points that prevent us from discovering the true density of the data?

In this paper we first provide a brief introduction to SOMs, followed by two extensions of the SOM and their application. We first present a method that combines the analytic SOM (non-Euclidean representation) with the scatter plot (Euclidean

CP755, *ISIS: International Symposium on Interdisciplinary Science*
edited by A. Ludu, N.R. Hutchings and D.R. Fry
© 2005 American Institute of Physics 0-7354-0240-X/05/$22.50

representation) into an interpolated model, describe our model, and provide an example interpolation of a microarray yeast expression data set. Second, we introduce an interactive technique that logically organizes occluded or overlapped points, a self-organizing map-based Smart Jittering algorithm. Large and high-dimensional data sets mapped to low-dimensional visualizations often result in perceptual ambiguities. We address this ambiguity and present a method to systematically organize the occlusion.

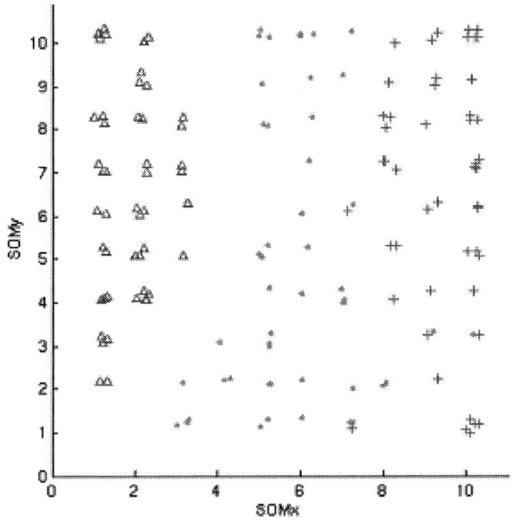

FIGURE 1. 10 x 10 SOM of the Fisher Iris flower data set.

SELF-ORGANIZING MAP (SOM)

The SOM is an unsupervised neural network that maps a set of n-dimensional vectors to a two-dimensional topographic map [2]. The training of an unsupervised neural network is data-driven, without a target condition that would have to be satisfied (as in a supervised neural network). The SOM combines an analytic and graphical technique to group data onto a low-dimensional (typically 2-dimensional) display and organize the data into clusters by this projection. The Kohonen SOM is similar to a k-means clustering algorithm, extending it by providing a topology-preserving mapping and placing similar objects in neighboring clusters. Numerous SOM algorithms and extensions have been developed in a multitude of fields which include engineering, military, and biomedical applications. These extensions include self-adaptive and incremental learning neural networks (SANN) that replace the static topology networks [2, 3] tree-structured SOM network architectures [4], and alternate neural-network based projections. Some of these approaches aim to determine the shape and size of the self-organizing structure during the learning process and some are targeted towards specific domains. The most common SOM uses a rectangular lattice display (Figure 1), although hexagonal and irregular examples have been studied. SOMs have a number of parameters which affect their success and repeatability (e.g., initial weight values, choice of neighborhood function). But perhaps the key drawback of the traditional SOM is that it tends to over-represent regions of low input density and under-represent regions of high input density.

We now briefly review the simplest SOM (see Kohonen [5] for details). We consider a data set with n dimensions. The learning of the SOM is the process in which we form a nonlinear projection of the records onto a map. The self-organizing grid or map consists of an array of output nodes (called neurons), each associated with an n-dimensional weight vector m_i (corresponding to the n dimensions of the input data set). The initial values of the m_i may be randomly selected, preferably from the data set. Each record is positioned on the map, one by one, until the data set is exhausted. The assignment of weight vectors occurs as part of the unsupervised learning process. The records are randomly drawn from the input distribution and presented to the network one at a time.

We map a record onto the SOM by calculating the similarity between the input vector and node i's weight vector m_i. Each node i receives the same input vector and produces a single similarity value. The input record maps onto the best-matching (winning) node c, based on the largest similarity or the smallest distance (depending on the implementation) as shown in Figure 2.

The weight vectors of nodes physically close to node c (up to a certain geometric threshold) adjust their weight vectors, "learning" something about the input. The adjustment depends on the size of the neighborhood, the value of the neighboring function and the learning function. This results in a local relaxation or smoothing of the neighborhood, which with continued learning leads to a global ordering.

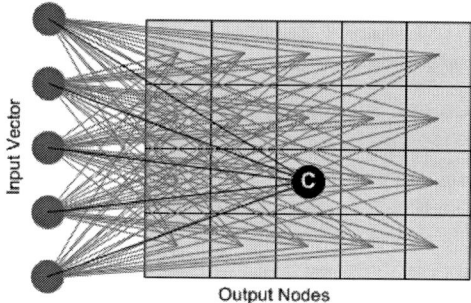

FIGURE 2. Mapping of an n-dimensional input vector to a 5 x 4 SOM rectangular grid. Winning output node c.

This process is repeated until the output map converges to a stable or organized state when the average error falls below a pre-specified value or a certain number of iterations have been reached. The self-organizing process works by repeated refinement and progressively smaller values of the learning function.

A variety of similarity metrics, neighborhood selection criteria and associated weight change alternatives exist. The similarity between a record and a node on the map is most commonly measured by the Euclidean distance, with the neighborhood consisting of a circular area around the maximally responding unit, whose weight changes are proportional to the Euclidean difference. The adjustment of neighborhood weights under a Gaussian function guarantees convergence. The neighborhood function with convex properties over a large range (around the winning node c) provides a defect-free environment.

VISUALIZATION INTERPOLATION

We first review some related work in visualization modeling. We describe the vector space model of visualization and the two layers in which we support visualization interpolation: the logical and the physical layers, followed by the description of our interpolation model.

Background

We briefly review the modern literature as a number of texts cover the earlier research [6, 7]. A number of commercial and academic systems combine visualization with data mining, most commonly as a presentation tool for process description and control. Becker [8, 9] described techniques for displaying decision trees, Bayesian classifiers and decision table classifiers. The loose coupling of visualization and data mining in specific domain exploration environments often starts off with visualization to gain an initial understanding of the data set and is followed by algorithmic analysis, most often terminating with some visualizations used to present the results [10, 11]. Healy [12], and Welling and Derthick [13] utilized data mining techniques to reduce the size and dimensionality of the data, followed by visualization tools to explore the results.

Model building or formalization of graphic techniques has a long history, with the graphical language of marks, points, lines, and areas, with Bertin [14] being one of the earliest attempts. Mackinlay's APT system [15] first applied graphical specifications to computer-generated displays, using a set of graphical languages and composition rules to generate 2-dimensional relational displays, using expressiveness and effectiveness as encoding parameters. The Sage system [16] extended this model by providing a richer set of data characterization tools and visual displays. Wilkinson [17] developed nViZn, a formal model of graphics describing statistical graphs using a subset of elements in the language of quantitative graphics. Wilkinson's work was extended by Stolte [18] into a relational data model using a specification directly tied to an interactive interface. Their Polaris system supports the interactive exploration of data warehouses based on hierarchically structured data cubes, generating visualizations built out of building blocks.

Analytical Interpolation: SOMn

We interpolate a SOM with a scatter plot and provide a link between the analytic visualization (the SOM) of the data and the Euclidean representation of that data (scatter plot), providing for a transition and intermediary views between a purely exploratory and purely modeling (analytical) views.

We modified the original SOM algorithm to create more locally based SOM spaces within a grid of two axes that are divided into bins and named the method SOMn (for SOM neighborhood). This algorithm facilitates the gradual transition from the SOM to the scatter plot, emphasizing the role of self-organization. Our model has the following added parameters: x and y axes of the output matrix, primary output nodes, secondary output nodes, and the SOMn neighborhood of qualifying output nodes.

The **SOMn topology** consists of two types of output nodes: primary and secondary output nodes. These two types of nodes are intuitively related to the coarse and fine mapping of the input vectors to primary and secondary output nodes, respectively. The number of primary output nodes P and secondary output nodes S specifies the output matrix, as shown in Figure 3. The number of output nodes along the x axis can be different from the number of nodes along the y axis, and we thus use P_x and P_y, S_x and S_y notation. For simplicity here each primary output node contains the same number of secondary output nodes.

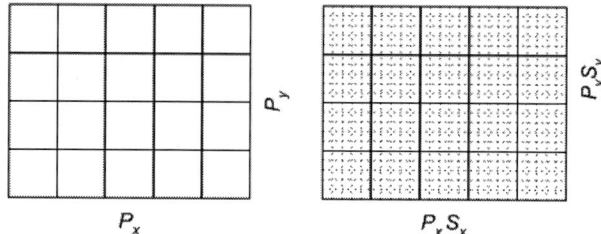

FIGURE 3. Example output grids, marked with the number of primary output nodes (left) and total output nodes (right). The primary output grid is 5 x 4, the secondary grid is 5 x 5.

The **primary output nodes** are used for the coarse mapping of input vectors and are arranged in a 2-dimensional primary output map, as in Figure 3. This is a "localized" SOM approach for each of the nodes in the grid of selected two dimensions, x and y. When the two SOMn layout dimensions are two variables from the data set, the grid resembles a scatter plot with bins along the axis. The assignment of a record is performed in two steps. Two data dimensions selected represent the x and y axes, and are used for the primary mapping, binning the data into the appropriate node on the output map.

After the primary binning of a record, the same input vector is then mapped to all qualifying **secondary output nodes** on the map. Secondary qualifying nodes are all secondary nodes within the primary winning node, as well as additional secondary nodes belonging to the SOMn neighborhood.

The **SOMn neighborhood** is a set of qualifying secondary output nodes. The size of the SOMn neighborhood may vary between 0 and a maximum, the number of all secondary output nodes outside the calculated primary output node along that axis, either $(P_x - 1) \cdot S_x$ or $(P_y - 1) \cdot S_y$, whichever is greater. The SOMn neighborhood is fixed throughout the process of mapping and independent of the SOM neighborhood. The size of SOMn neighborhood depends on the position of the winning primary output node.

Example SOMn Application

Simple example of the interpolation model, describing the approach and its merits is presented using the Fisher [2] Iris data set from the UC Irvine Machine Learning Repository. The Fisher data is quite familiar and simple. It contains fifty specimens from each of three species of Iris flowers: Iris *setosa*, Iris *versicolor* and Iris *virginica*. Its four dimensions are the length and width of sepal and petal leaves, measured in millimeters. We map this data set onto a 10x10 primary node grid on two dimensions with a 5x5 secondary output node grid within each primary node, creating effectively a 50x50 grid of output nodes, with sepal length and sepal width as the primary dimensions x and y. Varying the SOMn neighborhood from zero to the maximum generates a series of SOMn's which are visualizations of the interpolation space between the SOM and the scatter plot. These interpolations reveal additional information about the data.

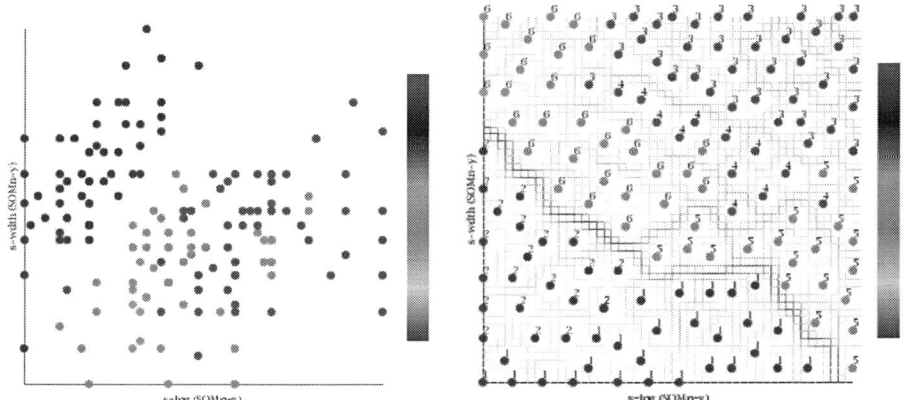

FIGURE 4. a. SOMn of the Iris flower data set, SOMn neighborhood set to 0. **b.** SOMn neighborhood size 33, groups 1 and 2 *setosa* flowers (selected at SOMn neighborhood size 16), groups 3 and 4 *virginica* flowers (selected at SOMn neighborhood size 24), groups 5 and 6 *versicolor* flowers (selected at SOMn neighborhood size 33).

We can see that the Iris *setosa* flowers are separable from the other two subspecies exhibiting small petals and short but wide sepals, as shown in Figure 4a in the top left location (with wide sepals). *Versicolor* and *virginica* subspecies are less clearly separable from each other, although *virginica* flowers have bigger leaves. Each SOMn can also denote the relationships between neighboring pairs of nodes using "lines of separation" that map the distances between neighboring weight vectors onto a gray color scale. We selected six groups of records (two for each flower type) that we identified in the SOMn neighborhood environments of size 16, 24 and 33, the groups are shown in Figure 4b. We identified them by requiring that at least one dark line of separation divides at least two distant weight vectors in order to indicate a dissimilarity of two areas and the existence of at least two groups of records. We notice that the records change position, when the neighborhood size changes. This is due to the adjustment of the neighboring nodes depending on the SOMn neighborhood value. These are not the only groups of records that can be observed in these displays, but we limited these for clarity. The six subgroups can only be identified in the SOMn neighborhoods varying between the two extremes (zero and max) and have not been observed before. We confirm the validity of these separate groups by using learning algorithms on the selected groups of records.

RESOLVING OCCLUSION

Overlap or occlusion occurs when a data set of n-dimensional records is mapped to an m-dimensional visualization space, where $m < n$ or even $m \ll n$. Overlap Ω occurs when the number of records r in a data set exceeds the number of physical points in a visualization of size v_x by v_y (in case of a rectangular two-dimensional visualization). Overlap may also occur when there are at least two records r_i and r_j that are not unique with respect to their dimensional values x and y or their non-linearly projected x and y positions. Overlap can also occur in scatter plots when the two presented dimensions have identical values. Moreover, visualizations based on non-linear projections from an n-dimensional space to an m-dimensional display most often result in multiple overlapping records regardless of the fact that their dimensional values are unique. All visualizations exhibit similar behavior when analyzing large, high-dimensional data sets. In general, they fail to handle overlapping records and crowding, when there are a lot of points displayed in a limited display space.

Smart Jitter

The goal of the Smart Jitter algorithm provides self-organization within classic visualizations. These may be linear or nonlinear projections of multi-dimensional data, such as scatter plot, polar charts and others. It is important to emphasize that Smart Jitter is not a technique that would determine the x and y coordinates for each input vector, but is rather a refinement method for spatial organization of overplotted input vectors. In general, our jittering approach can be applied to reduce occlusion given any record placement strategy (method to find the x, y position) on a two-dimensional display. We could use any dimension pair or mapping onto the x and y dimensions, or any other mapping onto a two-dimensional surface. The algorithm harnesses dimensional information of an input vector to provide local spatial organization while maintaining a relatively accurate x and y location on the low-dimensional surface of a matrix visualization.

The algorithm maps input vectors with similar properties to the same or neighboring output nodes of the display. Input vectors located in proximity of each other are likely to correlate more than vectors located farther apart. The correlation factor depends on the weight vectors of neighboring output nodes.

Displacement around an overplotted (x, y) position is driven by the dimensions of the overplotted data using a self-organizing map algorithm. The display surface is replaced with two grids; a secondary output grid within a primary output grid. The grid sizes (or resolution) are specifiable, or can be data-driven, based on the distribution and number of overplotted points, or on the overall number of displayed records. In Figure 5 we show how we grid the surface of a scatter plot, where the x and *y* dimensions correspond to dimensions 1 and 4 of the data set, respectively. Each primary output node contains a grid of secondary output nodes, creating a grid within a grid. In this example each primary output node contains a secondary output grid of 25 nodes, or 5x5 secondary nodes. We map a record onto the primary output node W_p as determined by the record's values of dimensions 1 and 4.

Mapping is first performed into the primary output nodes on the original projection of the data. Self-organization is repeated for every primary output node within the grid display. The secondary mapping first randomly initializes weight vectors of each secondary output node in the primary output node. The distance between an input vector and each output weight vector in this secondary grid is calculated, and the winning secondary output node W_s is determined based on the smallest Euclidean distance.

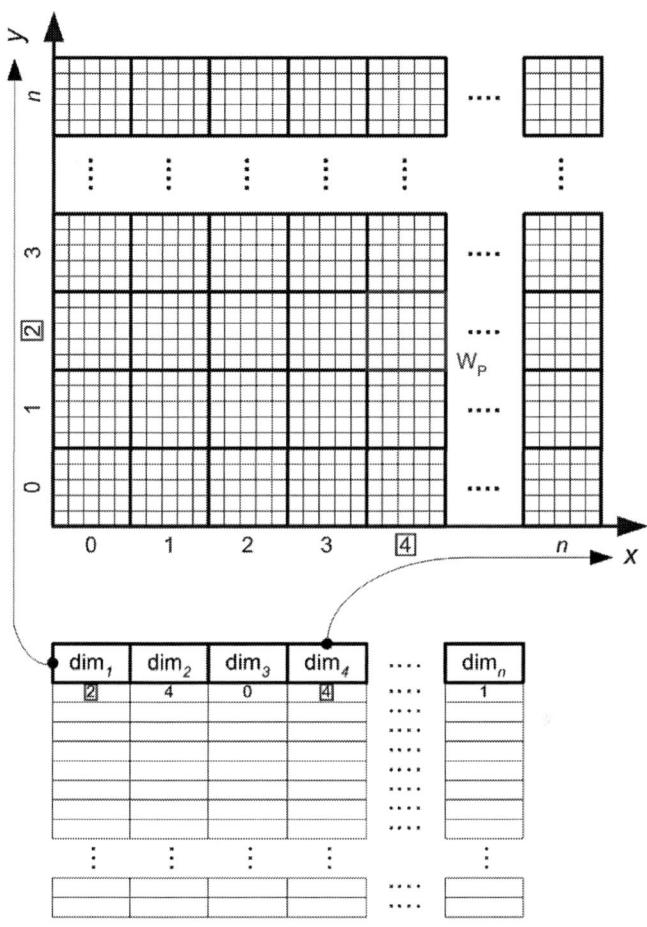

FIGURE 5. Secondary output grid within a primary output grid. W_p is one primary output node.

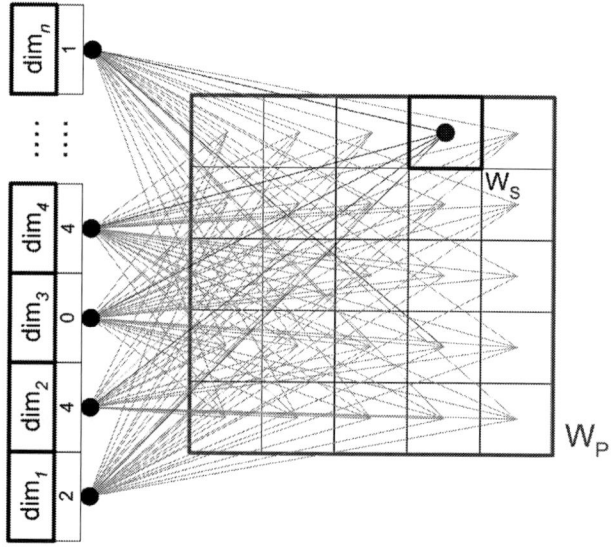

FIGURE 6. Mapping of a record into a secondary output node.

The record is mapped into that winning node W_s (Figure 6) and the weight vector of the winning node adjusted, in addition to limited functional adjustment of the neighboring secondary weight vectors, depending on the neighborhood function. Adjustment of weights and self-organization is not limited to a single primary grid, but is rather driven by the neighborhood function and the properties of the records. This process repeats for every primary output node, in successive training passes through the input data set.

Application of Smart Jitter

To demonstrate our Smart Jitter algorithm, we use the modified Fisher Iris flower data set as described above. We let the output of a classic scatter plot self-organize as shown in Figure 7.

Figure 7 shows a 5x5 primary and 10x10 secondary (5x5x10x10) Smart Jitter grid of the modified Iris data set. Each type of the Iris flower (*setosa*, *versicolor*, and *virginica*) is colored with its own shade of gray. Each input vector is first placed on the primary grid, as determined by the value of sepal length and sepal width dimensions and boundaries of the primary grid of the range of values. Each primary output node contains a 10x10 secondary grid. These 100 secondary output nodes are initialized with random weight vectors which are adjusted with every input vector mapping. The records are mapped to the closest secondary output node as determined by the shortest Euclidean distance between an input vector and the weight vectors of the secondary output nodes within a single primary output node (in our case 100 secondary weight vectors). The algorithm first maps every input vector until the data

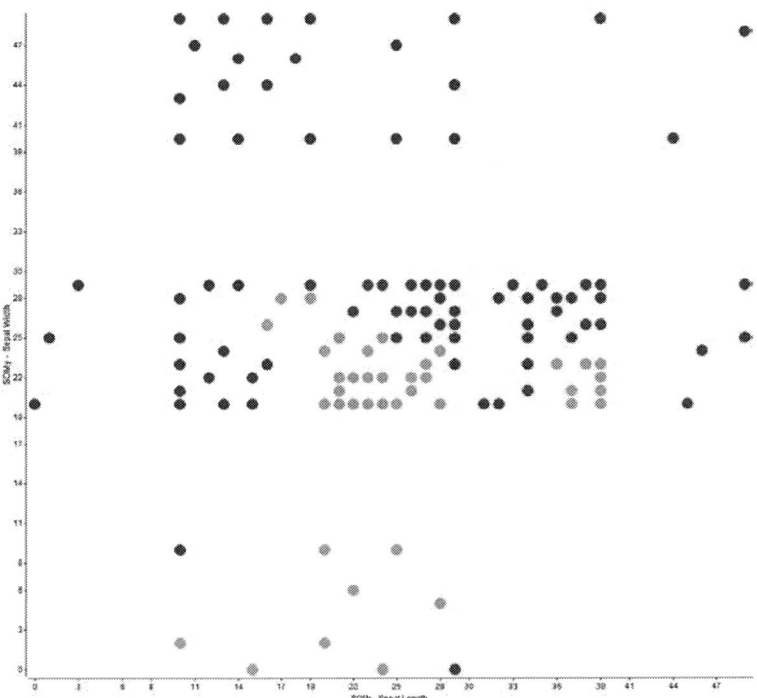

FIGURE 7. Smart Jitter plot of the modified Iris flower data set with 5x5 primary and 10x10 secondary self-organizing matrix.

set is depleted, and then repeats the training process until the target, self-organized, state is reached.

CONCLUSION

We presented extensions to the Self-Organizing Map, linking the SOM to a scatter plot visualization of the data. SOMn allows the intuitive exploration of a SOM through the interpolations to the scatter plot. We showed two applications of this approach, including the Smart Jitter technique that provides an increased visual scalability and offers higher intrinsic dimension than the classic visualization, as showcased on the scatter plot. The resulting mapping more efficiently alleviates crowding and overlaps, and emphasizes the relationships among the neighboring multi-dimensional records.

REFERENCES

1. Kohane, I.S., A.T. Kho, and A.J. Butte, *Microarrays for an Integrative Genomics*. 2002: MIT Press.
2. Nour, M.A. and G.R. Madey, *Heuristic and optimization approaches to extending the Kohonen self organizing algorithm.* European Journal of Operational Research, 1996. **93**(2): p. 428-448.

3. Fritzke, B., *Growing cell structure-a self-organizing network for unsupervised and supervised learning*. Neural Networks., 1994. **7**: p. 1441-1460.
4. Koikkalainen, P. and E. Oja. *Self-Organizing Hierarchical Feature Maps*. in *Intl. Joint Conf. on Neural Networks (IJCNN'90)*. 1990.
5. Kohonen, T., *Self-Organizing Maps*. 2nd Edition ed. Information Sciences. Vol. 30. 1997: Springer.
6. Card, S., J. Mackinlay, and B. Shneiderman, *Readings in Information Visualization: Using Vision to Think*. 1999, San Francisco: Morgan Kaufmann.
7. Spence, R., *Information Visualization*. 2000: ACM Press.
8. Becker, B.G. *Visualizing the Simple Bayesian Classifier*. in *KDD Workshop on Issues in the Integration of Data Mining and Data Visualization*. 1997.
9. Becker, B.G., *Visualizing Decision Table Classifiers*. 1998, Silicon Graphics, Inc.
10. Thearling, K., et al., *Visualizing Data Mining Models*, in *Information Visualization in Data Mining and Knowledge Discovery*, U. Fayyad, G. Grinstein, and A. Wierse, Editors. 2001, Morgan Kauffman.
11. Kohavi, R. *Data Mining and Visualization*. in *Invited talk at the National Academy of Engineering US Frontiers of Engineers. Available in Frontiers of Engineering: Reports on Leading-Edge Engineering From the 2000 NAE Symposium on Frontiers on Engineering*. 2000: National Academy Press.
12. Healey, C.G. *On the Use of Perceptual Cues and Data Mining for Effective Visualization of Scientific Datasets*. in *Graphics Interface '98*. 1998.
13. Welling, J. and M. Derthick. *Visualization of Large Multi-dimensional Datasets*. in *Virtual Observatories of the Future*. 2000.
14. Bertin, J., *Semiology of Graphics*. 1983: University of Wisconsin Press.
15. Mackinlay, J.D., *Automating the Design of Graphical Presentations of Relational Information*. ACM Transactions on Graphics, 1986. **5**(2): p. 110-141.
16. Roth, S.F., et al. *Interactive Graphic Design Using Automatic Presentation Knowledge*. in *SIGCHI '94*. 1994.
17. Wilkinson, L., *The Grammar of Graphics*. 1999, New York: Springer.
18. Stolte, C., D. Tang, and P. Hanrahan. *Query, Analysis and Visualization of Hierarchically Structured Data using Polaris*. in *8th ACM SIGKDD International Conference on Knowledge Discovery and Data Mining*. 2002.

Cognitive Radio – Genetic Algorithm Approach

Y. B. Reddy

Grambling State University, Grambling, LA 71245
Email: ybreddy@gram.edu

Abstract: Cognitive Radio (CR) is relatively a new technology, which intelligently detects a particular segment of the radio spectrum currently in use and selects unused spectrum quickly without interfering the transmission of authorized users. Cognitive Radios can learn about current use of spectrum in their operating area, make intelligent decisions, and react to immediate changes in the use of spectrum by other authorized users. The goal of CR technology is to relieve radio spectrum overcrowding, which actually translates to a lack of access to full radio spectrum utilization. Due to this adaptive behavior, the CR can easily avoid the interference of signals in a crowded radio frequency spectrum. In this research, we discuss the possible application of genetic algorithms (GA) to create a CR that can respond intelligently in changing and unanticipated circumstances and in the presence of hostile jammers and interferers. Genetic algorithms are problem solving techniques based on evolution and natural selection. GA models adapt Charles Darwin's evolutionary theory for analysis of data and interchanging design elements in hundreds of thousands of different combinations. Only the best-performing combinations are permitted to survive, and those combinations "reproduce" further, progressively yielding better and better results.

INTRODUCTION

A **cognitive radio** (CR) is a radio frequency transmitter/receiver [1-3] that is designed to intelligently detect whether a particular segment of the radio spectrum is currently in use, and to jump into (and out of, as necessary) the temporarily-unused spectrum very rapidly, without interfering with the transmissions of other authorized users. We view the CR as the evolutionary process from fully hardware based equipment to purely software-based equipment [4-9]. The process can be viewed in three stages:

- **Hardware Driven Radios**: Transmit frequencies, modulation type, and other radio frequency (RF) are determined by hardware and can be changed only by replacing the hardware.
- **Digital Radios**: The technology performs part of the signal processing or transmission digitally but is not programmable.
- **Software Defined Radio (SDR):** All models, functions, and applications can be defined, configured, reconfigured, and upgraded with new capabilities.

SDR is a flexible and proven technology and is in production. The next generation of SDR is *Cognitive Radio* [9-10]. CR is a software radio that has sensors and software that allow it to perceive the world around it and learn from experience. CR is a hard research topic within the realm of software radio. The functions of CR are:

- Know its functions and available services
- Know what services are interested to its users
- Know current degree of needs and future likelihood of needs to its users
- Learn and recognize usage patterns from the user
- Apply Model-based-reasoning about user needs, local content, and environmental content

CP755, *ISIS: International Symposium on Interdisciplinary Science*
edited by A. Ludu, N.R. Hutchings and D.R. Fry
© 2005 American Institute of Physics 0-7354-0240-X/05/$22.50

- Respond intelligently to an unanticipated event (i.e. a channel that it has never encountered before)

The term CR means different things to different persons [1-3]:

Mitola:

The term CR identifies the point at which wireless personal digital assistants (PDAs) and related networks are sufficiently computationally intelligent about radio resources and related computer-to-computer communications to:

- Detect user communications needs as a function of use context, and
- Provide radio resources and wireless services most appropriate to those needs
- Increase awareness that computational entities in radio have of their locations, users, networks, and the larger environment
- Cognition tasks that might be performed range in difficulty from the goal-driven choice of RF band, air interface, or protocol to higher-level tasks of planning, learning, and evolving new protocols.

FCC View (Federal Communications Commission):

- CR – a cognition radio is a radio that can change its transmitter parameters based on interaction with the environment in which it operates.
- SDR – a transmitter in which the operating parameters can be altered by making a change in software that controls the operation of the device without changes in the hardware components that affect the radio frequency emissions.
- Majority of cognitive radio will probably be SDRs, but neither having software nor being field reprogrammable are requirements of a cognitive radio.

Cognitive radio technologies are coming of age, supported by the growth in processing power in radio chipsets. IEEE 802 is moving with the regulatory process to bring cognitive techniques into new networking standards. The FCC [3] is beginning to open up the regulatory landscape for more extensive applications of cognitive radio technologies. The FCC is looking for innovative ways to enable "the next new thing" in spectrum management and commercial wireless activity. The innovative research applications including data mining, genetic algorithms, neural networks, fuzzy sets, and decision making are some of the recent attempts. In this research, we explore the possible applications of genetic algorithm models for improving the spectrum usage in CR technology.

Genetic Algorithm

Genetic Algorithm (GA) is a learning model [19-21] that uses formulas, rules, or arrangements to be optimized to find a solution. Genetic Algorithms allow us to solve large and complex problems. They are useful particularly to solve tricky nonlinear problems. A GA approaches the problem by using the principles of natural selection. First, a number of solutions (a population) are created by setting the parameters randomly throughout the search space. The parameters that contribute most toward solution are called the axis attributes [14-16] or spectrum activity parameters. The entity schema, the number of bits that an entity status (Center Frequency, Band Rate, Synchronous Certainty, Minimum Correlation, Maximum Correlation, Band Width, Peak Power, etc.) must match at least one of the search spaces. Otherwise, it does not belong to the entry group. The schema will be modified depending upon the role of axis attributes. The modified schema helps to decide the eligibility of receiving the information (the best fit otherwise generate deviation). The basic outline of the genetic algorithm approach includes the following steps [15-18]:

- Generate the initial population
  Individual nodes of the axis attributes are represented as strings of bits
- Evaluate all individuals of initial population (node values)
  A pair wise alignment helps to recognize the user on the network and then a fitness function must be defined that takes as input an individual and returns a number or vector as fitness of that input.
- Generate an individual that is relevant to receive the information.
  The set of bits called schema possesses the information and includes user entity state. It also contains type of user (some one surfing the net or penetrating the network), deviation from known region of normal user. Using the schema, we calculate the fitness of individual (Appendix). The operations that produce new generations include (a) mutation, (b) validation, and (c) crossover.

The above process will continue until the desired fitness value reached. The feedback of the ratio of relevant nodes helps the learning process to improve the performance. The paper discusses how cognitive radios learn about current use of spectrum (fitness of channel utilization) in their environment, make intelligent decisions to avoid the interference, and react to immediate changes in the use of spectrum by other users. The goal is to test whether the genetic algorithms are a better option or one of the options.

THE DATASET

In this study, we used the random samples of communication data for selecting the channels based on priority at each access point (AP) in a given spectrum. The priority list that is prepared depends upon the utilization of a particular channel with minimum interference and this interference should be below the defined threshold. If all the channels in priority are busy, then the lowest interference channel will be selected. The sample access point, where each access point is designed with fixed number of channels showing highest priority from left to right is provided below:

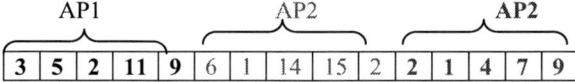

The above data set will be converted to chromosome as:

Ports	Value- Priority Left to Right	conversion
AP1	3 5 2 11 9	3,5,2,11,9,6,1,14,15,2,2,1,4,7,9
AP2	6 1 14 15 2	Convert each decimal number to 4 bit hexa number.
AP3	2 1 4 7 9	

217

The sample data for three access points is given in the following table:

Access Points	Boundary	Channels	Bandwidth (Mhz)	Priority	Success (Ns)	Trials (Nt)	Genetic Priority
A	880	10	877.4	1	500	512	9.77
A	880	12	885.7	3	450	520	8.65
A	880	13	843.77	3	670	800	8.38
A	880	8	810.98	6	345	700	4.93
A	880	2	888	5	245	500	4.90
A	880	4	854.98	2	780	900	8.67
B	895	10	877.4	4	200	400	5.00
B	895	14	900	6	122	345	3.54
B	895	9	885	1	900	902	9.98
B	895	6	779	5	234	500	4.68
B	895	7	784	2	800	910	8.79
B	895	5	789	3	445	678	6.56
C	990	4	854.98	3	344	567	6.07
C	990	3	987	1	400	400	10.00
C	990	2	888	4	235	677	3.47
C	990	1	878.9	2	876	956	9.16
C	990	11	743	6	200	789	2.53
C	990	13	978.98	5	234	900	2.60

The population of individuals (chromosome) is the group of access points and the each access point contains channel priority list. The left most channel number in the AP has highest priority. We created a population of 30 chromosomes to run our simulations. Each channel in the AP has weight. The fitness function of each individual in AP is calculated as shown in the next section.

Fitness Function

The information related to each individual is represented by string of bits called schema. Each schema has group of blocks. Each block represents specific information. The schemata are the collection of schemas (Appendix). With fitness proportionate replication, the number of individuals (m) in the population belongs to particular schemata (H) at time t+1 is related to the same number at time t by [16-18]:

$$M(H, t+1) = m(H, t)(fH(t)/f(t))$$

Where $f_H(t)$ is the average fitness value of the strings representing schema H, while f' is the average fitness value over all strings in the population. With the effects of reproduction the above equation can be written as:

$$m(H,t+1) \geq m(H,t)\frac{f_h(t)}{f'(t)}[1 - p_c\frac{\delta(H)}{l-1} - o(H)p_m]$$

Where
p_c crossover rate
δ length of schema
l length of string
p_m bit mutation probability
$o(H)$ length of string – do not care symbols in string

This result essentially says that the number of short, low order, above average schemata grows exponentially in subsequent generations of a GA. The building block hypothesis has been found reliable in many cases, but it also depends on representation and genetic operators and it is easy to find or to construct problems for which it is not verified.

The patterns presented to the genetic algorithm correspond to prioritized channel values. The set of attributes form a chromosome are of 15 (60 bits) genes length. The coding done with number of genes as: 5 hexadecimal digits per each AP. Each gene is a specific numeric value and coded as 4-bits. We have total of 15 digits or 60 bits of chromosome. Once we know the structure of chromosome, then we work on the fitness function on individual population as shown above.

MODELING WITH GENETIC ALGORITHM

The problem of interest is to find the fitness of specific individual channel in a given spectrum. The fitness of an individual based upon how many times that particular channel was selected with minimum interference. The weights are increased based upon the channel quality. Using the data analysis and genetic algorithm fitness function (Appendix) for random samples, the essential ingredients of the GA are as follows:
- A constant size population of individuals
- Each individual represents a point in the search space for a given problem through a suitable coding
- A fitness value is assigned to each individual in the population (Appendix)
- Individuals are ranked and selected according to their fitness in such a way that more fit individuals are more likely to enter the relevancy group.
- Genetic operators such as crossover and mutation (Appendix) are applied to pairs of individuals or single individual in order to produce new individuals

In the channel allocation or selection genetic priority in a given window time is highly important. For each selection and utilization rate, the priority of the channel is modified. We can generate all individuals using the table, but for simulation purpose only priority is used. The initial population of individuals may be produced as:
While termination condition not met do
 Evaluate the fitness of all individuals
 Select the fitter individuals and do
  {
    find the hamming distance of fitter individuals and relevant string
    if the hamming distance is not zero ignore the entity
     else the entity is relevant;
  }
Modification of schema of individual is based upon the priority of the channel. If the same priority is producing continuously, use the mutation operator at channel level. Remember that to produce a new individual, we always need two individuals:

- Present individual entity schema
- Old individual entity schema

DISCUSSION OF RESULTS

The following Figures (1 and 2) demonstrate the selection of channels in each AP. Figure 1 provides the number of trials of each channel in a particular AP and success rate of the channel. In Figure 2, we can see the generic priority of each channel in a given AP. If we correlate these two figures, we can conclude that generic priority depends upon the success rate of a channel [11-14]. This is a known factor that higher priority of a channel is depends upon the clarity (less interference) and utilization factor.

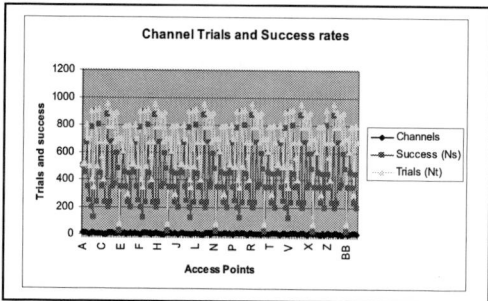
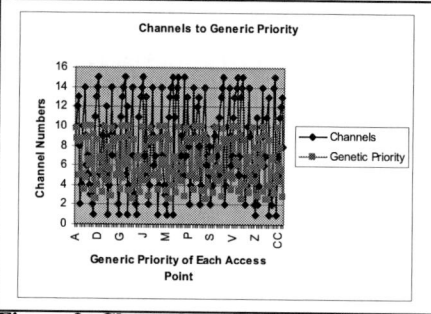

Figure 1: Channel trials and success rate   **Figure 2: Channels and Generic Priority**

We tested with genetic algorithm for selection and usage of a particular channel. So we created a database with random samples having the following data structure:
- population size of =30
- chromosome length = 60 (dataset of length 60 bits)
- Number of generations = 20
- Crossover probability=0.6
- Mutation probability=1.0/30=0.0333

Figure 3 and Figure 4 demonstrate the results of individual chromosome fitness. The program was run for 20 generations and the results show that the fitness function was not fully converged. Figure 3 shows that the convergence reached at 13th generation. The mean fitness of the channels is shown in series 2 of Figure 3 and convergence is seen after 13th generation. Figure 4 shows the convergence of all 20 channels and their priority. The channels 3, 9, 12, and 17 (series 3, 9, 12, and 17) have higher priority for selection. The results show that the genetic algorithm applications help to select higher priority channel (in a spectrum) with minimum interference and provide fruitful results.

The time factor is very important in our channel selection because the number of successful attempts out of number of attempts by a customer in a particular spectrum within a given time helps to signal the appropriate channel [12-14].

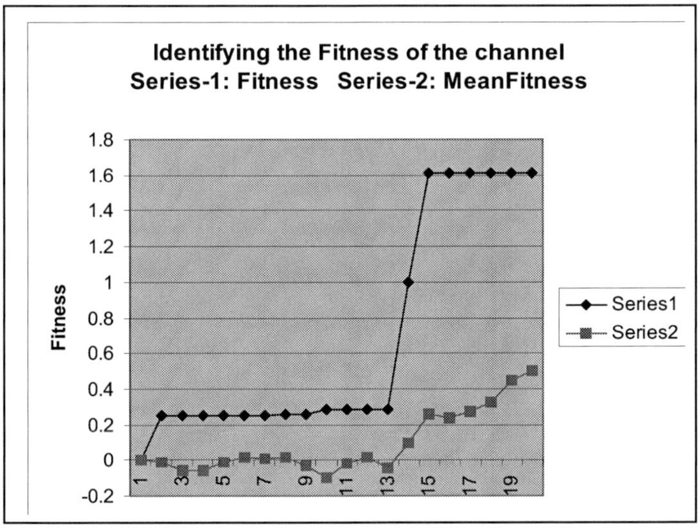

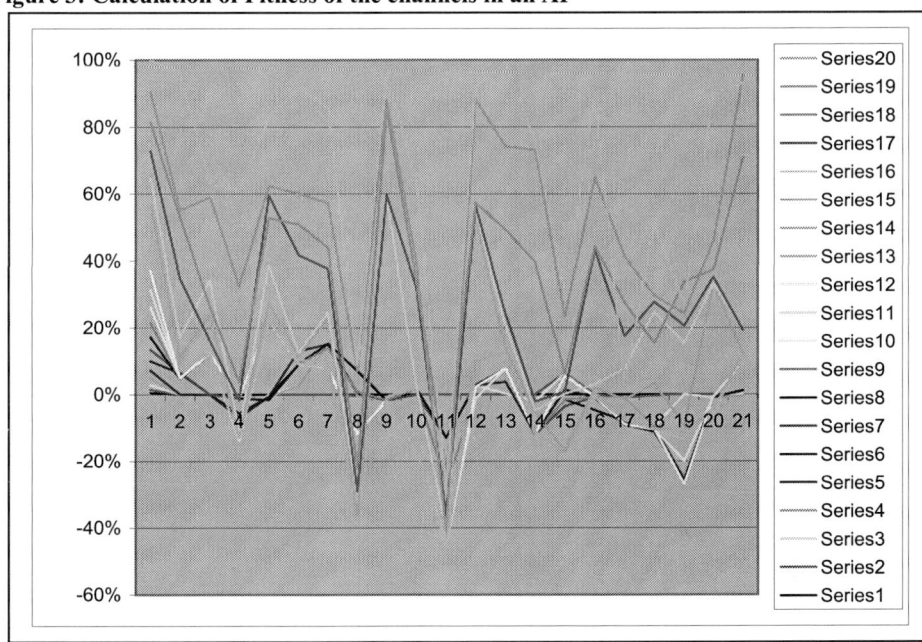

Figure 3: Calculation of Fitness of the channels in an AP

Figure 4: Fitness of Various channels in 20 generations

CONCLUSIONS AND FUTURE RESEARCH

In this paper, we have outlined genetic algorithm approach for spectrum utilization. We concluded that genetic algorithms can be applied to minimize signal interference and channel selection at each access point of a spectrum. The application of genetic algorithm for signal interference and channel selection is new. This bioinformatics learning process will give us new direction for better spectrum utilization and channel selection. The weight factor for channel utilization is very important in the dynamic selection of channels.

Remember that lot of unwanted data is generated during data collection. Therefore classification of data and selection of data related to access attributes helps us eliminate the processing of unwanted data; otherwise handling the interferences (hundreds of millions of records) would be an impossible task.

APPENDIX: GA – TERMINOLOGY

A Genetic Algorithm (GA) starts with an initial random population [24 - 27], and allocates more trials to regions of search space found to have high fitness. Genetic algorithms are adaptive methods which may be used to solve search and optimization problems, based on the genetic processes of biological organisms. Over many generations, natural populations evolve according to the principles of natural selection and "Survival of the Fittest". By mimicking this process, genetic algorithms are able to "evolve" solutions to real world problems, if they have been suitably encoded.

The genetic algorithms deal with parameters of finite length, which are coded with finite alphabet. The search starts from a population of many points. This process helps the candidate from trapping on local maxima. The transition rules used by genetic algorithms are probabilistic, not deterministic.

alleles: The chromosomes are composed of genes, which may take on some number of values

cardinality: The cardinality is the number of alphabet characters in a scheme. For example, in the binary alphabet $\{0, 1\}$ cardinality is 2 and the number of elements are 3, because we can add another symbol $*$ - do not care (can take 0 or 1), If the length of the string is 5, that is represented as $= \{10101\}$, there are 3^5 different similarity templates (schemata) can be created.

crossover: Crossover is a recombinant operator that takes two individuals, and cuts their chromosome strings at some randomly-chosen position. This produces two head segments and "tail" segments. The tail segments are then swapped over to produce two new full length chromosomes.

```
xxxxxxxx        xxx00000
00000000        000xxxxx
```

The two offspring each inherit some genes from each parent. This is *single point crossover*. Crossover is not necessarily applied to all pairs of individuals selected for mating. A choice is made depending on a probability specified by the user, this is typically between 0.6 and 1.0. If the crossover is not applied, the offsprings are simply duplications of the parents.

defining length: This is the distance between the first and last specific string positions. $d(011*1**) = 4$ because the last specific position is 5, and the first is 1.

fitness function: A fitness function must be devised for each problem; given a particular chromosome, the fitness function returns a single numerical fitness value, which is proportional to the ability, or utility, of the individual represented by that chromosome.

genotype: In natural systems, one or more chromosomes combine to form the total generic prescription for the construction and operation of some organism. The total genetic package is called the genotype.

hamming distance: The simplest distance calculation between two sequences. If two sequences of characters are equal length (number of characters in each sequence), then the hamming distance is the number of characters differ in their relative positions. For example:

Sequence s   SMITH JONE
Sequence t   SMELL JOHN
hamming distance (s,t)   3   2

mutation: Substitute one or more bits of an individual randomly by a new value (0 or 1)

10010010 1001010101
$\downarrow$
10010010 **0**001010101

phenotype: The organism formed by the interaction of the total genetic package with its environment (*product of the interaction of all genes*)

population size: The population of size n means the number of schemas considered. For example, the population of size 5 is:

 10011
 10001
 00111
 11100
 01010

schema: Schema H of length 7:
      String H $(b_1\ b_2 b_3 b_4 b_5 b_6 b_7)$
            = *11*0**
    or
    string A $(a_1\ a_2 a_3 a_4 a_5 a_6 a_7)$
           = 0110000

schema length: The length of schema H i.e. $\delta(H)$, is the distance between the first and last specific string position.
   The schema 011*1** has defining length $\delta = 4$ because the last position is 5 and the first specific position is 1, and the distance between them is $\delta(H) = 5 - 1 = 4$.
The schema 0****** has the length $\delta(H) = 0$.

schemata: A schema is a string over an extended alphabet, {0, 1, *}, where the 0 and the 1 retain their normal meaning and the * is a wild card or do not care symbol. This notational device greatly simplifies the analysis of the genetic algorithm method because it explicitly recognizes all the possible similarities in a population of strings.

Schema order: This is the number of fixed positions present in the similarity template.
e.g. o(011*1**) = 4, o(0*****)=5

variation: Change the bits in a way that the number encoded by them is slightly incremented or decremented.
  1001.0010.0101.1100

223

```
9      2      5      13

1001.0001.0101.1100
9      1      5      13
```

REFERENCES

1. Joseph Mitola III, "Cognitive INFOSEC", *IEEE MIT-S Digest*, 2003.
2. Joseph Mitola III and Gerald Q. Maguire, Jr, "Cognitive Radio: Making Software Radios More Personal", *IEEE Personal Communications*, August 1999.
3. FCC 03-322 Report available at http://hraunfoss.fcc.gov/edocs_public/attachmatch/FCC-03-322A1.pdf
4. Walter Tuttlebee., "Software Defined Radio: Baseband Technologies for 3G Handsets and Basestations*", Wiley & Sons;* 2004, ISBN: 0470867701.
5. Walter Tuttlebee (Editor)., "Software Defined Radio: Enabling Technologies", *John Wiley & Sons; 1st edition*, July 15, 2002, ISBN: 0470843187.
6. Markus Dillinger, Kambiz Madani, Nancy Alonistioti., "Software Defined Radio: Architectures, Systems and Functions", *John Wiley & Sons*, June 20, 2003, ISBN: 0470851643.
7. Joseph Mitola., "Software Radio Architecture: Object-Oriented Approaches to Wireless Systems Engineering", *John Wiley & Sons*, January 2000, ISBN: 0471384925.
8. Jeffrey H. Reed., "Software Radio: A Modern Approach to Radio Engineering", *Prentice Hall,* May 2002, ISBN: 0130811580.
9. Christian Rieser., "Biologically Inspired Cognitive Wireless", *IREAN Research workshop*, 2003 (genetic algorithm approach – Introductory).
10. Position paper Approved *by IEEE-Board of Directors*, "Improving Spectrum Usage through Cognitive Radio Technology", Nov 13, 2003.
11. Mitola, J., "Signal Processing Technology Challenges of Cognitive Radio", *Sixth Baiona Workshop on Signal Processing in Communications Advance Program*, September 8th, 2003
12. Dennis L. Chao, Justin Balthrop, and Stephanie Forrest., "Adaptive Radio: Achieving Consensus Using Negative Preferences", *Technical Report TR-CS-2004-08*, the university of New Mexico, Dept. of CS, Albuquerque, NM, 2004
13. Shin Horng Wong, Ian J. Wassell., "Dynamic Channel Allocation for Interference Avoidance in a Broadband Fixed Wireless Access Network",*3rd International Symposium on Communication Systems Networks and Digital Signal Processing (CSNDSP02),* Staffordshire, UK, PP 352-355, July 2002.
14. Shin Horng Wong, Ian J. Wassell., "Dynamic Channel Allocation Using a Genetic Algorithm for a TDD Broadband Fixed Wireless Access Network*", IASTED International Conference in Wireless and Optical Communications*, Banff, Canada, Pages 521-526, July 2002.
15. Bartee, Jon A., "Genetic Algorithms as a tool for phased array radar design", *Thesis, Naval Postgraduate School, Monterrey, Ca*, June 2002.
16. G F. Sahin, G. Abbate., "Genetic Algorithms For Parameter Optimization Applied To Sensor Coverage Problem*", Digital Wireless Communications VI*, April 2004.
17. Y. B. Reddy.," Royal Road Fitness Function for Relevance Filtering in Distributed Interactive Simulations", *CSMA'98*, Orlando, Florida November 1 – 3, 1998
18. Y. B. Reddy., "Genetic Algorithm Approach for Relevance Filtering in Distributed Interactive Simulations:, *Simulation Interoperability Workshop (SIW'98F)*, Orlando, Florida September 14 – 18, 1998
19. David E. Goldberg, "Genetic Algorithms in Search, Optimization, and Machine learning", *Addition-Wesley publishing Company*, INC., 1989.
20. David A Coley "An Introduction to Genetic Algorithms for Scientists and Engineers", *World Scientific*, 2003, ISBN: 981-02-3602-6.
21. Holland, J. H., "Adoption in Natural and Artificial Systems ", *An Arbor, University of Michigan Press*, 1975.

Medical Imaging and the Human Brain: Being Warped is Not Always a Bad Thing

James C. Patterson II, MD, PhD

Assistant Professor of Psychiatry,
Louisiana State University Health Science Center
Director of PET Neuroimaging Research, PET Imaging Center
Biomedical Research Foundation of Northwest Louisiana
1501 Kings Highway,
Shreveport, Louisiana, 71103

Abstract. The capacity to look inside the living human brain and image its function has been present since the early 1980s. There are some clinicians who use functional brain imaging for diagnostic or prognostic purposes, but much of the work done still relates to research evaluation of brain function. There is a striking dichotomy in the use of functional brain imaging between these two fields. Clinical evaluation of a brain PET or SPECT scan is subjective; that is, a Nuclear Medicine physician examines the brain image, and states whether the brain image looks normal or abnormal. On the other hand, modern research evaluation of functional brain images is almost always objective. Brain images are processed and analyzed with advanced software tools, and a mathematical result that relates to regional changes in brain activity is provided. The potential for this research methodology to provide a more accurate and reliable answer to clinical questions about brain function and pathology are immense, but there are still obstacles to overcome. Foremost in this regard is the use of a standardized normal control database for comparison of patient scan data. The tools and methods used in objective analysis of functional imaging data, as well as potential clinical applications will be the focus of my presentation.

Keywords: Positron Emission Tomography, neuroimaging, Nuclear Medicine, Alzheimer's disease.
PACS: 07.05.Pj Image processing

What is Positron Emission Tomography (PET)?

PET is the process whereby an image of the body or brain's function is generated. PET requires a cyclotron nearby for radioisotope production for radioligands. The most common radioisotopes are oxygen-15 [15O], nitrogen-13 [13N], carbon-11 [11C], and fluorine-18 [18F]. In very simple terms, a PET scan is generated when the radioligand molecule generates a positron, which quickly collides with an electron, and this collision subsequently results in the generation of dual photons at 180o to each other. These are detected simultaneously on opposite sides of the PET scanner by rings of photomultiplier tubes surrounding the scanner. The scanner acquires data about where and how many photons were generated, and produces a PET scan that

CP755, *ISIS: International Symposium on Interdisciplinary Science*
edited by A. Ludu, N.R. Hutchings and D.R. Fry
© 2005 American Institute of Physics 0-7354-0240-X/05/$22.50

represents this data. Function can be represented in many ways, with the most widely available radioligand being [18F]-fluorodeoxyglucose (FDG), the uptake of which provides a sensitive marker of glucose metabolism[1]. Well over 95% of clinical PET is done with FDG, and is related to oncology. This is because cancerous growths consume large amounts of energy - including glucose - and thus take up large quantities of FDG. This provides a very sensitive mechanism for the detection of cancer.

Another important use of FDG is to examine cerebral metabolism. The brain also uses large amounts of glucose. In fact, the brain only takes up 5% of the body's weight, but consumes 20% of the body's glucose[2]. PET neuroimaging is an important way for brain activity to be examined in vivo, and stretches well beyond just metabolism. PET is very suited to brain analysis, as there have been many radioligands made to label the many neurotransmitter receptors in the brain. Beyond glucose metabolism with FDG, the other two most common radioligands are H2150 (water PET - used to examine cerebral blood flow), and FDOPA, used to examine the dopaminergic nervous system. These are only a few of the many radioligands. The website for the Society for Noninvasive Imaging & Drug Development lists over 100 PET radioligands, many for brain imaging.

What is an FDG PET image?

An FDG PET image is a 3 dimensional matrix of "voxels". Whereas a "pixel" is a term used to refer to the smallest element of a 2 dimensional image (picture element >> "pixel"), a voxel is the term used to refer to the smallest element of a 3 dimensional image (volume element >> "voxel"). When discussing PET brain images that have been transformed into a standardized space, each voxel has a location in space (XYZ coordinates) and an intensity value, which represents the rate of metabolism in that location. A processed PET image of the brain has more than 200,000 voxels, each 2 mm3 in size. See Figure 1 for an example of an FDG PET image of a normal subject as well as a subject with Alzheimer's disease.

PET Imaging is a Multidisciplinary Field

PET neuroimaging research is a field of interest that would not be possible without input from a broad array of other areas of knowledge. These include Neurophysiology, Cognitive Neuroscience, Nuclear Medicine, Nuclear Physics, Mathematics, Neurology, Psychiatry, and of course Computer Science, Information Technology, and Bioinformatics. The capacity to generate digital images of the brain is critical, and computers provide this ability via 3D visualization software, digital image processing capabilities, high-speed networks, large-scale data storage and transfer, and software for analysis.

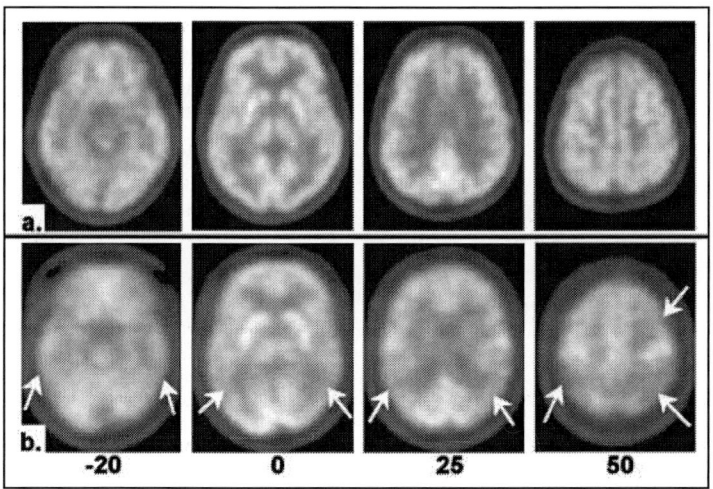

FIGURE 1. PET scans of a normal subject (a.) vs a patient with Alzheimer's disease (b). The patient with AD has characteristic bilateral posterior parietal hypometabolism, extending into the temporal lobes (double arrows). These axial image views are in radiological orientation (image left = subject right). The numbers along the bottom represent the z axis location (mm). The patient has more severe pathology in the left side of the brain, including left frontal hypometabolism (upper arrow in z=50 mm slice).

PET Neuroimaging Basics

PET images can be visualized in any number of ways. When the brain is visualized, often a color scale is used. These can become confusing, as different individuals may use different color scales, the color intensity can be varied in the software, and the same color can be used to represent different levels in different images. When PET images of the brain are looked at, it helps to have a basic understanding of brain orientation. Axial slices are slices through the brain from superior to inferior, sagittal slices are slices taken from left to right, and coronal slices are slices through the brain from anterior to posterior. Most PET scans of the brain used in the clinical arena are shown in the axial orientation (see Figure 1), with the patient's left side on the images right, as if you were looking at the patient's head from the foot of the patient's bed. This is called radiological orientation. However, it is not uncommon to see the opposite orientation in use as well, especially in presentations of a more academic bent, or from work originating from Europe. This orientation is called "neurological orientation", when the image has the patient's left side of the brain on the left side of the image.

Clinical Neuro-PET and the difference between subjective and objective analysis

PET is utilized for multiple different brain disorders, including Alzheimer's disease, cerebrovascular accidents, seizure disorders, cancer, and Parkinson's disease. Subjective evaluation of PET brain images involves reading of the scan by a nuclear medicine physician, who has been specially trained in this field. These doctors are very good at clinical diagnosis of various disorders, and are highly accurate for a given diagnosis. This is still the standard of care for nuclear medicine and radiology. Thus, when you get a clinical PET scan, there is no standard to compare to, because the results are qualitative and subjective, based on the nuclear medicine physician's interpretation of the PET image. In contrast, objective analyses compare unknown patient values to a set of known control values. A group analysis is an example of this. In this method, a group of patients with a disease are compared to a group of control subjects without the disease process. This allows objective evaluation to occur, however it has some drawbacks. To translate this type of analysis to clinical use is difficult, because patients do not walk into the clinic in groups, nor do they bring a group of controls with them. A test for a given problem thus requires two things: the ability to make a determination in a single individual and a standard for comparison that is readily available. The standard may be derived from a control group, but the control group is no longer necessarily present. An example of this is a white blood cell (WBC) count. If a patient had abnormally low or high numbers of WBCs, it can be indicative of an ongoing disease process. When you have a blood test of your WBCs, your value is compared to a "standard", with a range of normal values. The normal range for WBCs in the blood is 4,300 - 10,800 cells/µL. This was determined from measuring the number of WBCs in the blood of many healthy subjects who did not have any diseases. Once a normal range was determined, anyone can get a blood test to evaluate whether there WBC count was normal. There is no need for subjective evaluation, nor is there need for a group analysis to occur. To translate the objective analysis of a PET brain scan to the individual level, one must start with group analysis, and develop a normal brain "standard" for comparison. The optimal research design for an examination of a 41-year-old white male would be to compare his brain image to 100 brain images from 41-year-old males, preferable from the same region of the country and same heritage. However, for objective statistical analysis to become a reality, a database must be constructed with the broader population in mind. This requires obtaining a large distribution of a given variable such as age and gender and then correcting for these variables instead of matching.

PET Scan Processing: Spatial Normalization and Filtering

One of the problems faced with creating a normal standard for PET imaging is that humans, and thus human brains, have a significant degree of normal variability. Every head is somewhat different in size and shape, and thus to objectively compare brain images, certain steps need to be taken in order to facilitate across-subject comparison. This process in known as spatial normalization, or "brain warping". PET brain images are processed and analyzed with Statistical Parametric Mapping (SPM99)[3] implemented in Matlab (Mathworks, Natick, MA). In essence, processing with SPM results in brain scans that are brought into the same standardized shape and space as a template image. This is carried out through the use of a 12-parameter linear affine mathematical transformation. This involves four transformations for each of the three axes: translation, rotation, scale, and shear (see Figure 2). The end result is a brain image in the standardized stereotactic space knowns as Talairach space.[4] This image is then spatially filtered with a Gaussian filter to remove residual variation related to differences in gyri/sulci patterns between individual differences (see Figure 3). The size of the smoothing kernel is based upon the spatial resolution of the PET scanner from which the scan was derived, and is commonly set at 1.5 to 2 X the FWHM (full-width half-maximum) of the scanner. Spatial filtering also renders the data amenable to analysis via Gaussian Field Theory, and the General Linear Model.[5]

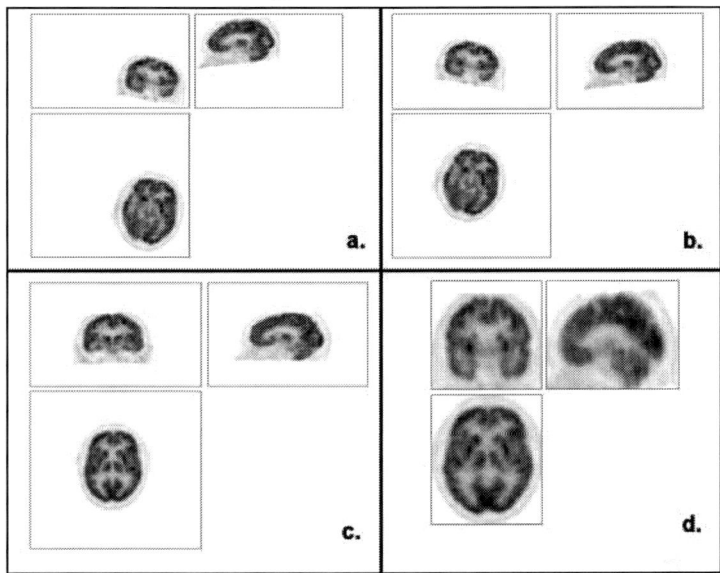

FIGURE 2. Spatial normalization of PET brain images. a.) Image in original orientation and location. b.) Image has been translated to center of image space. c.) Image has been rotated about the three axes. d.) Image has had scale and shear operations carried out to arrive at the final spatially normalized product.

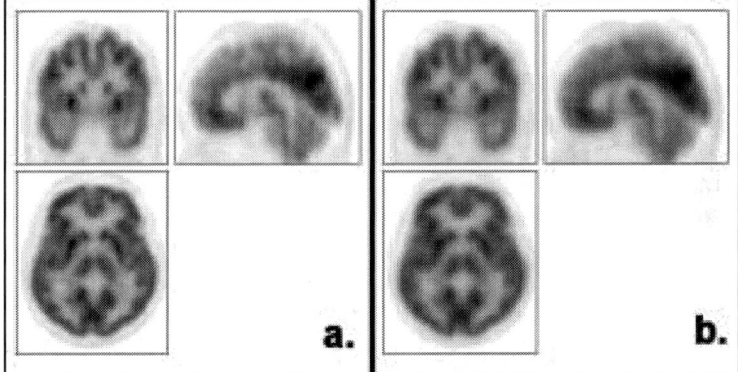

FIGURE 3. Spatial filtering of PET brain images. The spatially normalized scan in panel a. had a Gaussian spatial filter of 6 mm isotropic (6x6x6) applied. The resulting image is shown in panel b.

Clinical Application of Objective Statistical Analysis: Alzheimer's Disease

One of the most feared medical problems for an individual to face is Alzheimer's disease (AD). AD is the result of a chronic, pervasive process that causes degeneration and death of brain cells, and begins well before memory loss is noticed by the patient. AD is the most common cause of dementia in late life, present in approximately 10% in those 65 years and older, and almost 50% in those 85 and older.[6] While these numbers are fearsome enough in and of themselves, the prevalence in the aging population is increasing, and is projected to quadruple in the next half-century.[7] There have been numerous classification systems proposed for the objective description of early cognitive impairment.[8] The generally accepted standard in this regard has been adapted and published as a practice parameter by the American Academy of Neurology, and identifies Mild Cognitive Impairment (MCI) as a precursor to AD.[9] Current methodologies for early detection and diagnosis of MCI and AD take several approaches, including clinical evaluation, genetic analysis, neuropsychological tests, and functional neuroimaging. In AD, the PET brain image has a very characteristic pattern of abnormal metabolism which appears well before obvious atrophy on an MRI or CT scan. However, especially early on in the disease process, even the PET scan appears normal to the clinician. Previous results show that it is possible to find pathology in patients with early MCI using objective statistical analysis of PET brain images, before the scan appears abnormal to the NM physician. Three studies have shown PET can detect metabolic changes before cognitive decline.[10-12] All three were group studies: group of patients vs group of controls. However, patients rarely walk in to the clinic in groups, and rarely have their own control group in tow! Again, what is required is a method to bring objective analysis to the level of the individual patient. While individual patients with MCI can usually be detected with SPM analysis, often there is significant variability from person to person. Diseased areas are usually in a specific pattern just as in AD, but vary within that pattern. Some patients do not have the clearly defined pattern seen in AD. Another issue is what statistical threshold to use to determine abnormality. What is often found are clear patterns of abnormality that are consistent with MCI or early AD, but at a significance threshold that occasionally not even reach an arbitrary cutoff (p=0.001, uncorrected for multiple comparisons) for a given region. This brings back into play the power of clinical interpretation. While the numbers may not be right, the pattern may "look right". Another alternative to SPM analysis is the Cognitive Decline Index (CDI). The CDI is an index derived from sampling data from certain brain regions known to be commonly affected by AD, weighting the sampled data to stratify in regard to magnitude, and forming a ratio of a mean of some of the regions divided by the mean of the other regions. This number, when calculated for the control group, provides a standard normal range (0.949 - 1.042) against which other individuals can be compared, with complete objectivity (see Figure 4).

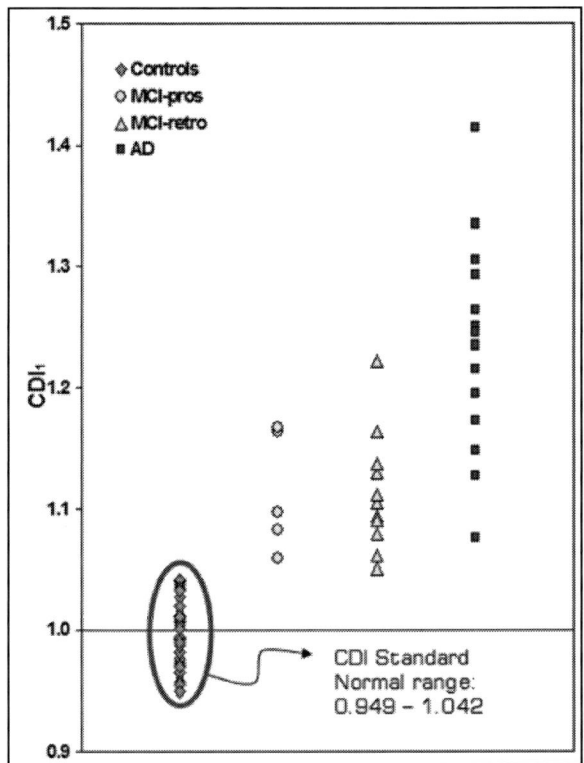

FIGURE 4. Graph of the Cognitive Decline Index, including CDI Standard Normal range. MCI-pros: Patients with MCI identified prospectively, MCI-retro: patients with MCI identified retrospectively from clinical cases, AD: Patients with AD identified retrospectively from clinical cases.

Conclusion

This type of project is made possible by the interaction of multiple areas of research to provide the necessary tools and paradigms in a multidisciplinary setting. Spatial normalization of brain images provides a system-critical methodology that is fundamental to this objective analysis. Without spatial normalization, PET brain images could not be compared across subjects. Without cross-subject comparison, objective analysis would not be possible. Without objective analysis, early detection of MCI via PET imaging could not occur. Objective statistical analyses of FDG PET brain scans appear to extend the Nuclear Medicine physician's diagnostic "reach" - further back in time along the continuum of the Alzheimer's disease process. With the technological advances discussed here, we hope to improve diagnostic accuracy at an earlier stage, and improve patient care in so doing.

ACKNOWLEDGMENTS

Parts of this material are patent pending. Research related to early detection of Alzheimer's disease was made possible by a research grant from the Kinsey Foundation and the Biomedical Research Foundation.

REFERENCES

1. Grafton, S.T., *Adv. Neurol.,* **83,** 87-103 (2000).
2. Hart, L., *How the Brain Works,* New York: Basic Books Publishers, 1975.
3. Friston, K.J., Ashburner, J., et.al., *Human Brain Mapping,* **2,** 165-168 (1995).
4. Talairach, J., and Tournoux, P., *Co-Planar Stereotactic Atlas Of The Human Brain.* Stuttgart: Georg Thieme Verlag, 1998.
5. Friston, K.J., Holmes, A.P., et.al., *Human Brain Mapping,* **2,**189-210 (1995).
6. Evans, D.A., Funkenstein, H.H., et.al., *JAMA,* **262,** 2551-2556 (1989).
7. Brookmeyer, R., Gray, S., and Kawas, C., *Am. J. Public Health,* **88,** 1337-1342 (1998).
8. Collie. A., and Maruff, P., *Aust N Z J Psychiatry,* **36,** 133-140 (2002).
9. Petersen, R.C., Stevens, J.C., et.al., *Neurology,* **56,**1133-1142 (2001).
10. De Leon, M.J., Convit, A., et.al., *Proc. Natl. Acad. Sci. USA,* **98,** 10966-10971 (2001).
11. Reiman EM, Caselli RJ, et.al., *Proc. Natl. Acad. Sci. USA,* **98,** 3334-3339 (2001).
12. Small GW, Ercoli LM,et.al., *Proc. Natl. Acad. Sci. USA,* **97,** 6037-6042 (2000):

Do Ecosystems Ever Converge? Evidence From Faunal Size Distributions Of Late Miocene North American Mammals

W. David Lambert

The Louisiana School for Math, Science, and the Arts, Department of Science, Natchitoches, LA 71457

Abstract. There is disagreement among ecologists as to whether ecosystem system behavior in general is the net result of all of the complex internal system interactions (bottom-driven) or if the behavior is driven by a limited number of key processes (top-driven). If ecosystems are primarily bottom-driven in nature, then it is unlikely that any two complex ecosystems will ever behaviorally converge as a simple matter of probability, or that their behaviors will ever be predictable. Conversely, evidence of ecosystem convergence would suggest that the systems are top-driven, with the corollary that their behaviors can be understood (and therefore in principle predicted) without a complete understanding of their internal workings.

Research has demonstrated that body mass distributions of terrestrial animals broadly reflect ecosystem function. Thus, comparable but causally disconnected terrestrial ecosystems that demonstrate similar body mass distributions would suggest ecosystem convergence. To look for this possible convergence, I generated body mass distributions in a time series for late Miocene North American mammal faunas from the Gulf Coastal Plains, Great Plains, and Pacific coastal region, and compared them with data from the modern Serengeti savanna region. The data show that during the early late Miocene Gulf Coastal Plain faunas resembled each other but were distinctly different from that of the Serengeti, the Great Plains fauna resembled the Serengeti, while the Pacific Coast fauna showed no resemblance to any of the others. However, during the latest Miocene the Gulf Coastal Plain faunas were transformed so as to strongly resemble the Serengeti fauna. The resemblance with the Serengeti was maintained by the Great Plains faunas until at least the end of the Miocene, while the Pacific Coast faunas remained distinctly different from the others. These findings suggest that the late Miocene ecosystems of the Gulf Coastal Plain and Great Plains regions (but not the Pacific Coast region) converged with that of the Serengeti savanna fauna and thus that these ecosystems were/are top-driven rather than bottom-driven in nature.

Keywords: landscape, mammals, thermodynamics, Miocene, ecosystem, body mass.
PACS: 05.70.

INTRODUCTION

Non-linear (i.e., self-perpetuating) non-equilibrium systems, which in this paper will henceforth be referred to as complex systems, can be broadly divided into two categories: bottom-driven and top-driven. Bottom-driven complex systems are those systems whose overall behavior is the net result of numerous interacting

CP755, *ISIS: International Symposium on Interdisciplinary Science*
edited by A. Ludu, N.R. Hutchings and D.R. Fry
© 2005 American Institute of Physics 0-7354-0240-X/05/$22.50

internal processes, all of which are important (although not necessarily equally important). The term bottom-driven is a direct reference to the importance of these lower level processes in the system hierarchy. Such systems show a high degree of connectivity and thus sensitivity, with the result that it is extremely difficult even in principle to forecast their responses to external disturbances (1). In contrast, the overall behavior of top-driven complex systems is controlled by a relatively small number of driving processes, the lesser processes in the system creating mostly inconsequential behavioral "noise." In contrast to bottom-driven systems, at least rough predictions can be made in principle about the response of top-driven systems to a disturbance, assuming that the effects of that disturbance on the key driving processes are understood (2, 3, 4).

This division of complex systems into top- and bottom-driven types has important ramifications for ecosystems and their conservation. Because of direct and indirect human activity, many if not most of the Earth's ecosystems are currently subject to disturbance, with many of them severely threatened. Conservation of these ecosystems requires that they be managed in the face of this disturbance, and this in turn requires knowledge of whether they are mostly top- or bottom-driven systems. If these systems are mostly top-driven, then in principle they can be managed as long as the relationship between their key driving processes and the disturbance(s) are understood. On the other hand, if they are mostly bottom-driven then their response to disturbance could be considered unpredictable, a fact that would be of great importance to managers and policy-makers.

One way to investigate the question of whether ecosystems are top- or bottom-driven systems is to search for examples of ecosystem convergence. Because only a relatively small number of factors would have to independently become congruent, as a simple matter of probability such convergence almost certainly must be restricted to top-driven systems. Thus, a finding of convergence between ecosystems would strongly indicate that they are top-driven in nature. However, this leads to the practical problem of demonstrating such convergence. One way to do this would be to look for geographically disparate but at least superficially similar modern ecosystems, and investigate whether they function similarly. However, a theoretical problem with this approach is that such modern ecosystems may not actually be causally disconnected, since a large number of biotic and abiotic factors have the potential to operate over hemispheric or even global scales (e.g. El Nino-related weather phenomena, bird migration, pelagic fish migration, and so on) (5, 6, 7). Thus, a finding of functional similarity in modern ecosystems might not clearly indicate independent convergence. A different approach that would circumvent the above problem would involve making a comparison between a modern ecosystem and a comparable but biotically and geographically distinct paleoecosystem. While such an approach as much as possible eliminates any ambiguities about causal connection, it faces a seemingly insurmountable logistical hurdle, namely observing processes and interactions in a currently non-existent ecosystem.

To a degree, the abovementioned problem could be overcome if an indirect "shadow" of key ecosystem processes existed that could be discerned in the fossil record. One possible example of such a factor is faunal body size distribution. Work by Holling and a number of colleagues (8, 9, 10, 11, 12) has shown that body sizes (masses) of terrestrial faunas are discontinuous, that is organized into clumps separated by well defined gaps. This distribution is explained by what is known as the Textural Discontinuity Hypothesis (TDH), which states that terrestrial landscapes are inherently discontinuous structurally, and that animals must adopt body sizes compatible with these discontinuities if they are to be successful in a given habitat. This phenomenon can be seen in human architecture, which is unimodally discontinuous in favor of individuals of average body size. Thus, individuals who are either too large or too small can experience significant problems in a wide range of common situations. The basis for important ecosystem processes casting a shadow in the form of faunal body size distributions lies in the source of the structural discontinuities. As stated by Holling (8), in terrestrial ecosystems key ecological processes operate on discrete spatial and temporal scales, and leave their signature imprinted on the landscape in the form of discontinuous structure. Thus, as products of landscape structure faunal body size distributions reflect important ecosystem structuring processes, and comparison of such distributions from different ecosystems should allow one to indirectly compare their functional modes.

One set of likely paleoecosystems that this methodology could be applied to is North America during the late Miocene (15-5 million years ago [mya]), with emphasis on the mammals as the best preserved aspect of the faunas. For a wide variety of reasons including mammals with similar adaptations (e.g., hippo-like rhinoceroses, giraffe-like camels, hyena-like dogs, a variety of horses including some presumably like zebras among many examples), the late Miocene ecosystems of North America has been compared with the savannas of modern East Africa (where savanna is defined as a mixed landscape consisting of grass and trees, with the tree density never becoming great enough to form a distinct canopy) (13, 14, 15, 16). The question of whether these temporally and geographically disparate ecosystems actually functioned in a similar manner or show only a superficial likeness remains unanswered, almost certainly in part due to logistical difficulties.

The purpose of this study is addressing the question of whether the late Miocene North American and modern East African ecosystems were/are functionally convergent. To achieve this goal, body mass distributions of mammal faunas, which serve as indirect reflections ecosystem function, were generated for each spatio-temporal region and compared for signs of congruence. In an attempt to look for historical and geographical patterns in the evolution of the late Miocene North American ecosystems, a time series of mammal faunas was created from 9-5 million years for three geographical regions, the Great Plains, the Gulf Coastal Plain, and the Pacific Coast. The data will show that the Great Plains converged with the savanna ecosystem of modern East Africa between 12 and 9 mya, and that this functional similarity lasted through the rest of the Miocene. The Gulf Coastal Plain converged with the modern East African savanna as well, but not until approximately 7 mya. The Pacific Coast faunas showed no evidence of functional convergence with the East African savanna during the late Miocene, a finding consistent with independent evidence that this region precocially developed a desert/chaparral landscape during the late Miocene rather than a savanna. The discovery of plausible convergence between late Miocene North American and modern East African ecosystems strongly suggests that these ecosystems are top-driven in nature. Assuming that these ecosystems do not represent special cases, this implies that top-driven ecosystems are common if not ubiquitous.

METHODS

As explained in detail elsewhere (10), in order to be utilized in this type of analysis a paleofauna must satisfy at least two requirements. 1) It must be large enough to reasonably represent the original fauna, and 2) it must have a reasonably wide size range, though not necessarily complete from mouse to elephant. Because of problems with making reliable body mass estimates (see below) and scarcity, both very small (less than 3 kg) and large (greater than 2,000 kg) species were excluded from the data sets. Generally, 20 was considered the minimum acceptable faunal size for inclusion in this study. However, for some faunas where the distribution was robust in the central region (a section observed to be particularly important for characterizing the body mass distributions of these faunas), sizes as low as 18 were utilized. There were some cases in which no one fauna in a critical geographical region and/or time period was diverse enough for inclusion in this study. In this situation where possible, faunal lists from localities that were reasonably close in geography and age were combined to form regional faunas that were treated as if they represented a single locality (locality faunal lists used in this study can be obtained from the author upon request).

The late Miocene mammal faunas were chosen from three geographical regions of North America: the Great Plains, Gulf Coastal Plain, and Pacific Coast (see Table 1). The choice of the first two regions was based on the fact that late Miocene mammal faunas from them are more commonly compared with modern African savannas than any other region of North America (e.g., 13). The Pacific Coast region was chosen as a test case. Evidence indicates that this region possessed a different landscape than the Great and Gulf Coastal Plains (see below), and thus, following the principles of the TDH, it should show faunal body mass distributions different from the aforementioned regions.

The modern East African mammal fauna used for comparison consisted of the Serengeti National Park (faunal data from [17]). This fauna was chosen because the Serengeti savanna ecosystem has been extensively studied (e.g. 18, 19), and because it encompasses an expansive flat landscape that is likely to be comparable to those of the Great Plains and the Gulf Coastal Plain.

Body mass estimates for extinct mammals were generated using linear regression equations derived from mass-correlated osteological/dental features in modern mammals (measurements of fossil mammals and regression equations used can be obtained from the author upon request). Though the only practical means for generating body mass estimates for extinct mammals, this methodology has limitations. The chief problem is consistently high intra-sample average prediction error for these regressions, with 20% being a relatively low value (no regression with an average predictor error greater than 30% was used in this study). Inputting measurements from extinct animals outside the original regressed sample can only unquantifiably magnify the error levels. Thus, body mass estimates generated for the extinct mammals in this study must be considered rough approximations at best. The faunal body mass distributions generated during this study are available at the following web address: www.lsmsa.edu/dlambert/miocenemammaldata.htm.

In the large majority of cases the gaps between clumps in body mass distributions are obvious to the eye in a casual perusal of the data, with the clumps commonly separated by a factor of 2 or more. However, this detection method obviously lacks statistical rigor. To overcome this problem, the body mass values in each distribution were log-transformed, with the resulting unweighted data sets subjected to a univariate cluster analysis with increasing sum of squares used as the clustering algorithm (20). The species (or their functional equivalents in cases where species could not be clearly identified) in each distribution were grouped according to the similarity trees produced by the analysis.

RESULTS

Body mass distributions of the early late Miocene faunas of the Great and Gulf Coastal Plains, circa 12-8 mya, are presented graphically in Figure 1. In each case, the late Miocene distribution is superimposed on the distribution from the modern Serengeti fauna for comparison. Examination of the body mass distribution of the Blue Jay/North Shore fauna (Great Plains, 9 mya) (Figure 1A) reveals a considerable congruence with that of the modern Serengeti, though there are differences in clump and location that can plausibly be explained in part by missing species and body mass estimation errors in the fossil fauna. The appearance of a late Miocene fauna that is

TABLE 1. Late Miocene Mammal Faunas Examined In This Study

Fauna	Age (mya)	Location
Valentine Railroad Quarry	12	Great Plains, western Nebraska
Blue Jay/North Shore	9	Great Plains, western Nebraska
Cambridge Quarry	7	Great Plains, western Nebraska
Santee/Devil's Nest	5	Great Plains, western Nebraska
Love Bone Bed	9	Gulf Coastal Plain, north Florida
McGehee Bone Bed/Mixon's Farm	8	Gulf Coastal Plain, north Florida
Moss Acres Racetrack/Withlacoochee River 4A	7	Gulf Coastal Plain, north Florida
Palmetto	5	Gulf Coastal Plain, central Florida
Ricardo	9	Pacific Coast, southern California
Merhten Formation/ Peace Valley/ Mt. Eden	6	Pacific Coast, central California
Schutler/Rome	5	Pacific Coast, western Oregon

similar to the modern Serengeti at 9 mya begs the question of when this congruence first occurred. The youngest Great Plains fauna older than this one that was both available and suitable for this type of analysis was the Valentine Railroad Quarry fauna (12 mya), whose body mass distribution is presented in Figure 1B. The clump pattern for the Valentine Railroad Quarry fauna is significantly different from that of the modern Serengeti fauna in that its second clump unambiguously straddles the space between the second and third Serengeti clumps. Thus, the body mass distributions of the Great Plains faunas converged with that of the modern Serengeti between 12 and 9 mya. Unfortunately, no suitable Great Plains faunas from the 8 mya interval were available for analysis.

The body mass distributions of the early late Miocene Gulf Coastal Plain faunas (9 and 8 mya) are considerably different from those of the contemporaneous Great Plains fauna, the 12 million year old Valentine Railroad Quarry fauna, and the modern Serengeti fauna (Figures 1C and 1D). The most noteworthy feature of these two faunal distributions is a shared large clump that largely encompasses the third and fourth clumps of the modern Serengeti fauna.

Body mass distributions for the middle late Miocene faunas of the Great and Gulf Coastal Plains, circa 7 mya, are presented graphically in Figure 2. Examination of the plot of the Cambridge faunal data from the Great Plains reveals that, like the early late Miocene faunas discussed above, there is a reasonable congruence with the pattern for the Serengeti. Examination of the data plot for the Moss Acres Racetrack/Withlacoochee River 4A fauna reveals a similar congruence with the modern Serengeti, which is significant because the Gulf Coastal Plains faunas from 9 and 8 mya consistently show a very different clump pattern. Thus, it would appear that the Gulf Coastal Plains faunas experienced a transformation between 8 and 7 mya that paralleled that which occurred to the Great Plains faunas between 12 and 9 mya.

Body mass distributions for the latest Miocene faunas of the Great and Gulf Coastal Plains, circa 7 mya, are presented graphically in Figure 3. The pattern observed in the body mass distributions of both the Great Plains and Gulf Coastal Plain during the middle late Miocene, namely congruence with the modern Serengeti fauna, continued to the end of the Miocene. Thus there was a consistent, overlapping trend in the body mass distributions for both regions in the late Miocene, the Great Plains from at least 9 mya to 5 mya and the Gulf Coastal Plain from 7 to 5 mya.

Plots of the faunal body mass distributions from the three Pacific Coast faunas examined in this study, representing ages of 9, 6, and 5 mya accordingly, are presented in Figure 4. The body mass distribution of the 9 million year old Ricardo fauna shows very little congruence with the modern Serengeti fauna, or for that matter any of the late Miocene faunas from the Great or Gulf Coastal Plains. Similarly, the body mass distribution of the 6 million year old Merhten Formation/Peace Valley/Mount Eden fauna shows no clear congruence with any of the other modern or late Miocene faunas examined in this study, including the older Ricardo fauna from the Pacific Coast. The body mass distribution of the 5 million year old Schutler/Rome fauna shows some similarity to the modern Serengeti, but its wide second clump broadly overlaps the second and third of the of the modern clumps. Thus, unlike the Great and Gulf Coastal Plains there is no evidence of convergence between the body mass distributions of the late Miocene Pacific Coast and the savanna faunas of modern East Africa.

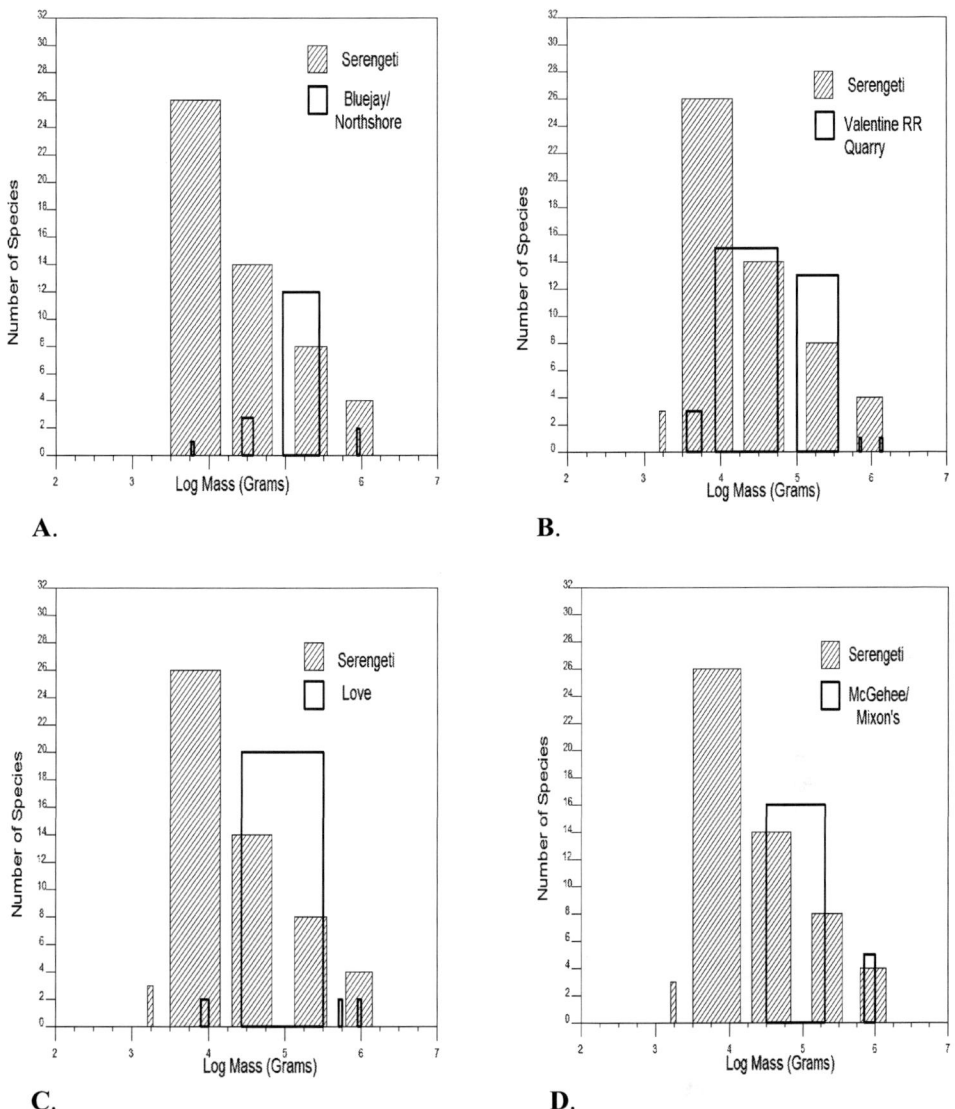

FIGURE 1. Plots of early Late Miocene Great Plains and Gulf Coastal Plain faunal body mass (grams, log-transformed) distributions against a reference plot from the modern Serengeti fauna. A. Blue Jay/North Shore fauna, Nebraska, 9 mya; B. Valentine fauna, Nebraska, 12 mya; C. Love fauna, Florida, 9 mya; D. McGehee Bone Bed/Mixon's Farm fauna, Florida, 8 mya.

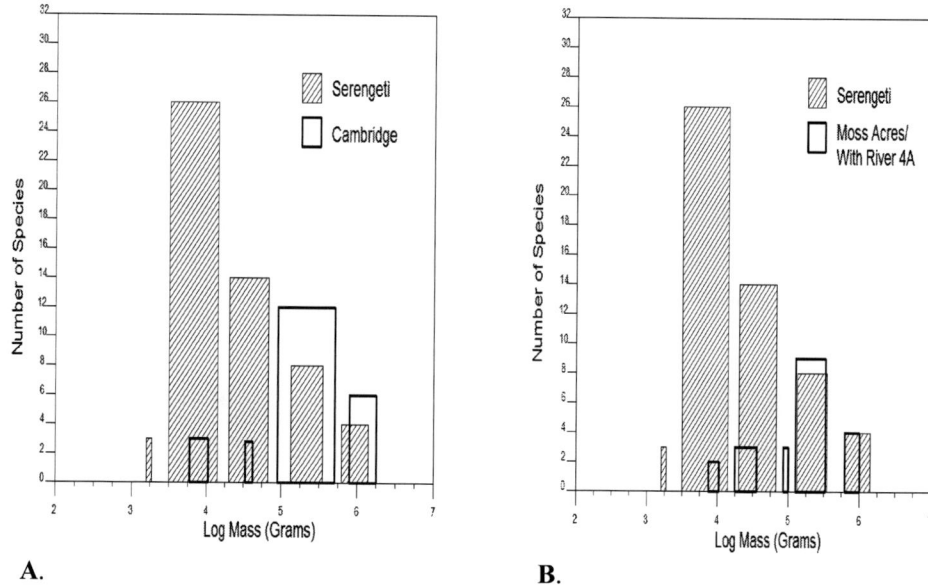

FIGURE 2. Plots of middle Late Miocene Great Plains and Gulf Coastal Plain faunal body mass (grams, log-transformed) distributions against a reference plot from the modern Serengeti fauna. A. Cambridge fauna, Nebraska, 7 mya; B. Moss Acres Racetrack/Withlacoochee River 4A fauna, Florida, 7 mya.

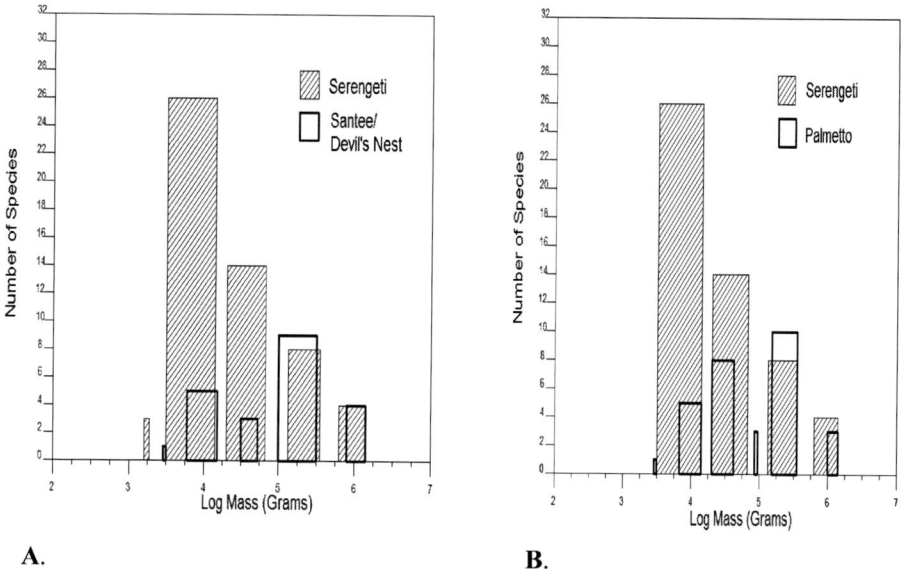

FIGURE 3. Plots of latest Miocene Great Plains and Gulf Coastal Plain faunal body mass (grams, log-transformed) distributions against a reference plot from the modern Serengeti fauna. A. Santee/Devil's Nest fauna, Nebraska, 5 mya; B. Palmetto fauna, Florida, 5 mya.

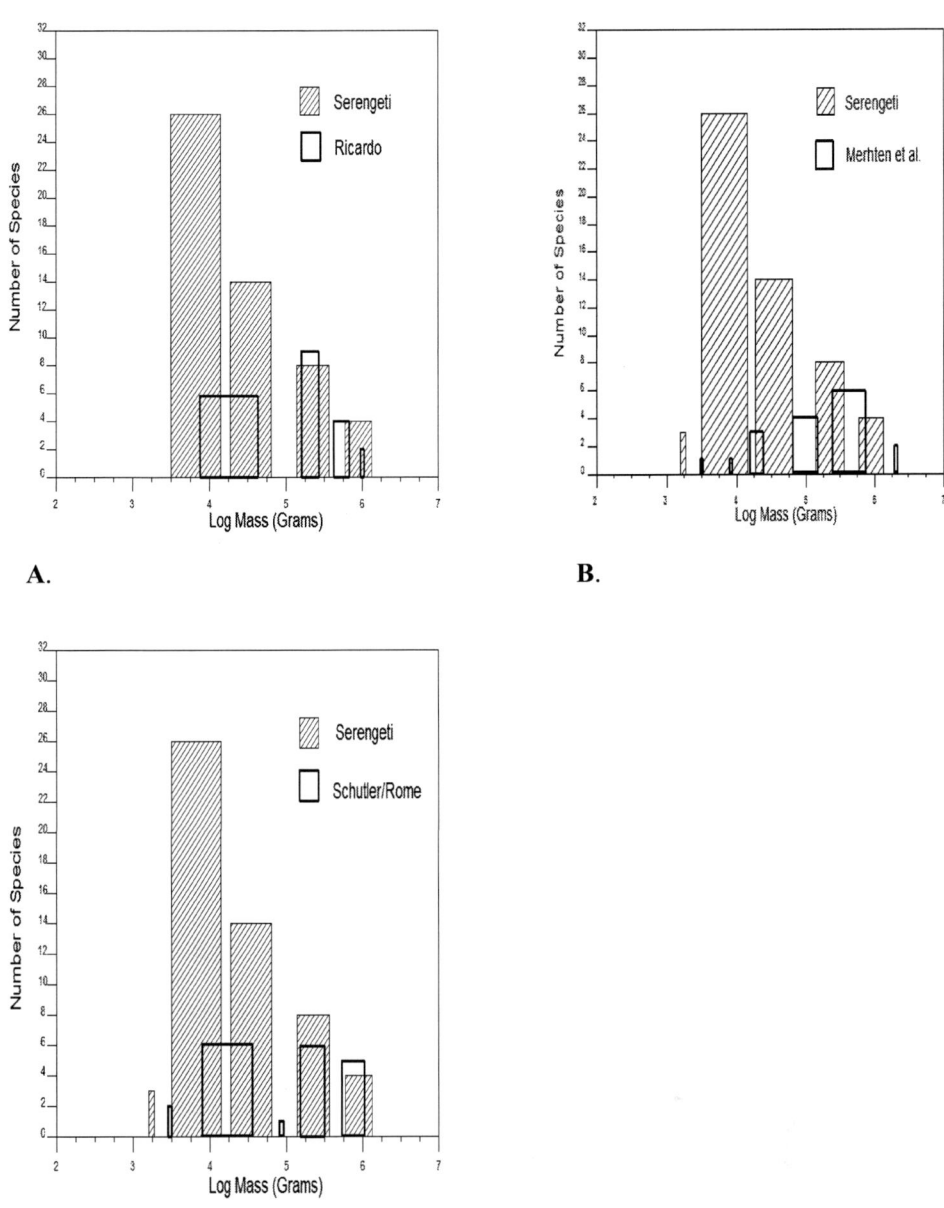

FIGURE 4. Plots of late Miocene Pacific Coast faunal body mass (grams, log-transformed) distributions against a reference plot from the modern Serengeti fauna. A. Ricardo fauna, California, 9 mya; B. Merhten Formation/Peace Valley/Mount Eden fauna, California, 6 mya; C. Schutler/Rome fauna, Oregon, 5 mya.

DISCUSSION

The fundamental question addressed in this study is whether ecosystems can functionally converge, which has a direct bearing on the issue of whether they are top- or bottom-driven. Thus, this leads to the following specific question: Do the data support convergence between late Miocene ecosystems of North America and the modern savanna ecosystem of East Africa?

From 9-5 mya, the mammal faunas of the Great Plains consistently showed clump patterns congruent with those of the modern Serengeti mammal fauna. In contrast, from 9-8 mya the faunas of the Gulf Coastal Plain showed a regionally consistent pattern that is significantly different from the modern Serengeti fauna. However, around 7 mya a sudden transformation occurred, with the Gulf Coastal Plains faunas following the Great Plains faunas in converging with the modern Serengeti. As in the case of the 9-8 mya interval, this new pattern remained regionally consistent until at least the end of the Miocene at 5 mya. The Textural Discontinuity Hypothesis and its corollaries argue that critical, structuring ecological processes leave their shadow in the faunal body mass distributions of terrestrial ecosystems. If the TDH is correct, then the patterns in the aforementioned faunal body mass distributions indicate that on at least some level the ecosystems of the late Miocene Great Plains and later the Gulf Coastal Plain functionally converged with the modern savanna ecosystems of East Africa.

It is noteworthy that the pattern trend observed in the Great Plains and Gulf Coastal Plain did not apply to the Pacific Coast. Evidence from a variety of sources including palynofloral data indicate that during the late Miocene the Pacific Coast experienced a precocial drying trend, producing regional chaparral and desert landscapes much like those found in the region today rather than savanna (14, 21). Assuming that these lines of evidence are correct regarding the structure of these regional landscapes, then indeed the Pacific Coast pattern differences from the Great and Gulf Coastal Plains are consistent with the principles of the TDH, though they do not definitively prove them.

As discussed above, the fundamental reason for investigating the phenomenon of functional convergence between ecosystems was trying to answer the question of whether ecosystems are top-driven or bottom-driven in nature. The behaviors of top-driven ecosystems are in principle easier to understand and therefore manage than bottom-driven ones, since far fewer driving factors need to be considered. The convergence in body mass distribution patterns observed between certain late Miocene North American mammal faunas and that of modern East Africa plausibly suggests that these causally disconnected ecosystems functionally converged, which therefore implies that they were/are top-driven in nature. Unfortunately, considering the small sample size involved, it is impossible to draw any meaningful conclusions about whether this condition is widespread or rare among ecosystems from this study. However, it is noteworthy that the examined ecosystems are not exceptional in any obvious way. Should this prove to be the case, then it is plausible that future studies of ecosystems will reveal that top-driven behavior is common among them if not outright ubiquitous.

ACKNOWLEDGEMENTS

The Florida Museum of Natural History, Nebraska State Museum of Natural History, and Los Angeles County Museum of Natural History kindly gave me access to their collections. S. David Webb, Buzz Holling, Paul Marples, Craig Allen, and Jan Zenzimir provided guidance and valuable suggestions during the conduction of this study. Craig Allen provided invaluable criticism of this manuscript. This project received financial support from NASA Earth Observation Systems and Terrestrial Ecosystems grants, and a generous grant from the Richard Brown Foundation of the Louisiana School for Math, Science, and the Arts.

REFERENCES

1. Holling, C. S., G. Peterson, P. Marples, J. Sendsimir, K. Redford, L. Gunderson, and W. Lambert, "Self-organization in ecosystems: lumpy geometries, periodicities, and morphologies," in *Global Change and Terrestrial Ecosystems*, edited by W. Self and B. Walker, Oxford, U. K.: Oxford University Press, 1996, pp: 346-384.
2. Perry, D. A., *Forest Ecosystems*, Baltimore: Johns Hopkins University Press, 1994, pp. 18-19.
3. Jorgensen, S. E., *Integration of Ecosystems: A Pattern*, Dordrecht, Germany: Kluwer Academic Publishers, 1992, pp. 100-102.
4. Chaisson, E. J., *Cosmic Evolution: The Rise of Complexity in Nature*, Cambridge, MA: Harvard University Press, 2001, pp. 10-34.
5. McHugh, M. J. and J. C. Rogers, *J. of Climate* **14**, 3631-3642 (2001).
6. Zolkevich, S., *Conservator* **18**, 4 (1997).
7. Block, B. A, H. Dewar, S. B. Blackwell, T. Williams, E. D. Prince, C. J.Farwell, A. Boustany, S. L.H. Teo, A. Seitz, A. Walli, and D. Fudge. 2001. *Science* **293**, 1310-1314 (2001).
8. Holling, C. S., *Ecol. Mono.* **62**, 447-502 (1992).
9. Holling, C. S. "What barriers? What bridges?" in *Barriers and Bridges to the Renewal of Ecosystems and Institutions*, edited by L. H. Gunderson, C. S. Holling, and S. Light, New York: Columbia University Press, 1995, pp. 3-36.
10. Lambert, W. D. and C. S. Holling, *Ecosystems* **1**, 157-175 (1998).
11. Allen, C. R., E. A. Forys, and C. S. Holling, *Ecosystems* **2**, 114-121 (1999).
12. Holling, C. S., and C. R. Allen, *Ecosystems* **5**, 319-328 (2002).
13. Webb, S. D., *Ann. Rev. Ecol. and Syst.* **8**, 355-380 (1977).
14. Webb, S. D., "The rise and fall of the late Miocene ungulate fauna in North America," in *Coevolution*, edited by N. H. Nitecki, Chicago: University of Chicago Press, 1983, pp. 267-306.
15. Webb, S. D., "Ten million years of mammal extinctions in North America," in *Quaternary Extinctions: A Prehistoric Revolution*, edited by P. S. Martin and R. G. Klein, Tucson, AZ: University of Arizona Press, 1984, pp. 189-210.
16. Webb, S. D., R. C. Hulbert, W. D. Lambert, "Climatic implications of large herbivore distributions in the Miocene of North America," in *Paleoclimate and Evolution: With Emphasis on Human Origins*, edited by E. S. Vrba, G. H. Denton, T. C. Partridge, and L. H. Burckle, New Haven, CT: Yale University Press, 1995, pp. 91-108.
17. Swynnerton, G. H., *Mammalia* **22**, 435-450 (1963).
18. Sinclair, A. R. E., and M. Norton-Griffiths, *Serengeti*, Chicago: University of Chicago Press, 1979, pp. 1-379.
19. Sinclair, A. R. E., and P. Arcese, *Serengeti II: Dynamics, Management, and Conservation of an Ecosystem*, Chicago: University of Chicago Press, 1995, pp. 1-673.
20. Wishart, D., *Clustan* (software), Edinburgh, U.K.: Clustan Limited , 2003.
21. Wing, S. L., "Tertiary vegetation of North America as a context for mammal evolution," in *Tertiary Mammals of North America*, edited by C. M. Janis, K. Scott, and L. Jacobs, Cambridge, U. K.: Cambridge University Press, 1998, pp. 37-65.

Interdisciplinary Social Science: An Example of Vertical and Horizontal Integrative Strategies

Subhash Durlabhji

College of Business, Northwestern State University, Natchitoches, LA 71497, USA

Abstract. A "Concept-Centered" strategy for Integrative Studies was proposed and implemented in the creation of the book *Power in Focus: Perspectives from Multiple Disciplines*. Essays on the ubiquitous concept of Power were solicited internationally and a final cut of ten essays from ten different disciplines, written specifically for this project, were included. This provides an example of what might be called Horizontal Integration, as it cut across *multiple disciplines*. One of the essays in the volume provides an example of Vertical Integration, as it applies a psychodynamic hypothesis concerning the development of Power relations among humans *across hierarchical levels*, from the child to the family to other groups and institutions in society, including finally entire nations and regions of the world.

Keywords: Social Science, Interdisciplinary, Power, Concept-centered investigation..
PACS: 89.65.-s

INTRODUCTION

In his landmark book *Consilience: The Unity of Knowledge* [1], Edward O. Wilson addressed the need for interdisciplinary research not only in the natural sciences but also within the social sciences: "If the social and natural sciences are to be united, the disciplines of both need to be defined by the scales of time and space they individually encompass and not just by subject matter as in past practice, and then they need to be connected" [pp. 208]. While admittedly in a forum where it must seem out of place, this paper seeks to advance Wilson's project by describing a strategy for embarking on *both* the tasks enumerated above.

A concept-centered strategy for Integrative Studies was proposed and implemented in the creation of the book *Power in Focus: Perspectives from Multiple Disciplines* [2]. This strategy has the potential to cut across disciplinary boundaries both within and between the social and natural sciences. The book consists of essays on the concept of Power from ten different social science disciplines. Viewing these essays collectively yielded four generalized hypotheses about the nature and dynamics of Power that are shown to be valid in the context of different disciplines. This could thus be called a "horizontal" integrative exercise. One specific essay utilized a concept from Psychology to link hierarchical levels of analysis within the behavioral sciences, from the individual to groups to societies to humanity as a whole. This is presented in the paper as an example of a "vertical" integrative effort.

CP755, *ISIS: International Symposium on Interdisciplinary Science*
edited by A. Ludu, N.R. Hutchings and D.R. Fry

POWER IN FOCUS

The premise of this book, and the strategy for integrative exploration it represents, is as follows:

"From the eternal dance of unimaginably huge galaxies and stars to the most microscopic constituents of all matter, from the twists and turns of geo-political affairs involving the destinies of nations to the most mundane activities of ordinary humans, power is invariably implicated in one way or another, either as an explicit, central organizing principle or as part of subtle, often unconscious or hidden processes. Reflecting this ubiquity of the phenomenon of power is an equally pervasive occurrence of power as a key concept in the whole range of disciplines of study devoted to the study of the material and social world, from Astronomy to Zoology and from Anthropology to Zooarcheology [3]."

If power is indeed a key concept in all physical and social sciences, it is either the case that the same concept has unique meanings in different fields or that there is some commonality in the meaning of power across the range of disciplines in which it makes an appearance.

An invitation was posted through various mailing lists on the Internet to scholars in the whole range of physical and social science disciplines, asking for essays on Power from the perspective of their own field of interest. Power was defined very generally, as *an ability to act or produce an effect*. All the more specific and precise definitions seem to not contradict this formulation, which came essentially from the dictionary.

TABLE 1. *Power in Focus* Table of Contents.

Title	Author
Introduction	Subhash Durlabhji
Fusion Power And Rhetorical Power: A Communication Perspective On Nuclear Energy Research	William J. Kinsella
Political Economy And The Tax State: The Power Of The Purse In Historical Perspective	Matias Vernengo
Whose Voices Matter? Feminists Stretch The Boundaries Of International Relations Discipline	Elina Penttinen
Power And The Definition Of "Order": The Misuse Of The Diagnosis Of Attention Deficit Hyperactivity Disorder	Ken Jacobson
The Psychodynamics Of Dominance And Submission	Paul Ziolo
Power In Interpersonal And Family Relationships	Hendrika Vande Kemp
The Power Of The Proposition	Tahir Wood
All My Relations: An Anthropological Reflection On The Faces Of Power	Maire Anderson-McLean
Power In The Framework Of Cognitive Neurosciences	Alexandre Castro-Caldas
Power And The Emerging Dynamic SuperGlobe	Crockett Grabbe

It took much too long but in any case, this book is the result. In spite of the centrality of the concept of Power in many natural sciences, none of the essays finally included in the collection are from the natural sciences. The Table of Contents displayed in Table 1 indicates the range of social sciences represented.

Multiple disciplines are represented here and this is useful enough by itself. It falls short of being inter-disciplinary or integrative, however. For this some additional work has to be done, a beginning for which is indicated in the Introduction to the book. The effort here is to try to extract from these ten essays generalizations that apply to power across disciplinary boundaries, which could then be subjected to further investigations to establish their validity, perhaps in the end to result in a "Unified Field Theory" of Power.

HORIZONTAL INTEGRATION

The following sections outline four generalizations that were extracted from an analysis of the ten essays in this book. These are presented here as tentative and fragmentary elements of a general theory of Power. The purpose is not to present these generalizations as valid but to demonstrate the potential utility of a horizontal integrative strategy.

Source

From the primordial source of power (the sun) to the corporations and institutions that wield enormous economic and/or political power in the social world, *the structure of the powerful entities appears to be the source of their power*. The structure of an entity -- whether it is inorganic or organic, microscopic or global, or individual or collective -- may be defined as the system of relationships and roles among its constituent elements. Fusion and fission, for example, release power through modifications of the structure of (nuclear) molecules. The different powers of various organic compounds are similarly associated with their structures. Life itself is distinguished from inorganic matter not by its constituent elements but by the way the elements are arranged or structured. Even slight modifications to the structure result in the disappearance of its powers, emphasizing again the link between organization and power.

The Introduction provides numerous other examples in which structure gives rise to power. Cognitive powers are in fact a hierarchically organized collection of conceptual structures or *schema* built up over time on a foundation of, and constrained by the properties of, biological structures. The more complete and integrated – well-organized – these *schema* are the more "intelligent" is the individual, precisely in the sense of having a greater ability to act or produce effects in the service of solving problems. Similarly, when the organization of the elements that collectively constitute personality is weak or somehow disturbed, the result is a deterioration of the individual's powers. All social organizations, from corner grocery stores and country clubs to the mightiest multinational corporations and national governments, derive their power from organization, a fact readily recognized through reference to terms like "organized labor", "organized crime", "grass-roots organization", and so on.

Kinsella's essay [all essays from the book are identified in this paper by the author's last name. Page numbers in the text below refer to the edition of the book cited in the Reference section] concerns the "big science" project involved with generating fusion in the laboratory. Fusion provides one illustration – from Physics – of the relationship between structure and power. In a different context, the power of science itself rests on the structure of the scientific method – "rules of argumentation, standards of evidence and proof, and methods of appealing to the classical principles of logic and authority" [p. 18] – which must be adhered to for the results of experiments to be accepted as valid. Science is an *organized* (i.e. structured) activity. Furthermore, "contemporary big science is a network of organizational relationships" [p. 20]. Scientists today participate in a multitude of communities in addition to their own narrow disciplines, including for example research teams, laboratories, universities, and private and public funding sources.

Pentinnen made much the same point in her discussion of the dynamics of power within the discipline of International Relations. In fact, both Kinsella and Pentinnen paid homage to the seminal ideas of Michael Foucault [4] in their essays. Foucault's entire project can be viewed an analysis of the structure of discourse and its relationship to power. Here is one example in which the established curriculum – the accepted structure of a program of study – was used to wield power: "When my colleague and I proposed a course on women and politics, including Feminist International Relations Theory, to the Political Science department, the proposal was initially declined. The Professors politely told us that in our department the feminist perspective had never been included in the study program, and because there had never been a feminist course offered this meant that the feminist perspective was not necessary!" [p. 76].

Jacobson demonstrated that the source of the power that adults wield over children in the context of schools is the school's structure, specifically the rules by which disciplinary measures are employed. Ziolo's essay, on the "psychodynamics of dominance and submission", is at its core a structural analysis, though the entity whose structure is implicated – the unconscious mind – is impossible to validate empirically. Castro-Caldas focused on not the mind but the physical organ called the brain, which is undoubtedly the seat of human cognitive power. He described the brain as a "community of elements that are organized according to complex rules." Castro-Caldas outlined the current knowledge about how this neurobiological structure results in the production of power. An example of the link between the structure of the brain and power is provided by the loss of power that results from injury to specific structures within the brain.

These examples provide support for the proposition that *the structure of powerful entities appears to be the source of their power.* This proposition is integrative in that it appears to operate in the context of power as it is understood and utilized in many different social science disciplines. These examples thus demonstrate the utility of the concept-centered interdisciplinary approach embodied in the book *Power in Focus: Perspectives from Multiple Disciplines.*

Metamorphosis

Perhaps no other aspect of power establishes it as being a unitary phenomenon as much as the great variety of ways in which *one form of power transforms into another*, and then another, in an endless, eternal dance. The multiple forms of power found in the material and social domains turn out to be like the waves of an ocean that rise and fall and, in doing so, fuel the birth of another wave. Solar power becomes fuel, which in turn is converted into various forms of mechanical and physical power. In the social context, personal qualities can translate into political power, which in turn can become organizational or corporate power, which can be utilized to produce vast quantities of material power.

A most convincing demonstration of this quality of power is made by Kinsella. The quest for the ultimate source of material power, fusion, depends, even for its very existence, on a particular set of shared values, beliefs and visions of a future among the citizens of a society – that is, on cultural power. The quest materializes in the form of a "big science" project through the confluence of vast material, intellectual, organizational and institutional powers. The experiments subsequently require multiple transformations of physical power, from electrical to mechanical power, then again back to electrical power, which is converted to the massive quantity of heat required to produce fusion. The end result is the production of data, which in turn must be converted, through the application of rhetorical power (i.e. persuasive communications) among the scientists, into "scientific knowledge."

In turn, this production of knowledge enhances the credibility of the scientists, institutions and sponsors involved in the project, increasing the influence or power of their publications. This enhanced power then reappears as the power to attract more support and resources – that is, political power and material power. And here Kinsella comes back full-circle to cultural power: the transformation of the success of the Big Science project into political or material power may or may not take place, since cultural power may have shifted since the inception of the project many years earlier.

Vande Kemp described power relations between people as a nuanced dance among three dimensions: Focus on Other, Focus on Self, and "Introjects", that is, a power interaction internal to the person, such as "self-blame" or "self-affirm." Every interpersonal interaction is characterized by a series of interchanges in which each person is adjusting among these three dimensions, depending on the reactions of the other person.

Anthropology provides a number of examples of the shifting nature of power in cultural systems. Anderson-McLean's essay depicted the eons-old struggle between the powerful and the marginalized in cultures everywhere, with religious power becoming economic power, which in turn becomes political power, and so on "in a never-ending chain of articulation." Finally, Grabbe described the emergence of a new locus of power, wrought by the marriage of communication and technology – an army of satellites in the skies above us is rapidly transforming the power of societies and nations throughout the world. This "SuperGlobe" is another example of how the power unleashed by science and technology is transformed into military, economic, political and social power.

Enhancement

A key aspect of the process by which an entity increases or alters its power is *change of structure through differentiation and integration.* Increased power implies increased ability to act or have an effect. Differentiation refers to the emergence or invention of a new response or structural element. Integration in this context refers to the incorporation of new responses or structural elements into existing structures. For example, over the millennia less powerful organisms added to their powers through trial and error responses – differentiation – to environmental pressures and opportunities and eventual integration of successful new responses into their genetic make up, to yield a "more evolved" or more powerful creature. Similarly, the astounding powers of the adult human are built up, layer by layer, through a trial-and-error process involving responses to new challenges and opportunities presented by the environment. These responses involve the differentiation or elaboration of new neurobiological structures within the brain. After a "critical mass" of such experiences is accumulated a process of integration of these new structures into existing structures or *schema* takes place. Social organizations also grow in power by inventing new problem-solving tactics and strategies in response to environmental challenges and opportunities, and eventual integration of these new solutions or response patterns into the main body of the organization (through the creation of a new policy or a new department, for example).

In his essay on Big Science projects Kinsella outlined the familiar process by which the immense power of science and technology has accumulated over the past two or three centuries. An experiment is conducted, a few new facts are uncovered, new ideas for experiments emerge in response to new challenges. The scene is repeated in thousands of labs in many countries, each focused on a tiny part of the puzzle, until finally a "critical mass" of knowledge is accumulated. Then some scientist or scholar endowed with visionary imagination integrates all these differentiated bits and pieces into a brilliant new theoretical framework that produces a quantum leap in knowledge and understanding – and power.

Pentinnen's review of the emergence of the academic discipline of International Relations is the familiar story of the development of every academic department. A new course is proposed and offered in response to an emerging need. As other scholars publish more work in the area, a critical mass of issues, concepts, and theories emerges. Perhaps three or four courses focused on this narrow niche begin to be offered as a minor, a first level of integration. A degree program may be the next level of integration, followed by a department and eventually a separate college within the University. The very notion of "Interdisciplinary" or "Integrative" studies is illustrative of the differentiation-integration dynamics of the growth of the power of a field of study.

Neural organization provides another example. Castro-Caldas suggested that judgment and understanding, which may be viewed as more powerful than other brain functions such as perception, are based on a synthesis of memory traces of a wide array of experiences. This becomes clear in cases of brain damage such as dementia, in which organic damage to the brain cells lead to loss of long-term and even short-term memory, resulting in loss of judgment and understanding. In other words, the more powerful capacities of the brain such as judgment emerge from integration of a host of differentiated experiences stored in memory and neural traces (which is why age is a necessary, though not sufficient, pre-requisite to wisdom).

Dynamics

A final element shared by power in its various forms is the central role of *balance of power* in the way it operates. Balance of power is at the very core of the existence of all matter: the structure of the atom is in fact a dynamic interplay between its various constituents, held in place through a delicate balance between forces of attraction and repulsion; the same pattern is repeated in the much vaster arena of the entire cosmos. Perhaps one of the most awe-inspiring aspects of living creatures is the process by which hundreds of chemicals, each having different properties or powers, are kept within incredibly narrow tolerances for the whole to continue to function. Similarly, psychological health depends on a balance among the various powers contained within the personality – in Freudian parlance, the superego, ego, and id. The resilience and indeed greatness of American democracy flows directly from the ingenious system of checks and balances among the legislative, judicial and executive powers of government.

Almost every essay in the volume illustrates this phenomenon of the balance of power. Kinsella set up his discussion of big science projects by pointing out that this kind of science draws its power and sustenance from a favorable cultural climate, which in turn is dependent upon many powerful stakeholders being aligned in just the right mix. If the balance of power among these elements shifts, so does the resultant ability of scientists to obtain support for their projects.

Vernengo engaged the issue of balance of power with the very first sentence of his essay: "Political economy was developed at a time when the rising capitalist class was struggling to reduce the power of the aristocracy and of feudal institutions that constrained its ability to expand" [p. 42]. In fact of course the history of not only the economy but of human affairs as a whole can be viewed in terms of a never-ending struggle for power among individuals, groups, organizations, institutions, and ideas.

In all societies, the balance of power between adults and children clearly favors adults. Jacobson found this reality deserving of caution, especially in the context of the diagnosis and treatment of "attention deficit hyperactivity disorder" among schoolchildren. He also found balance of power dynamics operating in what he calls the "children's world", characterized by its own roles, statuses and sanctions. Ironically, a child's challenge toward adult authority may often have more to do with the balance of power *within* the children's world: "It is evident that the majority of [children's] behaviors are motivated to gain peer attention. Since challenging adults leads to peer attention, it could be argued that those behaviors are also motivated to gain peer recognition, not simply to challenge adult authority" [p. 106].

Vande Kemp's analysis of streams of interpersonal interactions reveal the often subtle ways in which balance of power determines the strategies used by the parties to the relationship, as well as the outcomes experienced by them. The categories of symmetrical, complementary and parallel relationships discussed by Vande Kemp are distinguished from each other in terms of balance of power: a symmetrical relationship is one between two people who behave as if they have equal status (equal power). In a complementary relationship, the two people are of unequal status, so that one person always initiates action while the other always submits. And in parallel relationships, the individuals alternate between symmetrical and complementary relationships in response to changing situations.

In the context of cognitive neuroscience, the most obvious example of balance of power relates to the two hemispheres of the brain. Castro-Caldas noted: "Every time we receive information from the outside world, both hemispheres become informed and the subsequent processing of the information has to be decided. *It is through the inter-hemispheric dialog of power that this becomes possible*" [p. 278, emphasis added]. In fact of course, the ability of all organic systems to function effectively depends on balance of power among hundreds of forces and processes.

VERTICAL INTEGRATION

The analysis above illustrated the integrative potential of concept-centered investigations across multiple disciplines. Paul Ziolo's essay on the "Psychodynamics of Dominance and Submission" utilizes a set of ideas from psychology and applies them to *hierarchical* levels of social science disciplines, and can be viewed as concept-centered vertical integration. Ziolo's essay is based on the premise that patterns of dominance-submission imprinted in the child's unconscious mind forms the template or *archtype* for his or her Power relations as an adult. This much is firmly rooted in the psychoanalytical tradition of Freud and his followers. Ziolo goes beyond the psychodynamics of the individual, however. He suggests that childhood experiences form the template for Power relations within the family, within and between groups in society, between "psychoclasses" – groups whose members share child-rearing patterns and early experiences – and within organizations and institutions like the school. Ziolo takes a process experienced intensely during childhood and employs it in explaining Power phenomena at different hierarchical levels of analysis, thereby linking together in one framework the fields of psychology, social psychology, family psychology, organization behavior, sociology, anthropology, political science and economics. A brief review of the suggestions made at each level is provided below.

Developmental Psychology: Dominance-submission relations between human beings originate in generic trauma, human reproductive strategies and childrearing modes, sexual dimorphism, culturally enhanced gender differentiation, continual sexual arousal and neoteny-induced dependence.

Social Psychology: These most basic dominance-submission roles are expanded and reinforced through family relationships, in which the individual is induced to assume these roles *vis-à-vis* parents and siblings.

Sociology, Organization Behaviour, Education: Further reinforcement takes place through schooling and education.

Psychohistory: Childrearing and socialisation modes shared by certain individuals give rise to "psychoclasses", whose mutual interaction and power relations account for the hidden dynamic of the historical process.

Social Psychology, Sociology, Economics: Social or economic classes may form the intersection point of two or more psychoclasses, so that the unconscious psychodrama of trauma re-enactment may be played out both within and between groups.

Anthropology: The total system of projection, introjection and psychological defence evolved by a particular society over time is called 'culture'.

Political Science: The greater the degree of environmental insecurity for a given society, the more elaborate and oppressive the power structures that evolved within it.

It is the more traumatised, less advanced (and more numerous) psychoclasses that are most drawn to the exercise of power. Historically, the conflicts engendered by this process have often been avoided through emigration. This is becoming less and less of an option in the densely populated global society of today.

Political Science, War: The economic benefits of technological advance provide a positive feedback for further advance and also encourage reduction in social inequality. At the same time however, technological advance becomes an instrument for the enforcement of power. When cultural advance is deemed to be too progressive -- i.e. when dominance-submission relations are felt to be weakening -- the society seeks 'rebirth' through a common act of sacrifice: war. War is therefore manufactured in the interests of political, social and economic power and arises from inner psychological fears of individuation -- of breaking away from ancient patterns of dominance.

CONCLUSION

There are many social scientists intensely committed to the Interdisciplinary approach. My purpose here was to interject into this conference of physical scientists an awareness of the parallel work being done in the social sciences. At higher levels of integration, the physical and social sciences will also be linked by interdisciplinary concepts. As Edward O. Wilson emphasized, "The social sciences are intrinsically compatible with the natural sciences. The two great branches of learning will benefit to the extent that their modes of causal explanation are made consistent" [5]. This paper suggested that concepts like Power can serve to unite not only the fields within social science but also serve as a bridge between the social and natural sciences. Perhaps one of you will undertake a similar project on Power in the many disciplines of the natural sciences.

REFERENCES

1. Wilson, E. O., *Consilience: The Unity of Knowledge*, New York, NY: Vintage Books, 1998.
2. Durlabhji, S., *Power in Focus: Perspectives from Multiple Disciplines,* Lima, OH: Wyndham Hall Press, 2004.
3. Durlabhji, S., *Op. Cit.,* pp. xiii.
4. See, for example, Michel Faucault, *Power/Knowledge: Selected Interviews and Other Writings*, Colin Gordon (Editor). London: Harvester Wheatsheaf Press, 1980.
5. Wilson, E. O., *Op. Cit.*, pp. 205

Conference Summary and Closing Remarks

Dr. Andrei Ludu

I have prepared a short summary of our conference. We had thirty-five presentations on approximately 70 problems, from at least seven fields of science. This table illustrates the representation from various fields.

TABLE 1) Percentage of Presentations from each fields.

Discipline	Percentage of Presentations
Biology	36
Physics	30
Chemistry	13
Computer/numerical or mathematically oriented	20
Social Sciences	1

As you can see, a lot of talks were biophysical in nature. Figure 1 contrasts the percentage between experimental and theoretical approach. I'd say we had a fairly balanced presentation.

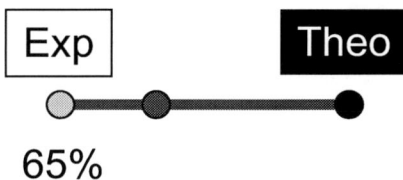

Figure 1: The percentage between experimental and theoretical (computer simulations or exact calculations) presented at ISIS.

The ratio between micro-scale to macro-scale systems being investigated is illustrated in figure 2. Micro-scale investigations are analogous to studying a single tree within a forest, while macro-scale investigations are analogous to studying the entire forest. There was a larger focus towards microscopic approaches in this conference. This makes sense, since a lot of presentations dealt with the dynamics and energetics than about kinematics and descriptive approaches.

CP755, *ISIS: International Symposium on Interdisciplinary Science*
edited by A. Ludu, N.R. Hutchings and D.R. Fry
© 2005 American Institute of Physics 0-7354-0240-X/05/$22.50

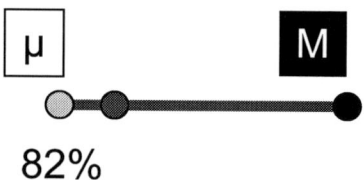

82%

Figure 2: The percentage between micro and macro scale presentations was 82%.

In field of physics, the ratio between classical and relativistic quantum mechanics approach is presented in next figure 3. One can notice a tendency towards classical physics, mainly because of the scale involved, that is systems above the micron scale.

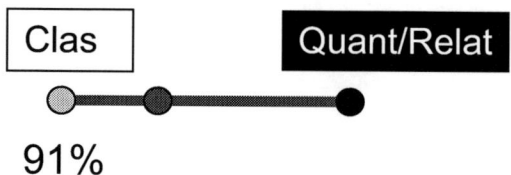

91%

Figure 3: The percentage between micro and macro scale presentations was 82%.

In the figure 4, illustrates how interdisciplinary the presentations were. Each edge represents connections between the different fields of science (represented here by vertexes, or corners). The thicker the connection is, the larger the frequency of presentations between those two fields. For example, the large majority of presentations used drew from both biology and physics. The next largest connection lies between biology and computer modeling.

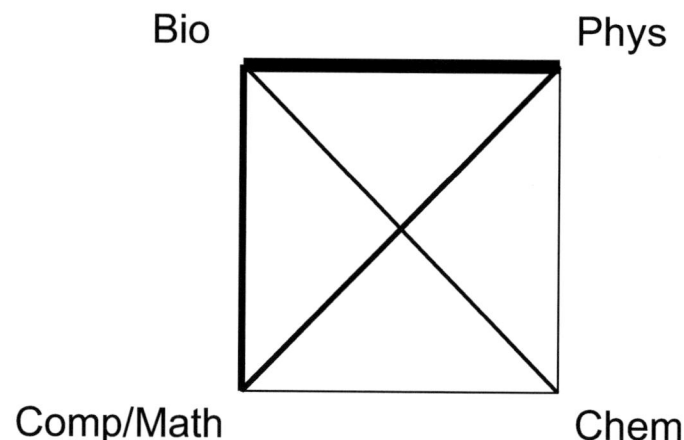

Figure 4: The interdisciplinary nature of the presentations. At each vertex, Biology, Physics, Chemistry and Computational/Mathematical approaches are represented. A darker line illustrates a higher frequency of presentations that drew from two fields.

Finally figure 5 is a 3-D representation of a *correlation function* between the presentations. The vertical axis counts number of presentations. The "Inter-D" axis measures, on a scale of 0 to 5, the number of presentations that connect different (disjointed) fields within the same discipline. For example, in the discipline of physics we could compare how many presentations relate classical mechanics of elastic bodies with optics. The "Multi-D" axis measures, on a scale of 0 to 5, the number of presentations that connect different disciplines of science. In a very focused single discipline conference, such a diagram would show one tall peak placed somewhere on the "Inter-D" axis; the remainder of the graph would show zero. A conference focused on a certain field of science, such as the The American Physical Society March Conferences, would show a distribution of different bars, but all lined up along the "Inter-D" axis, at some constant coordinate for the "Multi-D" axis. A broad conference, which contains talks from several disjoint fields of science which are not interrelated would provide a line of bars of different heights, all lined up along the "Multi-D" axis and placed at a constant value to the "Inter-D" axis.

On the contrary, a real interdisciplinary conference where different sciences and different fields within each science are interrelated and connected in different presentations, would show a general distribution of bars all over the horizontal plane. The closer the bars are to the horizontal diagonal, the more interdisciplinary and cross-disciplinary coherent content the conference has. In the case of the ISIS symposium, we have such an interrelated and connected character of the presentations, as one can see from the figure below. The large majority of the talks are placed along the main diagonal of the horizontal plane. In that, this symposium was a success, and has completely attained its goals.

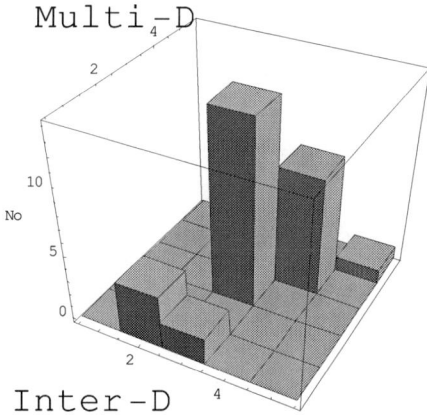

Figure 5: The interdisciplinary nature of the presentations shown as a three dimensional talk. A truly interdisciplinary conference should have a large number of presentations along the diagonal—much like ISIS had.

Dr. Nathan R. Hutchings

I have a few comments I would like to add. First, I'd like to thank all the participants. I believe my analysis is less complicated than Andy's. We had an excellent group of speakers who all gave excellent talks.

I like to thank the supporters who haven't yet been formally recognized. Ted Jones, was the principal fund raiser for this event. I would also like to thank Richardson Technologies, who made a very generous contribution of two thousand dollars and McGraw Hill who also made a generous contribution. I would also like to thank Northwestern for letting us use their facilities, ARAMARK for the catering, and the technical staff. The technical staff has done a very good job of putting out the fires, and making transition smooth. Finally, I would like to thank the other local entities that I will not mention by name.

We organized the conference around three objectives, and these three objectives were met. I think it's safe to say from Andy's analysis that we truly had an interdisciplinary collection of speakers here today—this was the first objective. Our second objective was to stimulate some interdisciplinary discussion. I believe that from the questions we had and from the discussions that have occurred during the conference that this objective was met as well. Indeed, we managed to open some dialog between some otherwise disconnected fields. The final objective was to challenge the boundaries for each of us. The conference impacted how we perceive our own research and the research of others. I believe that through this conference we were able to truly challenge those boundaries of each one of our research fields. We were forced to ask ourselves if there are elements from other disciplines that could enhance the way we do our work everyday in the lab.

So again I commend you all and I thank you all for your participation. If you have semantics you need ironed out, receipts or technical details of any kind, please don't hesitate to e-mail Andy, Darrell, or me, anytime. We would be glad to help you out with any questions you may have.

Thank you all.

Dr. Charles J. Brokaw

I was asked to give some remarks at the end. I'm going to make a very short, somewhat personal, summary of this conference.

My father graduated from college in 1922 and went into the field of radio communications, so I grew up in that environment learning a little bit about it. A little later I started studying flagellum in 1960 or so. There wasn't to much my dad and I could talk about the things we were interested in, they were just to different. Neither of us would have imagined at the end of the century that there would be methods being used that could be used to optimize radio communications and almost similar methods could be used for analyzing images of flagella. Things are really happening in terms of developing new methods. Of course a lot of the progress is dependent upon on digital computation; however, the computing power is available. Researchers have to develop methods on how to use them the computing power. Meetings like this one facilitate the fusion between the methods from one problem to another.

When I initially found out about this meeting, I was interested in attending. I expected to come and find a small group of other people interested in flagella and was sure I would have some interesting things going on to talk about. That has turned out to be very true, but the rest of the meeting has really been a surprise. I am really impressed by how interesting the content of the other presentation have been, but the fact that the speakers have really made an effort to stress the interdisciplinary importance, value of their contributions.

I don't have enough data to know how this developed. It could be a very deterministic result, resulted from the efforts of the organizers. Or it could be a very beautiful work example of self-organization—of course facilitated by the environment provided by the organizers. In any event, it's time for us as members of the audience and speakers to express our appreciation for this to Northwestern State University, to the organizers, to their staffs, to the sponsors who provided the money and these came together and created for me at least and hopefully for the rest of you as well a very worth wild three days and I hope that it will be very worth while to Darrell, Nathan, and Andy in their future endeavors to develop a new program here at Northwestern State. So let's thank Darrell, Nathan and Andy. I would like everybody to so how much we appreciate you people. This has really been worth the trip, thank you all very much.

Author Contact List

Andrei Ludu
ludua@nsula.edu

James C. Patterson II
JPatte@lsuhsc.edu

David R Mitchell
MitchelD@upstate.edu

Charles Wolgemuth
cwolgemuth@uchc.edu

Christophe Zimmer
czimmer@pasteur.fr

Christo I. Christov
christov@louisiana.edu

Charles J. Brokaw
brokawc@its.caltech.edu

Marios Christou
christou@ulm.edu

Yenumula B. Reddy
ybreddy@gram.edu

Young S. Kim
yskim@physics.umd.edu

Sorinel Adrian Oprisan
soprisan@uno.edu

Marjan Trutschl
mtrutsch@pilot.lsus.edu

W. David Lambert
DLambert@lsmsa.edu

Aghalaya S. Vatsala
vatsala@louisiana.edu

Armon Kargol
akargol@loyno.edu

Christian Cibert
cibert@ijm.jussieu.fr

Constance Schober
cschober@odu.edu

Darrell R. Fry
fryd@nsula.edu

Jayanta Choudhury
jxc9551@louisiana.edu

Joachim A. Maruhn
maruhn@th.physik.uni-frankfurt.de

Nathan R. Hutchings
Hutchingsn@nsula.edu

Nick Cogan
cogan@math.tulane.edu

Rui Zhu
rui.zhu@uleth.ca

Takahiro Harada
harada@chem.scphys.kyoto-u.ac.jp

Walter Greiner
greiner@th.physik.uni-frankfurt.de

Subhash Durlabhji
durlabhji@nsula.edu

Author Index

A

Ablowitz, M. J., 40
Archuleta, L., 185

B

Blazquez, S., 177
Brokaw, C. J., 107, 253
Bürvenich, T. J., 17

C

Choudhury, J., 85
Christou, M. A., 70
Christov, C. I., 46, 53, 70, 85
Cibert, C., 117
Cogan, N. G., 190
Cvek, U., 204

D

Dunham, A., 185
Durlabhji, S., 244

F

Frischknecht, F., 177
Fry, D., 185

G

Gell-Mann, M., 1
Greiner, W., 17
Guillén, N., 177

H

Harada, T., 165
Hutchings, N. R., 91, 137, 153, 253

K

Kargol, A., 159
Kim, Y. S., 61

L

Labruyère, E., 177
Lambert, W. D., 233
Ludu, A., ix, 91, 137, 253

M

Maruhn, J. A., 34
Ménard, R., 177
Mitchell, D. R., 130

N

Noz, M. E., 61

O

Olivo-Marin, J. C., 177
Oprisan, A., 198
Oprisan, S. A., 198

P

Patterson II, J. C., 225

R

Rains, J., 185
Reddy, Y. B., 215
Roussel, M. R., 172

S

Schober, C. M., 40

T

Trutschl, M., 204

V

Vatsala, A. S., 78

W

Westergard, A. M., 153
Wolgemuth, C. W., 145

Y

Yang, J., 78

Z

Zhang, B., 177
Zhu, R., 172
Zimmer, C., 177